APPLICATIONS
OF
NUCLEAR PHYSICS

J. H. FREMLIN

M.A., Ph.D(Camb.), D.Sc., F.Inst.P., M.I.E.E.
Professor of Applied Radioactivity, University of Birmingham

HART PUBLISHING COMPANY, INC.
NEW YORK CITY

APPLICATIONS OF NUCLEAR PHYSICS

CONTENTS

CONTENTS

CONTENTS

CONTENTS

PREFACE

Increasing numbers of workers in all branches of research are using the techniques of nuclear physics. Few of them wish, or have the time, to take a full course in the subject. Some understanding of the principles however is invaluable if a proper choice is to be made of the best means to solve a given problem.

This book is not intended to replace a full course in nuclear physics. Readers who wish to learn a serious amount of that subject and who wish to find out how our knowledge in the field was gained, should refer to one of the several excellent—and much larger—text-books now available. The names of two or three are given in Section 2 of the Bibliography.

For most of the applications described, the research-worker can be successful without such detailed knowledge, so long as the commoner types of behaviour and interaction of nuclei are known, in much the same way as a general can be successful without much detailed knowledge of his soldiers' physiology.

The basic nuclear information required here appears in Chapters I and II and is presented dogmatically (and often very approximately) although I have also tried to arouse an interest in the further study of nuclear physics for its own sake.

The main part of the book is concerned with instruments and techniques. I have dealt at some length with sources of error and limitations of accuracy. It has been my experience that the danger of a little learning is well demonstrated by the student who has just been told that for statistical reasons the probable error in a count of N is $0.67\sqrt{N}$. Too often he at once forgets all other sources of error and is happy to believe that all he has to do to get a final result good to 0.01% is to leave his equipment to count a hundred million particles.

The particular examples chosen to illustrate the methods described, and even the classes of examples represented by individual chapters, have necessarily been chosen from a much wider field according to my own knowledge and interests. This procedure is not the best, but it is much the quickest and I wishfully think that it is more valuable for the examples in a scientific text-book to be up to date than to have the perfect balance that can be obtained only in treating subjects that have been studied for a considerable time.

I would like to offer my particular thanks to Dr. J. N. Kudahl for his help with the presentation of several chapters; to my wife for the numerical computations required for many of the tables; to her and to my daughter Margaret for most of the illustrations and to nearly everyone of my colleagues in the Physics Department of the University of Birmingham for their willing, if sometimes unwitting, assistance.

J.H.F.

APPLICATIONS
OF
NUCLEAR PHYSICS

CHAPTER ONE

THE ATOM AND THE NUCLEUS

1.1 Composition of the Atom

Modern ideas of the atom and of the nucleus have remarkably little in common with the pictures evoked by their classical names. It is perhaps no coincidence that few people nowadays remember what are the pictures suggested by the classical names.

An atom is now regarded as consisting of a massive positively charged nucleus surrounded by a cloud of negatively charged electrons all of which together comprise less than 1/1000 of the mass of the atom. The electron-clouds have an effective diameter in the region of 10^{-8} to 10^{-7} cm while the diameters of the nuclei are some hundred thousand times smaller. The forces retaining the electrons close to the nucleus are almost entirely electrical. Magnetic forces produce some observable minor effects but gravitational forces are entirely negligible, being less than the electrical by a factor of about 10^{36}.

The electrons are distributed in a series of diffuse, overlapping "shells", the innermost being called the K-shell, the next L and after that M, N, O. . . . Each shell can take only a limited number of electrons, the successive maximum numbers being twice the squares of the natural numbers; thus the K-shell can take up to 2, the L-shell 8, the M-shell 18 and so on.

The electron-nucleus system has lowest potential energy when the two particles are closest together. Energy is therefore absorbed when an electron is moved from an inner shell to an outer one and is liberated when an electron falls from an outer shell to a vacancy in the inner one. When all electrons in an atom have the lowest possible potential energies consistent with the limitations described, the atom is said to be in the ground state. If an electron is raised to a higher energy, leaving a vacancy in an inner shell, the atom is said to be excited.

This picture originated with Bohr, who showed that the simpler optical emission spectra of radiating atoms in a gas could be explained quantitatively if (1) the electrons in each shell were supposed to be moving in orbits under the influence of the electrostatic attraction of the nucleus and (2) each electron were supposed able to stay only in those orbits in which its angular momentum was an integral multiple of a minimum value $h/2\pi$, where h is Planck's constant.

This limitation gave a discrete set of possible stable orbits with well-defined energy differences. Bohr thus explained also what had already been known for some time, that visible and other radiations were not emitted continuously but were emitted in well-defined amounts of energy called quanta; these amounts exactly corresponded to the energy-difference between the initial and final orbits of a radiating electron.

The frequency v of a quantum of radiation of any given energy E is always the same, E/h, where h is again Planck's constant. This is now known to be true for all radiation, from the longest-wavelength radio-waves to the shortest-wavelength X-rays. Radio- and other low-frequency waves appear to be a continuous flow of energy rather than a succession of separate quanta only because h is so small. Its numerical value is $6\cdot55 \times 10^{-27}$ erg seconds so that one quantum of radiation at a frequency of 1 Mc/s, an important region for broadcasting, has an energy hv of only $6\cdot55 \times 10^{-21}$ ergs. To detect radiation at this frequency requires an enormous number of quanta per second which are then effectively indistinguishable from a continuous flow of energy.

Bohr made no attempt to explain why electrons should be able to revolve indefinitely in his proposed "stable orbits" without losing energy continuously by radiating, as required by classical electromagnetic theory. This was left to the theory of wave mechanics, which considers an electron as being continuously distributed round a stable orbit rather than as a revolving particle.

Interpenetration of the electron clouds of two different atoms involves strong repulsive forces and in the ordinary course of events the nuclei of neighbouring atoms will not come within ten thousand diameters of each other. Although the reasons are different, one can see that, if each human being maintained a number of stones whirling around him at a fair fraction of the speed of light in unpredictable orbits on mile-long strings, social contacts would be rare.

The usual physical, chemical and biological interactions of atoms are determined entirely by the electrical interaction of the outermost regions of the electron clouds. In a sense, therefore, chemistry, biology and much of physics are all branches of electrodynamics. The rather feeble electrostatic forces shown by an elementary experiment with pith balls—when our climate allows it to work at all—leaves one unprepared to expect that the strength and rigidity of steel should be due to identical forces. The charge on an electron is only $1\cdot6 \times 10^{-19}$ coulombs, but a single electron may be within 10^{-8} cm of two or more nuclei in a solid, all pulling at it, and the inverse square law of force makes these pulls of impressive size for the tiny particles involved. The weak forces involved in the pith-ball experiments arise from the fact that the excess or deficit of electrons on each ball is exceedingly slight. It would probably be less than one electron per hundred thousand million atoms. If a pair of pith balls had one a deficit and the other an excess of as much as one electron per ten thousand atoms, they would attract each other with a force of a thousand tons from a metre apart and, if released, would strike each other with an energy equivalent to the explosion of about ten tons of T.N.T.

1.2 Composition of the Nucleus

The nucleus, like the atom, is a complex structure. It is built up of two kinds of particles of nearly equal mass. One, the proton, has a positive electric charge equal in magnitude but opposite in sign to that of the electron. The other, the neutron, has no electric charge at all. Like the electrons in the outer regions of the atom, both kinds of nuclear particle behave as though they were arranged in a series of "shells", the later ones being less tightly bound than the earlier. These shells, however, are much less distinct than the electron shells and show themselves mainly in a greater stability for the nuclei which have just completed a shell of either neutrons or protons. Thus, nuclei containing two, eight, twenty, fifty, eighty-two and a hundred and twenty-six of either neutrons or protons are more tightly bound than neighbouring ones and are called *magic number nuclei*. Nuclei such as that of ordinary helium, containing two protons and two neutrons, are "doubly magic", and outstandingly stable.

The protons in the nucleus are alone responsible for the electrical force which holds the electron-cloud forming the outer part of the atom. The neutrons contribute nothing to this, though they are very important in the nucleus as we shall see later. The simplest atom, that of hydrogen, consists of one proton, by itself constituting the nucleus, and one electron. All other atoms have nuclei containing both protons and neutrons.

One cannot reasonably expect to be able to visualise the properties of nuclear matter, as assemblages of protons and neutrons at normal nuclear densities may be called. The density of such matter is of the order of a hundred million tons per cubic centimetre. This is as much greater than the density of lead as the density of lead is greater than that of the residual gas in a high vacuum. Suppose we imagine intelligent beings, composed entirely of gases at very low pressures and having no direct acquaintance with matter at higher densities than that of air at 10^{-6} mm of mercury. They would not find it easy to visualise the properties of solids and liquids. The properties of nuclear matter may be just as different from anything we know, and are not simplified by the fact that its constituent particles are always in motion with velocities in the region of two-tenths of the velocity of light.

1.3 Nuclear Forces

The forces holding the protons and neutrons together clearly cannot be electrical. One or more neutrons might well be weakly bound to one proton by electrical forces, just as any neutral conductor is polarised and attracted to a charged body of either sign. Two protons, however, would repel each other strongly and stable nuclei with large numbers of protons would be impossible. In fact, lead has perfectly stable nuclei containing 82 protons and nuclei containing as many as 103 protons have been synthesised.

As already indicated, gravitational and magnetic forces are much too small to be responsible, and we must postulate the existence of some quite new, nuclear, attractive force to overcome the repulsive electrical one. A large part of modern nuclear research is concerned with the detailed nature of this force. It turns out to be a combined effect of several different forces all of them unfamiliar in ordinary life and all having a very short range of action (of the order 10^{-13} cm or less) so that it is very unlikely that they can ever be made to manifest themselves directly in macroscopic experiments.

That they have a very short range was clearly demonstrated by the original experiments in which Rutherford proved the existence of the nuclei. Helium ions (α-particles) were scattered from gold atoms exactly as they should have been if each were a point positive charge, twice and seventy-nine times respectively the charge of an electron. Thus at the distance of closest approach, perhaps 2×10^{-12} cm, the specifically nuclear forces were negligible compared to the electrical ones. Yet inside the nucleus, at distances only twenty times smaller, they must be *greater* than the electrical ones for the nucleus to stay together at all. This means that they are very large indeed. The electrical repulsive force acting on a single proton at the outside of the uranium nucleus due to the presence of the other 91 is easily calculated to be in the region of 20 kg wt. The nuclear attractive force must be more like 100 kg wt.

If we could make gross structures with the tensile strength of nuclear matter, a thread 1/1000 mm diameter could be made strong enough to lift the top 25 km of England and Wales out of the sea—thus offering a fresh solution to the controversy of Channel bridge versus Channel tunnel. Alternatively, if a ten-megaton H-bomb were exploded in a bag made of a sheet of nuclear matter only 10^{-12} cm thick, the bag would not burst.

With such forces involved it is not surprising that the energies of nuclear interactions are large. The usual range of such energies approaches a million times the usual range of energies involved in chemical reactions. It is thus possible, as we shall see later, to detect the interactions and changes of individual nuclei. Hence, instead of considering reaction-energies per gram or per gram-molecule as in normal chemical practice, it is usual to consider the energy of each individual event. The unit used is based on the amount of energy transferred to an electron in passing through a potential difference of one volt. This is called an *electron-volt* (eV), and is $1 \cdot 60 \times 10^{-12}$ ergs. One eV per molecule is equivalent to $23 \cdot 2$ kilocalories per gram-molecule. The electron-volt is thus of quite a suitable size for use in connection with chemical reactions. For the much higher energies of nuclear reactions, however, it is inconveniently small and the usual unit used is the MeV or million electron-volts.

An MeV is still not very large in c.g.s. or practical units, being $1 \cdot 6 \times 10^{-6}$ ergs or $1 \cdot 6 \times 10^{-13}$ joules. It is so large for the tiny particles concerned, however, that an energy change of a few MeV will not only produce drastic effects on nuclear structure but will also produce observable effects on the masses of the nuclei affected.

1.31 *Mass and energy*

Before the atomic nucleus was discovered, Einstein had shown that energy, in any form whatever, has a corresponding mass. Thus a wound-up watch is a little heavier than an unwound one and a mixture of hydrogen and oxygen is a little heavier than the water which can be formed from it—when weighed at the same temperature. The mass m corresponding to an energy E is given by Einstein's equation $E = mc^2$, where c is the velocity of light. Since c is very large, the mass-equivalent of the quantities of energy normally encountered is exceedingly small; one joule is $1 \cdot 11 \times 10^{-14}$ grams or, looking at it the other way round, electrical energy at 1d. per kilowatt hour would cost about £100,000 per gram. The mass-change in a chemical reaction is unobservably small, but a nuclear energy-change may well have a mass as high as $0 \cdot 1 \%$ of that of the interacting particles, which is easily measurable. The kinetic energy of an electron ejected in a nuclear transformation may be several MeV, which will have a mass several times the original (rest) mass of the electron itself. This mass, $9 \cdot 1 \times 10^{-28}$ g, corresponds to $0 \cdot 511$ MeV; the proton rest-mass $(1 \cdot 67 \times 10^{-24}$ g) to 938 MeV.

It is worth stressing that there is no qualitative difference, in their relation to mass, between nuclear energy and such familiar forms as gravitational potential energy or heat. In many early texts it was erroneously stated, or implied, that the equation $E=mc^2$ applied specifically to nuclear energy. It was often stated also that the energy of, for example, uranium fission "came from" the loss of mass of the uranium nucleus which occurred in the process, and that the discovery of Einstein's equation made the discovery of fission possible. The first point is of course true in a sense, just as the energy of a steam-engine "comes from" the loss of mass occurring when coal combines with oxygen. In neither case does the knowledge that $E=mc^2$ contribute much to our detailed knowledge of the processes involved. The theoretical importance of the equation can hardly be overemphasised, but its contribution to experimental nuclear physics up to the discovery of fission was quite small and it was no more *essential* to the discovery of fission than it was to the invention of the steam-engine—or, for that matter, to the discovery of fire.

The only further point which is here worth making about the specifically nuclear forces is that detailed investigation shows them to be even shorter in range than the diameter of a very moderate nucleus. One nucleon (proton or neutron) is in effect attracted only by its immediate neighbours. This gives a large nucleus some of the properties of a liquid drop whose form is determined by surface tension. This effect makes it much easier to understand the process of fission described in a later chapter. It also shows that the threads and sheets of nuclear matter imagined above could not be maintained, in spite of their fantastic strength, any more than could similar structures made of water.

1.4 Isotopes

It has been stated that the chemical properties of any atom depend upon the electron-clouds forming the outer parts of the atom. The number of negative electrons comprising these must be equal to Z, the number of positive protons, in the nucleus. This number Z is called the atomic number.

The number, N, of neutrons in the nucleus has no effect on the number of electrons. All the atoms of any one chemical element thus have the same characteristic atomic number but may have different neutron numbers.

Any nucleus may be defined by these two numbers Z, N. In practice Z and the quantity $A=Z+N$ are more usually used; thus an atom or nucleus with six protons and six neutrons will be called "carbon 12"; the name carbon tells us that $Z=6$, since all carbon atoms must have the same atomic number and $A=Z+N=12$ is conveniently close to the atomic weight on the (old) chemical scale. The atom or nucleus with six protons and seven neutrons will similarly be called "carbon 13". Each of these types of atom will then be called an *isotope* of carbon. In formal equations they may be written $^{12}_6C$, $^{13}_6C$ (usual British practice) or $_6C^{12}$, $_6C^{13}$ (usual American and international practice). A few elements, such as aluminium, have only one stable isotope ($_{13}Al^{27}$) while others such as tin may have a great many ($_{50}Sn^{112}$, $_{50}Sn^{114}$, $_{50}Sn^{115}$, $_{50}Sn^{116}$, $_{50}Sn^{118}$, $_{50}Sn^{119}$, $_{50}Sn^{120}$, $_{50}Sn^{122}$, $_{50}Sn^{124}$). (Note that tin has 50 protons, a "magic" number.)

In common parlance the term "isotope" is usually used to mean only the radiactive isotopes, with which we shall deal in the next chapter. Reports in the daily press often imply that all isotopes are deadly dangerous. In fact, every possible atom is an isotope of some element or other. For obvious reasons the far rarer radioactive ones support a much more voluminous literature.

1.5 Nuclear Interactions

We conclude this chapter with a statement of the principles involved in considering nuclear interactions. The first three are familiar in everyday life, the remaining ones being of practical importance only in determining the structure of atoms and nuclei.

The normal laws of mechanics: conservation of energy, conservation of momentum and conservation of angular momentum are all obeyed. This, perhaps, is not surprising; for forty years or so it has been known that some or all of these quantities disappear in β-radioactivity but an unobservable particle (the neutrino) was invented to be blamed for carrying away the missing energy, etc. Direct detection of this particle was claimed for the first time as recently as 1959, but the evidence was not nearly strong enough to have been accepted if there were not other good reasons for believing in its existence.

The next principle is novel; it was first enunciated by Heisenberg and is known as the *Uncertainty principle*. This states that the momentum and position of a particle cannot simultaneously be exactly known. If the uncertainty in position in any direction x is δx and the uncertainty in momentum in direction x is δp, then $\delta x \times \delta p \geqslant h$, where h is Planck's constant, $6 \cdot 55 \times 10^{-27}$ erg sec. This is not of much importance in ordinary life; if we know the velocity of a one-ton lorry at a particular instant to 1 cm/sec (about $0 \cdot 1 \%$ at 20 m.p.h.) it would not matter that it was theoretically impossible to determine its position at the same instant more closely than $6 \cdot 5 \times 10^{-33}$ cm. For an electron it is however of great importance. The single electron in the hydrogen atom, in the ground state, has a kinetic energy of motion about the nucleus of 13·6 eV. The velocity corresponding to this energy is over 2000 km/sec, but the electron mass is so small that the momentum is only $2·0 \times 10^{-19}\,g$ cm/sec and its position cannot be defined to better than 3×10^{-8} cm. This figure is close to the "size" of the hydrogen atom and in fact one can fairly say that the uncertainty principle defines the effective sizes of atoms and, by a similar argument, of nuclei.

The uncertainty principle can also be written in terms of energy and time: $\delta E \,.\, \delta t \geqslant h$. This means that if the energy of an event is known with great accuracy the instant of its occurrence cannot be well defined. The significance of this will become clear later on.

The principle of quantisation of angular momentum is also of little importance in ordinary affairs, but important on the atomic scale. According to this, it is impossible for the angular momentum of any system to change except by integral multiples of $h/2\pi$ often written $\hbar$. This limitation is always strictly obeyed. Thus both proton and electron have an angular momentum of $\tfrac{1}{2}h/2\pi$ as though they were tiny spinning tops. It is impossible for either of them just to change the direction of the axis of its spin in an arbitrary way. They can stay the same or they can exactly turn upside down so that $\tfrac{1}{2}h/2\pi$ becomes $-\tfrac{1}{2}h/2\pi$, i.e. a change of $h/2\pi$. An atom or a nucleus may have a "spin" of an odd number of times $\tfrac{1}{2}h/2\pi$ if it contains an odd number of nucleons, but if it is excited by absorption of energy can change its total spin only by an integral multiple of $h/2\pi$.

The last principle we shall mention here is Pauli's *exclusion principle*. This states that in any given system it will be impossible for any two identical particles to be in exactly the same state, i.e. to have identical energy, momentum and angular momentum. This is found not to apply to all particles but does apply to the electrons, protons and neutrons with which we are concerned. It is obedience to this principle that leads to the limitation of numbers of electrons and nucleons in their various "shells".

CHAPTER TWO

RADIOACTIVITY

2 Stable and Unstable Nuclei

The vast majority of atomic nuclei in the world of today have remained entirely unchanged since the material composing this planet was assembled some five thousand million years ago. Many *nuclides*, or kinds of nuclei, are now known however, which after a short or long period of time change spontaneously. A nucleus capable of such a spontaneous change is known as a *radioactive nucleus*. The initial nucleus is often referred to as the *parent* and the resulting nucleus as the *daughter*. The known radioactive nuclides now outnumber the stable nuclides three or four to one and more are discovered every year.

Five or six types of radioactive change can be distinguished.

2.1 Beta-emission

The commonest and most important of these is known as β-*particle emission* or β-*emission*. In many ways a neutron and a proton behave as though they were two different states of the same basic nuclear particle. A free neutron is not permanently stable but eventually changes into a proton giving off an electron and a *neutrino* in the process. The neutrino is an almost unobservable particle with no charge. Since it has no positive practical effects in ordinary work with radioactive materials we shall usually neglect it and speak only of the ejection of an electron. Much energy is released at the same time and the neutrino and electron will be ejected with an energy of 750,000 electron-volts to be shared between them. This process can also occur inside a nucleus. For example, in the nucleus of carbon 14, one of the eight neutrons will eventually change into a proton, ejecting an electron but itself staying within the nucleus. The nucleus resulting has then seven protons instead of the original six and is an isotope of nitrogen, $_7N^{14}$. $_6C^{14}$ is then said to be a β-*active nucleus*.

This behaviour might lead one to suppose that a neutron is not really as fundamental a particle as a proton but is some kind of close combination of a proton and an electron. This is not so. The inverse change from proton to neutron can also occur inside a nucleus. Thus oxygen, like carbon, has a radioactive isotope of mass 14. In this case we start with eight protons and six neutrons. One of the eight protons will in due course change into a neutron, ejecting a neutrino* and a *positron*, also known as a "β^+", or a *positive electron*. We have then seven of each kind of nucleon and have again produced $_7N^{14}$.

The positron is a particle exactly similar to an ordinary electron except that its electric charge is positive instead of negative. It is not normally found free in nature, being created only in radioactive transformations or in other high-energy interactions. Once created, it would exist indefinitely, just as do ordinary electrons, if left to itself. In ordinary matter, however, it will soon meet a negative electron; the positive and negative pair then mutually annihilate each other, giving rise in the process to a pair of *photons* or quanta of radiation which are emitted in opposite directions from the point of interaction. If both particles are initially at rest, and finally have vanished, the energy available for the quanta of radiation must be given by Einstein's relation $E = 2m_0c^2$, where m_0 is the rest mass of either electron. Each quantum has then an energy m_0c^2, or 0·511 MeV, with a mass exactly equal to that of one electron. This of course is its moving mass. The *rest-mass* of a photon is zero; in other words, if brought to rest it ceases to exist.

Since the energy of the two quanta, when absorbed in matter, will eventually appear as heat, each positron produced in a radioactive change (or *decay*) will, with an ordinary electron from its surroundings, always finish up as a small quantity of heat. ($2·5 \times 10^{13}$ pairs would have to be annihilated to give one calorie.)

This is much nearer the "turning of mass into energy", which Einstein's equation popularly represents, than is the small percentage change of total mass which occurs when a constant number of nucleons are rearranged in a nuclear change.

Einstein's law, however, gives no hint that such a mutual annihilation *will* occur, any more than it explains why a proton and an electron should *not* annihilate one another, with a thousand times larger release of energy.

Its value lies in its ability to tell us *how much* energy will be available if the process does occur and if, as we do, we know the masses of the electrons involved.

2.2 Electron-capture

Another way in which a nucleus with an excess of protons can transform one of these to a neutron is by *electron-capture*. This occurs

* For the purposes of this chapter we do not need to distinguish between the neutrino produced in β^+-emission and the "mirror-image" anti-neutrino produced in β^--emission.

mainly in the heavier elements, in which the density of electrons at the nucleus is large. If energy is released thereby, an electron in the same atom may be absorbed by a proton in the nucleus. The electron usually comes from the innermost, K, shell but sometimes from one of the higher shells. This process is described as *K-capture, L-capture*, etc., as well as, more generally, electron-capture. The proton absorbs the electron and changes into a neutron; for example, the radioactive cobalt isotope $_{27}Co^{57}$ changes entirely by this process to $_{26}Fe^{57}$. This process may easily be energetically possible (i.e. energy may be released) when there is not enough energy available for positron emission. In electron-capture, energy corresponding to the mass of the absorbed electron is given to the nucleus while in the other case 0·511 MeV has to be provided by the nucleus for the creation of the positron. Hence about 1 MeV must be released by electron-capture before positron-emission also becomes possible.

Where plenty of energy is available, either process may be possible for the same nucleus. Thus, when $_{30}Zn^{65}$ changes to $_{29}Cu^{65}$, 0·325 MeV is available above what is needed to create a positron and about 2% of nuclei transform themselves this way while 98% change by electron-capture. A large release of energy favours positron-emission so that in the nearby nucleus $_{30}Zn^{63}$, which gives 2·36 MeV over the minimum threshold value for positron-emission, 90% of nuclei decay this way and only 10% by electron-capture.

We shall now consider the source of the energy released in radioactive decay. The nucleus has a diffuse shell-structure and the protons and neutrons it contains have each a set of discrete possible states with differing binding energies. The binding energies of the protons are somewhat less than those of the neutrons because, though the nuclear attractive forces are the same, the protons repel each other electrically while the neutrons do not (see Fig. 1). It is most unlikely in any nucleus that the binding energy of the last neutron should be the same as that of the last proton. This does not necessarily matter, but if the last neutron is not as tightly bound as a proton would be in the *next*, empty, proton-level, energy would be liberated if the neutron were to change into a proton and transfer to this level. Sooner or later this process will occur and will be repeated if it is still energetically favoured. Similarly, if the last proton is less tightly bound than the next neutron would be, the reverse change can take place. For any given number of nucleons, there will be a particular most stable number of protons. In light nuclei, where the electrostatic forces are not yet large, this will be close to the number of neutrons. In heavier nuclei the neutrons are more seriously favoured and in the heaviest elements there will be some 60% more neutrons than protons in a "*β-stable*" nucleus.

As in the case of the atomic electrons, each nuclear level can take a pair of particles of identical energy but opposite spin. Each member of a completed pair, whether of neutrons or of protons, will be more tightly bound than will be a single particle in the corresponding level. Hence the nuclides with even numbers of protons and neu-

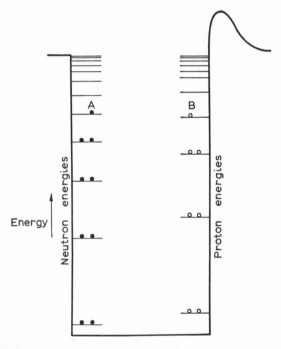

Fig. 1 Schematic arrangement of energy-levels in a *β*-active nucleus. Energy will be liberated if the odd neutron at level A emits an electron and becomes a proton in the lower energy-level B.

trons tend to be more stable than those with odd numbers. The isotope-chart shows that the elements with even numbers of protons have larger numbers of stable isotopes than those with odd numbers and that more of these stable isotopes have even numbers of neutrons. The total numbers are striking; of the *β*-stable isotopes listed up to bismuth (mid-1961), there are 162 even-even nuclei, 104 even-odd and only 4 odd-odd (all of which have 7 or less protons).

The energy released in a *β*-decay, beyond that required for *β*-

particle creation, escapes in several ways. In the β^- and β^+ processes, part appears simply as the kinetic energy of the ejected β-particle. In both β-emission and electron-capture, energy is given also to the other freshly created particle already mentioned, the neutrino. This is a particle like an uncharged electron, with the same spin of $\frac{1}{2}h/2\pi$, but with a much smaller, probably zero, rest mass. It is necessarily created simultaneously with a β-particle and the energy immediately available is shared between them, in what to the outside observer appears a completely random manner. Hence the β-particle may emerge with any kinetic energy from almost zero up to practically the whole of the surplus energy available from the transition. The β-particle alone is observed; it is a fair assumption, made in order to comply with the principle of the conservation of energy, that the rest of the energy is in each case taken away by the neutrino. It cannot however be said yet to be *experimentally* proven, though nuclei have been observed to recoil by an amount consistent with that to be expected from the ejection of high-energy neutrinos.

In many cases some part of the energy released is retained for a time by the nucleus. This occurs when, for example, a neutron, on changing to a proton, finds itself in an energy state above the lowest state available to the new proton (although, of course, of lower energy than the original state of the neutron itself). The nucleus is then left in an excited state and will in due course (usually in 10^{-13} sec or less) emit the energy of excitation as a quantum of γ-radiation. If there are several possible energy-states between that initially produced and the nuclear ground-state, a series of γ-quanta may be radiated as the system "steps down" from one state to another.

A nucleus that remains in an excited state for a measurable time is called an *isomer* of the same nuclide in its ground state. This ground state may itself be either stable or radioactive. Thus the stable $_{35}Br^{79}$ has a 4·8-second isomer emitting a 0·2 MeV γ-ray and the unstable $_{35}Br^{80}$ has a 4·5-hour isomer giving a ·037 γ-ray. Some isomers have expectations of life of many years.

The de-excitation of a nucleus need not always take place by γ-emission. The excitation energy may instead be transferred directly to one of the extranuclear electrons of the same atom. This electron is promptly ejected from the atom with a kinetic energy equal to the loss of excitation energy by the nucleus minus the electron's own binding energy in the atom. This process is known as *internal conversion* of the nuclear excitation-energy. Such an ejected electron could be mistaken for a β-particle directly created in a radioactive decay. In practice, when a reasonably strong source is available, the internal-conversion electrons are easily distinguished experimentally by the fact that they have all exactly the same energy, while the electrons resulting from true β-emission have a continuous distribution of energies up to a characteristic maximum, owing to the sharing between themselves and the simultaneously created neutrinos. At most, if two or three different γ-rays are simultaneously being replaced by conversion electrons, two or three discrete energies will appear.

By whichever mechanism the energy of excitation is lost, some nuclear excitation following β-emission might almost be said to be the rule rather than the exception. Only half a dozen or so β-emitters are known followed by none at all; for example the positron-emitter $_9F^{18}$ and the electron-emitter $_{15}P^{32}$.

Following electron-capture, the energy which is released immediately must all be taken by the created neutrino, since no electron is ejected. This neutrino cannot easily be detected, but evidence for its emission has been provided in several cases by observation of the recoil of the residual nucleus following the decay.

As in β-emission, electron-capture is usually followed by γ-emission, from the observation of which the decay may be inferred. If there are no γ-rays, the electron-capture can still be detected because it must necessarily be followed by X-ray emission. The electron absorbed will almost always be taken from one of the inner shells of the atom. When an electron from an outer shell falls in to fill the gap, a quantum of one of the characteristic X-radiations of the atom concerned is emitted. Note that the X-ray is characteristic of the resulting, not of the initial atom; thus when $_{31}Ga^{67}$ changes by K-capture to $_{30}Zn^{67}$ it is the K-X-ray of zinc which is subsequently emitted, not that of gallium.

Since the approximate wavelength of an X-ray is easily determined, and the characteristic X-rays of all elements are known and tabulated*, this gives us a powerful method for identifying the atomic number of an unknown electron-capturing nucleus.

2.3 Alpha-emission

We now consider an entirely different form of radioactivity which does not involve the creation of any new particles. This is α-*particle radioactivity*, or α-*emission*. In this, an α-particle, which is simply the nucleus of a helium atom ($_2He^4$), is ejected leaving a residual nucleus with two fewer protons and two fewer neutrons. α-emission does not occur at all in the first half of the table of elements, but it is common in the heavier elements above lead. It is of historical importance both because most of the naturally occurring radioactive nuclides are

* See Appendix V

α-active and because α-emission was understood much earlier than β-emission.

As the number of particles in a nucleus increases the individual nucleons are at first more and more tightly bound since each new particle can be attracted by an increasing number of others. The range of nuclear forces is so short, that the forces binding each particle increase very little, after the diameter of the (roughly spherical) nucleus has reached even four nucleons. The disruptive electrostatic forces, however, continue to increase more and more rapidly as the

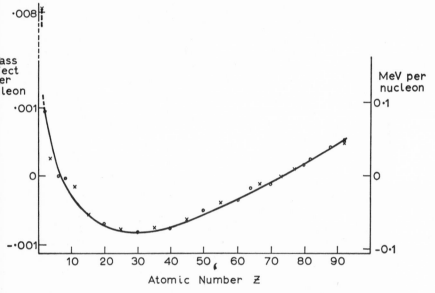

Fig. 2 Variation of mass-defect with atomic number. The nucleus $_6C^{12}$ is taken as the standard on both mass- and energy-scales.
O—Even-even nuclei.
X—Even-odd nuclei.

number of protons in the nucleus increases. Hence the total binding energy per nucleon passes through a flat maximum in the region of iron (element 26) and then begins to fall, increasingly rapidly as the atomic number rises.

In Fig. 2 is shown the variation of binding energy per nucleon with atomic number for the known elements.

The binding energy (negative, since energy would have to be supplied to take a nucleus apart into its constituent particles) may also be given in terms of its equivalent mass and this is also shown, in *milli-mass-units* in the same figure. One *mass unit* is one-twelfth of the mass of the C^{12} atom, which recently displaced O^{16} as the basis of chemical and physical atomic weights. It is equivalent to 931 MeV so that one milli-mass-unit is a little less than 1 MeV. The difference between the mass of a given nucleus and the sum of the masses of the free nucleons from which it is made is sometimes known as the *packing fraction*, but the term is not often used now.

From Fig. 2 it is clear that if two light nuclei were put together into a larger one, energy would be liberated. On the other hand, energy might be liberated by the breaking up of very large nuclei.

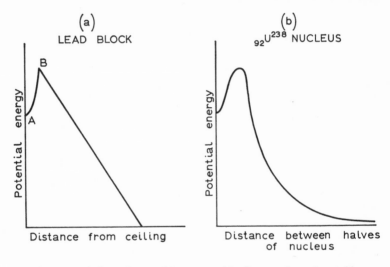

Fig. 3 Variation of potential energy with distance from the stable position for (*a*) a block glued to the ceiling, (*b*) a nucleus of U^{238}. The distance corresponding to A–B in Fig. 3a, which represents the range of elastic extension of the glue, is much exaggerated for clarity.

Apart from $_4Be^8$, the β-stable light nuclei represent completely stable assemblages of particles, while the heavy nuclei, above about rhodium (element 45) are only metastable. Thus, if for example an $_8O^{16}$ nucleus could be extracted from $_{47}Ag^{107}$ leaving $_{39}Y^{91}$, nearly 3 MeV would be liberated.

On the basis of classical mechanics, not even the heaviest nuclei would be actually unstable, although if a nucleus of uranium were broken in halves, some 200 MeV would be released.

The short-range nuclear forces holding such a nucleus together are still far stronger than the electrostatic forces pushing it apart. A similar picture would be shown by a lead block firmly glued to the ceiling. The glue might be capable of exerting an adhesive force many times the weight of the block, but this force would remain in being through less than 1/10 mm of downward displacement of the block before the glue came unstuck so that the energy absorbed by the glue during such a displacement would be negligible compared with the gravitational energy subsequently transformed into kinetic energy as the block fell to the floor.

The energetics of these two situations may be represented by very similar diagrams such as are shown in Fig. 3. To pass over the brief initial rise of potential energy requires an initial supply of extra energy from outside; the rise is appropriately described as a *potential barrier*.

At this point the analogy between the heavy nucleus and a lead block stuck to the ceiling breaks down. In the absence of an external impetus and assuming a permanent glue (and a permanent ceiling, not usually put up by builders) the lead block would remain in position for ever—the nucleus will not. According to our best picture, the wave-mechanical one, a nucleus, or an associated group of nucleons such as an α-particle, does not have an exactly defined position in space but is distributed over a finite region. This distribution is not cut off sharply by the potential barrier although it is very rapidly attenuated by it. This is schematically illustrated for an α-particle in Fig. 4. (α-particles may not necessarily have a permanent existence in a large nucleus, but it is clear that such sub-groups of nucleons are frequently formed and persist for significant times; during such times they behave as though they constitute a single particle.) The distribution-function in the lower part of the figure represents the calculated solution of the wave-equation for an α-particle in a nucleus. The potential-energy distribution due to the rest of the nucleus is shown in the upper part of the figure. The potential energy, E, of the α-particle, is positive with respect to the outside world but negative with respect to the top of the barrier. The height of the distribution function may be taken to represent the probability that the α-particle may be found at the corresponding distance from the nuclear centre. Classical mechanics would allow it to be anywhere inside the limits shown but would give zero probability anywhere outside this region as it would be utterly impossible for it to "climb" the barrier without some external supply of energy. Wave-mechanics, however, predicts a finite probability of its being within the thickness of or even outside the barrier. If it does find itself outside, it will be repelled away. From the point of view of the external observer, it will simply have been shot out of the nucleus with kinetic energy E.

This is the fundamental basis of α-particle radioactivity. The *possibility* of such radioactivity exists for most nuclides from the region of barium (element 56) upwards, and for practically all nuclides above the middle of the rare earths. The *probability* of emission, however, is exceedingly small for most nuclides below polonium (atomic number 84). The wave-mechanical attenuation through a high and wide barrier is very great. The shape of the potential barrier does not vary fast from one element to the next, so that the main factor determining the height and width to be penetrated is the height

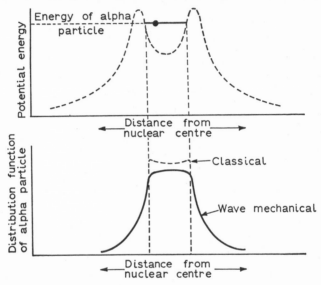

Fig. 4 The upper figure shows the potential energy of a stationary α-particle as a function of distance from the centre of an α-active nucleus. The heavy line shows the energy-level of an α-particle which has also, at the centre of the nucleus, appreciable kinetic energy.

The lower figure shows the probability that the particle should find itself at any particular distance from the centre.

of the energy-level of the α-particle. This also represents the kinetic energy which the particle will have after emission. Consequently the probability of emission increases very rapidly with the energy of emission. This energy, by the conservation principle, clearly depends on the difference of binding energy between the original and final nuclei. It thus increases with the steepness of the binding-energy curve of Fig. 2 and, correspondingly, so does the probability of α-emission.

Similar considerations determine the kinds of particle which have a reasonable chance of being emitted. Most nuclei above rhodium (45) are unstable against $_8O^{16}$-emission, i.e. energy would be released if an oxygen nucleus were ejected. With sixteen nucleons to hold on to however, the potential barrier against $_8O^{16}$-emission is so high and so wide that such emission is exceedingly improbable. For many nuclides the improbability is such that it is 10^{20} or so to 1 that no such atom in the known universe has yet broken down in this way. Even if we don't then say that the breakdown is impossible, it will do very well until something impossible comes along. Although the energy released by oxygen- or carbon-emission would often be greater than that released in α-emission, the latter, with its smaller potential barrier, is far more probable in all practical cases; in fact, neither of the former has ever been observed.

2.4 Spontaneous Fission

In another form of natural radioactivity, *spontaneous fission*, a large nucleus, such as that of $_{92}U^{238}$ or of the still heavier artificial elements, breaks apart into two parts of comparable size. The potential-barrier against such a breakdown is high, but the energy available for release is enormous—some 200 MeV—and hence the probability is not undetectably small. It rises with atomic number even faster than does the probability of α-emission. From being unobservable for elements below about thorium (90) it becomes one of the most probable forms of decay by the time we reach fermium (100) and for still heavier elements becomes so rapid that it is likely to form the main practical limit to the production of elements much above 103. Further discussion of fission, which is a much more complex process than indicated here, will be deferred until Chapter 8.

2.5 Quantitative Aspects of Radioactive Decay

We have so far discussed the probability of any form of radioactive decay only in a general way. For any one radioactive species or nuclide the probability is a characteristic quantity best expressed in terms of the *decay constant* λ. This is defined by the equation

$$\frac{dN}{dt} = -\lambda N \qquad 2.1$$

where N is the number of nuclei of the substance concerned at time t and dN/dt is the rate of change of N, i.e. the number of breakdowns per second. The negative sign represents the fact that N gets smaller as t increases. It must be noted that the equation, like any other concerned with probability, is truly valid only for indefinitely large values of N. The practical importance of this limitation will be discussed later.

A more fundamental point to note is that the equation involves only the number of radioactive atoms, taking no account of their age or history. This corresponds with the experimental fact that the probability of decay of any one atom is exactly the same whether it is newly formed or has existed for a long time. This is quite different from most large-scale structures. A man who has already lasted eighty years is much less likely to survive for a further eighty than is a man who has so far lasted only ten.

The decay-constant is not only unaffected by the age of the surviving atoms at any time but is unchangeable by any ordinary chemical or physical agency. This is the natural result of the lonely isolation of the nucleus of an atom from the outside world by its entourage of electrons. The only exception to this is decay by electron-capture. In the light atom $_4Be^7$ the electron distribution in the L-shell can be altered drastically by the state of chemical combination and even the K-shell may be secondarily distorted. As the total concentration of electrons at the nucleus is changed, the probability of electron-capture is changed. Even in this very favourable case the effect is small. In no other case has any effect been observed at all, though it must in principle exist for other electron-capturing nuclides. Except for the solitary case of $_4Be^7$ therefore, the quantity λ in equation 2.1 is for practical purposes absolutely constant for any radioactive substance.

By integration of 2.1 we can find the number of nuclei which will survive at any later time. Suppose that at time t_0 we have N_0 nuclei. Then from equation 2.1,

$$t = -1/\lambda \int_{N_0}^{N} \frac{dN}{N}$$

$$t - t_0 = -\frac{1}{\lambda}[\ln N]_{N_0}^{N} \; *$$

$$t - t_0 = -(1/\lambda) \ln (N/N_0)$$

$$N = N_0 \exp [-\lambda(t - t_0)]$$

* Writing $\ln N$ for $\log_e N$.

If we measure time from the instant when there were N_0 nuclei, i.e. if we put $t_0 = 0$.

$$N = N_0 \exp (-\lambda t). \qquad 2.2$$

This shows that the same time will be taken for the same *proportion* of nuclei to decay whatever the absolute number present may be. The variation of N/N_0 with t can be measured and this enables us to calculate λ.

The probability of decay of a given nuclear species is usually given in terms of the *half-life* rather than the decay-constant. This is simply the time it takes for half of any given initial number of nuclei to decay. The relationship of half-life to decay-constant is found from equation 2.2 by putting $N = \frac{1}{2}N_0$. Then if we call the half-life $\tau_{\frac{1}{2}}$

$$\frac{1}{2} = \exp(-\lambda\tau_{\frac{1}{2}})$$

or

$$\tau_{\frac{1}{2}} = (\ln 2)/\lambda \qquad 2.3$$

The *average life* $\bar{t}$ of a nuclide is also used occasionally. This can be found from equations 2.2 and 2.3 to be given by

$$\bar{t} = 1/\lambda \qquad 2.4$$

Hence $\tau_{\frac{1}{2}} = \bar{t} \ln 2$, and the half-life is $0.693\ldots$ times the average life.

2.51 *Alternative modes of decay*

It is possible for a given kind of nucleus to decay in more than one way. Thus $_{29}Cu^{64}$ can decay in three ways; by β^--emission (38 %) to give $_{30}Zn^{64}$ or by β^+-emission (19 %) or electron-capture (43 %) to give $_{28}Ni^{64}$. This does not affect the definitions of decay-constant or half-life which we have given as they depend only on the number of the original nuclei which survive and not at all on what happens to those which do not survive. It is clearly of interest, however, to know the separate decay-probabilities for the three modes of decay. We can describe these in terms of three *partial decay-constants*,

$$\lambda^-, \; \lambda^+ \text{ and } \lambda^e$$

such that $\lambda = \lambda^- + \lambda^+ + \lambda^e$.

We can also define *partial half-lives* as

$$\tau_{\frac{1}{2}}^- = (\ln 2)/\lambda^-, \; \tau_{\frac{1}{2}}^+ = (\ln 2)/\lambda^+, \; \tau_{\frac{1}{2}}^e = (\ln 2)/\lambda^e.$$

These are the half-lives which the Cu^{64} would have had if only one mode of decay were possible.

Clearly $1/\tau_{\frac{1}{2}}^- + 1/\tau_{\frac{1}{2}}^+ + 1/\tau_{\frac{1}{2}}^e = 1/\tau_{\frac{1}{2}}$. \qquad 2.5

2.52 *Factors which determine half-lives. Beta-decay*

Experimental methods of measuring half-lives are described in Chapter 7.

The range of observed half-lives for β-decay extends from 22 milliseconds for $_5B^{12}$ to 6×10^{14} years for $_{49}In^{115}$, the ratio being about 10^{24} to one. The reasons for this enormous range are interesting.

B

2.521 *Energy*. The first factor is simply the amount of energy released in the change. Other things being equal, the half-life becomes rapidly shorter as the energy available increases. Thus the change $_5B^{12\beta-} \rightarrow {_6}C^{12}$ releases 13·4 MeV while the long-lived β-emitters will usually release only a fraction of one MeV. This, however, is not the only factor. One of the lowest-energy β-emitters is the naturally occurring $_{82}Pb^{210}$

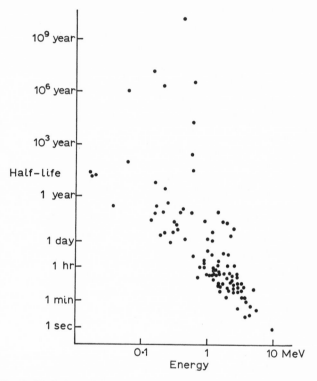

Fig. 5 Variation of half-life of β-emitters with energy of emitted particle. Only those emitters have been shown which go to a single level of the daughter-nucleus.

which releases only 0·017 MeV and has a half-life of 19·4 years. On the other hand the indium isotope $_{49}In^{115}$ already mentioned, with half-life 6×10^{14} years, releases some 30 times as much; about 0·5 MeV.

If we plot half-life on a logarithmic scale against energy as in Fig. 5, the β-emitting nuclides mostly lie in a fairly well-defined band. A number of nuclides, however, lie off this band on the side representing a higher half-life, some of them a long way off.

2.522 *Spin*. This divergence from a simple pattern arises from changes of angular momentum in the decay. The initial and final nuclei will each have a perfectly definite total angular momentum which will, as we saw in Chapter 1, be an integral multiple of $h/4\pi$. If these spins are the same, no difficulty of conservation arises because, although the β-particle and the neutrino have each a spin of $h/4\pi$, these can be in opposite directions (*antiparallel*) with no net effect on the system remaining behind. Again, if the spins of the initial and final nuclei differ by 1 unit, $h/2\pi$, no difficulty occurs, since the β-particle and neutrino can leave with parallel spins in the appropriate direction to compensate the change. If, however, there is a spin-change of two or more units this simple method of compensation does not work. Such a transition is said to be *forbidden*.

The created particles must then leave with some orbital momentum about the centre of the nucleus as well as with parallel intrinsic spins if total angular momentum is to be conserved. In other words, instead of being ejected directly from the nucleus along a line through its centre, at least one must leave along a line which does *not* pass through the nucleus so that its linear momentum has a moment about this. There is then in effect a torque on the nucleus which can transfer the angular momentum required. The probability of β-emission is much reduced, or the half-life increased, by this requirement. The amount of the effect depends on how much angular momentum has to be transferred. If the transition has a spin-change of two units, one unit must come in this way from orbital momentum. The transition is then called *once forbidden*, and the half-life is increased about a hundred times. If the total spin-change is three units, two units must be derived from the displacement of the effective flight-paths of the particles; the transitions are then said to be *twice forbidden*, and so on. Each extra unit of spin to be found decreases the emission probability by about one extra factor of a hundred. The very long half-lives characteristic of the group of nuclides in Fig. 5, furthest from the main band, correspond to a spin-change of 4 units. Thus for example $_{49}In^{115}$ has a spin of 9/2 units while $_{50}Sn^{115}$, to which it decays, has a spin of only 1/2.

It will be seen from Fig. 5 that the energy-release and the half-life of a new isotope provide between them a strong hint as to the spin-change involved in the transition. Similar considerations apply to the half-lives of positron-emitters.

2.53 *Electron-capture*

In the case of electron-capturers the mechanism is slightly different. Here only one particle, the neutrino, can be emitted "off-centre", but on the other hand in a heavy atom there is a choice of electrons to capture which already possess a series of different angular momenta.

We saw earlier that, where there is enough energy available, the K-electrons are more likely to be captured, being concentrated more closely round the nucleus. For a spin-change of 0 or 1, no other factor comes in. If there is a spin-change of two or more in the transition, however, one unit of spin can be obtained by capture of a suitable member of the L-shell. The relative likelihood of this is then greatly enhanced. Since only a neutrino is ejected, the energy-release may be difficult to determine. The ratio of L-capture to K-capture may then take the place of the half-life-energy relation as a guide to the spin-change which occurs.

Both positron-emission and electron-capture occur in many nuclides. As we have seen, when positron-emission is energetically possible, electron-capture must always be possible as well, since 0·511 MeV of energy are then added from the electron rest-mass to that available for release while in β^+-emission an *extra* 0·511 MeV must be found for the positron rest-mass.

Above the threshold, high energy favours β^+-emission but a large spin-change may favour electron-capture. With an extra 1·02 MeV available, a larger number of excited states in the daughter nucleus may be accessible, some of which may have spins closer to that of the parent nucleus. In this case particular γ-rays may appear, following electron capture, which cannot follow β^+-emission from the same nuclide. A more important factor in determining the relative probability of β^+-emission and electron-capture, when plenty of energy is available, is the atomic number of the radioactive nucleus. The K-electron density at the uranium nucleus is twelve thousand times as great as it is at the nucleus of beryllium—the lowest element with an observable neutron-deficient isotope. Thus electron-capture is much favoured among the heavier elements. This is illustrated in Fig. 6.

2.54 *Alpha-decay*

In α-emission there is a rapid increase of the probability of leakage through the nuclear potential barrier as the initial energy of the α-particle increases. This leads to an even more rapid variation of half-life with energy than in β-emission. Thus half-lives from 3×10^{-7} sec for $_{84}Po^{212}$ to 2×10^{15} years for $_{60}Nd^{144}$ are already known for a range of from 8·8 to 1·9 MeV. The longer half-life given is certainly not the greatest in existence. It represents only the limits of our present capacity for measurement. As indicated earlier, *all* nuclides in the higher part of the periodic table are unstable to α-emission. Many longer half-lives will no doubt be measured quite soon, but it will be a long time before activities as low as, say, one α-emission per ton per century can be detected or measured, though they may easily exist. The effect of energy on half-life is much less obscured by the influence of spin than is the same effect in β-emission. The momentum of a 5 MeV α-particle is nearly 40 times that of a 5 MeV β-particle so the displacement of its line of flight from the centre of the nucleus, to transfer a given twist, will be correspondingly smaller. The variation of half-life with energy is nearly exponential as shown in Fig. 7. This exponential relation is known as the *Geiger-Nuttall law*.

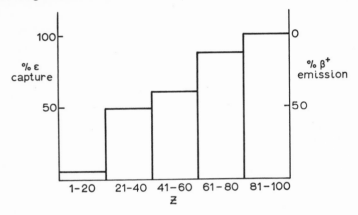

Fig. 6   Block diagram to show the increase in probability of electron-capture, as compared to β+-emission, with increase in atomic number of the active nucleus. All neutron-deficient nuclei have been included for which data were available up to the end of 1961.

2.55   *Radioactive series*

As has been shown, all radioactive materials decay according to a simple exponential law. A mixture of different activities may often be sorted out by analysis of the complex decay curve into a series of separate exponential curves as described in Chapter 7. If this is difficult, the different nuclides can be separated chemically or methods of production devised which will make them separately in the first place.

A special case in which this procedure is difficult, or even useless, arises when one activity grows out of the other. Many active nuclides are so far from stability that the daughter- as well as the parent-nuclei are radioactive. There may be a whole series of successive decays before a stable end-product is reached. Thus some fission-product nuclei have so large an excess of neutrons that half a dozen

β^--decays may occur in succession. The activities need not all be of the same kind. In the natural radioactive series starting with $_{92}U^{236}$, eight α-decays and six β^--decays occur before the stable nucleus $_{82}Pb^{208}$ is reached. In an old uranium ore the whole series of products will be found together in various quantities.

We shall here investigate the behaviour only of a series consisting of two active substances. If we begin with N_{10} atoms of the parent

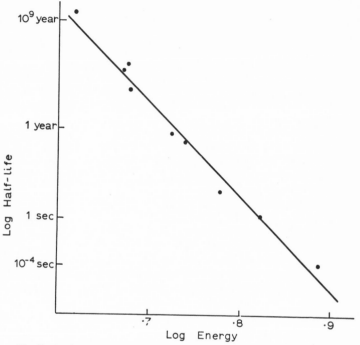

Fig. 7 Variation of α-particle energy with half-life for the U^{238} decay-series. The nearly linear relation forms the Geiger-Nuttall Law.

nuclide, with decay constant λ_1, after time t, the number N_1 of atoms surviving will be

$$N_1 = N_{10} \exp (-\lambda_1 t). \qquad 2.6$$

If the number of atoms of the daughter nuclide, decay constant λ_2, at time t is N_2, in a time δt, $\lambda_1 N_1 \delta t$ new atoms are produced from the parent but $\lambda_2 N_2 \delta t$ are lost by decay. Hence

$$\delta N_2 = \lambda_1 N_1 \delta t - \lambda_2 N_2 \delta t$$
$$= \lambda_1 N_{10} \exp (-\lambda_1 t) \delta t - \lambda_2 N_2 \delta t$$

whence $\qquad dN_2/dt = \lambda_1 N_{10} \exp (-\lambda_1 t) - \lambda_2 N_2$

The solution of this is, if we suppose that at time $t=0$ there were no daughter-atoms,

$$N_2 = \lambda_1 N_{10} \exp(-\lambda_2 t)\{1 - \exp(\lambda_2 t) \exp(-\lambda_1 t)\}/(\lambda_1 - \lambda_2) \qquad 2.7$$

(see for example Milne-Thomson and Comrie, Standard Four Figure Tables—differential equations).

What is observed experimentally may be either $\lambda_1 N_1$ (the rate of breakdown of the parent) $\lambda_2 N_2$ (the rate of breakdown of the daughter) or $\lambda_1 N_1 + \lambda_2 N_2$ according to whether one or both of the decays are recorded by the detecting equipment used. We shall consider all these cases.

2.551 If the daughter-decay is not detectable, observation becomes that of a simple exponential decay of a single nuclide, which has already been examined.

2.552 If we observe *only* the decay the daughter (as might happen, for example, if the parent was an electron-capturer and the daughter a positron-emitter), equation 2·7 represents directly the variation of activity with time.

It will be noticed that of the two terms involving t, the first represents a *fall* of activity with time and the second a *rise*. If t is sufficiently small, the second term will always be in control. Initially therefore, the observed activity will rise. This is physically obvious; at first there are no daughter-atoms to decay, but as soon as some parent atoms have decayed some will be created.

The subsequent shape of the decay curve depends very much on the values of λ_1 and λ_2. Some special cases will be considered.

2.5521 $\lambda_2 \gg \lambda_1$ (i.e. the half-life of the daughter is much shorter than that of the parent).

Thus, while $\lambda_1 t$ is small, so that we can write $\exp(-\lambda_1 t) \simeq 1$, equation 2·7 reduces to

$$N_2 = \lambda_1 N_{10}\{\exp(-\lambda_2 t) - 1\}/(-\lambda_2)$$

$$\therefore \qquad \lambda_2 N_2 = \lambda_1 N_{10}\{1 - \exp(-\lambda_2 t)\}$$

This shows a continuous rise of the observed quantity $\lambda_2 N_2$ to the limiting value $\lambda_1 N_{10}$ with a corresponding rise of the actual number N_2 of daughter atoms to the limiting value $(\lambda_1/\lambda_2)N_{10}$. In other words, the rate of decay of the daughter goes steadily up till it settles down to a value identical with the rate of decay of the parent. The daughter is then said to be in *secular equilibrium* with the parent. Physically, one daughter-nucleus on the average is being produced for each daughter-decay, keeping the total amount constant. Such cases are common among the natural radioactive materials. Clearly, observation of the daughter alone from the beginning of the experiment can give both the half-life of the daughter (from the rate of build-up to equilibrium) and the steady decay-rate of the (possibly unobservable) parent.

By the time that $\lambda_1 t$ is no longer small, the initial build-up of daughter-activity will be over and exp $(-\lambda_2 t)$ will be very small (since $\lambda_2 \gg \lambda_1$). Then equation 2·7 reduces to

$$\lambda_2 N_2 = \lambda_1 N_{10} \cdot \exp(-\lambda_1 t)$$

so that the daughter finally appears to decay at a rate which is actually that of the parent. The daughter is then said to be in *transient equilibrium* with the parent. This is of extreme practical importance as, if this result is forgotten, it is very easy to mis-identify an activity on the assumption that the decay-rate measured is that of the actual substance observed.

2.5522 The next special case is that when $\lambda_1 \gg \lambda_2$. Examination of equation 2·7 shows at once that in this case we get a very rapid rise of the number of daughter-atoms to N_{10}, followed by the decay given by

$$\lambda_2 N_2 = \lambda_2 N_{10} \exp(-\lambda_2 t).$$

Physically this means that if the half-life of the parent is relatively very short, then all parent atoms will have changed to daughters before any appreciable number of the latter have time to change, after which they decay according to the ordinary law for a single substance.

2.5523 The case of $\lambda_2 = \lambda_1$ is of mathematical interest as, for this, equation 2·7 gives an indeterminate result. By finding N_2 as $(\lambda_1 - \lambda_2)$ is allowed to tend to zero in equation 2·7, or by going back to the derivation of the equation itself, it can be shown that for this case, where $\lambda_1 = \lambda_2 = \lambda$

$$N_2 = N_{10} \lambda t \exp(-\lambda t)$$

so that $\qquad \lambda N_2 = \lambda^2 N_{10} t \exp(-\lambda t),$

which gives a decay which never becomes exponential although for large values of λt it approaches this. Such a case could easily be mistaken for one in which a mixture was being examined of a large number of nuclides with slightly different half-lives.

2.553 When both parent and daughter are simultaneously observed, experiments on the ejected particles will show $\lambda_1 N_1 + \lambda_2 N_2$ where

$\lambda_1 N_1 = \lambda_1 N_{10} \exp(-\lambda_1 t)$ and $\lambda_2 N_2$ is found from equation 2·7. The practical analysis of the resulting decay curves will be left to a later chapter. Here we will consider only one rather odd special case* which again could lead to confusion if it were not suspected. This occurs where the half-life of the parent material is just half that of the daughter. Thus $\lambda_1 = 2\lambda_2$ and on substituting for λ_1 we find

$$\lambda_1 N_1 + \lambda_2 N_2 = 2\lambda_2 N_{10} \exp(-\lambda_2 t),$$

the two terms in $\exp(-2\lambda_2 t)$ cancelling out. Thus the combined system gives a decay with exactly the same rate of fall as would be given by the daughter alone but appearing to begin with twice as many daughter-nuclei as were actually present of the parent-nuclei at the start of the experiment. Since the decay which is experimentally observed would look throughout like a perfect exponential, characteristic of a single radioactive species, there would be nothing to arouse the suspicion that two might be concerned. Once suspicion was aroused it could be confirmed either by chemical separations done at different times after the beginning of the experiment or by setting up a detector-system which would discriminate—even if only partially—between the particles or radiations emitted by the two substances.

More complex cases occur where both parent and daughter activities are observed but with different efficiencies. The general equation for these cases gives no further insight into the physics involved and the few practical cases of interest will be dealt with in later chapters as they arise.

2.6 Units

We conclude this chapter with a note on the units in which radioactivity is measured. The quantity of a long-lived radioactive isotope can conveniently be given in grams. The quantities of short-lived radioisotopes which are usually used are, however, unweighably small. For example, a convenient sample of the widely used nuclide $_{15}P^{32}$ (half-life 14.3 days) might give ten thousand disintegrations a minute. Its weight would be $1\frac{1}{2} \times 10^{-14}$ gm. Since we cannot measure such a small weight and since in any case it is the rate of disintegration with which we are concerned a new unit, the *Curie*, has been defined to measure this. Originally it was defined as the quantity of any radioactive substance which had the same number of disintegrations per unit time as one gram of radium 226. More recently, to meet the need for a unit which would not change every time a new measurement was made of the half-life of radium 226,

*Drawn to my attention by Dr. K. F. Chackett.

B*

the Curie has been redefined as *the quantity of any radioactive nuclide in which the number of disintegrations per second is* $3 \cdot 700 \times 10^{10}$. This is rather large for most laboratory work; the millicurie (mCi) and microcurie (μCi), $3 \cdot 7 \times 10^7$ and $3 \cdot 7 \times 10^4$ disintegrations per second respectively, are more often used.

Another unit which is sometimes employed for the same purpose is the *Rutherford*, representing 10^6 disintegrations per second.

CHAPTER THREE

DETECTION OF PARTICLES
I. GAS DETECTORS

When a charged particle passes through a gas at high speed it will excite some of the gas atoms and ionise others. This ionisation is the basis of most of the devices used to detect the fast charged α- or β-particles emitted in radioactive decay.

The minimum energy required to produce an ion pair in a gas depends upon the nature of the gas but is usually in the region of 10 eV. The efficiency of ionisation by fast charged particles is however not very high, much of the energy going into atomic excitation and elastic collisions. The average energy dissipated per ion pair produced is therefore usually several times the minimum, being between 25 and 35 eV for most of the common gases. A particle with energy 1 MeV will thus produce some 30,000 ion pairs, the charge on which is much easier to observe than the one or two charges of the original particle.

3.1 Ionisation Chambers

The device which is simplest in principle for the observation of these ions is the ionisation-chamber. This is a vessel filled with air or other gas through which a particle to be detected can be made to pass. The ions produced by its passage can be separated by an electric field and those of one sign collected on an insulated plate. The resulting change of voltage on this plate can then be recorded. The principle is illustrated in Fig. 8. A difference of potential V is maintained between the case and the collector. If the collector is positive, electrons or negative ions produced in the gas will move to it, while positive ions will be driven to the case.

The collectors of the early ionisation chambers were simply connected up to an electrometer which recorded the total charge received. This system was barely sensitive enough to detect the entry of a single charged particle, even when this had an energy of several MeV. The capacitance of the collector plus electrometer might be, say, 50 e.s.u. (roughly 50 pF) and a good quadrant electrometer would need 10^{-6} e.s.u. (3×10^{-4} volts) of potential for a measurable deflection. The charge needed for such a deflection would then be 5×10^{-5} e.s.u. or 10^5 electron charges, requiring 2·5 to 3·5 MeV for the original particle. Such a system would have a response time of the order of a minute so that observation of individual particles would be exceedingly slow.

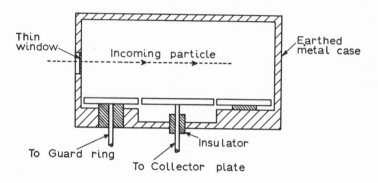

Fig. 8   Diagram of Ionisation Chamber.

3.11 *Ion chambers as monitors*

The simple ionisation chamber is still useful for the detection of large numbers of particles. Where the particles are entering at a low rate, the system can be sufficiently well insulated to indicate reliably the change of voltage after collecting for many hours. When they are entering at a rate of thousands a second, a relatively in-sensitive instrument with a much quicker response can be used to record them.

The ion chamber is often used as a monitor to show how long it is safe to work with radioactive materials. A typical modern instru-ment with quick response is shown in Plate I. The insulated central rod is charged to about 100 volts from a suitable D.C. supply. This deflects the fibre to one end of the scale, conveniently observed through a lens in the end of the main tube. Ionisation in the chamber at the end of the instrument discharges the central rod, causing the fibre to move across the scale by an amount proportional to the number of ions collected. A full-scale deflection is produced by less than 1 e.s.u. of charge (2.10^9 electronic charges) and insulation is so good that the fibre moves over only one per cent. or less of the scale per day in the absence of outside radiation sources.

3.12 *Ion chambers for single particles*

An ionisation chamber can be used satisfactorily for detection of individual particles if the small induced voltages are amplified. The gain required is high, a million times or more, if the output of the amplifier is required to operate a recording instrument with reasonably quick response. This is expensive and hence the ionisation chamber is not now used simply to count charged particles. It is still of great value however for the simultaneous recording of particles and measurement of their energy.

3.121 *Conditions of operation.* As was stated above, the number of ion-pairs produced by a fast particle in a gas is proportional to the energy which is dissipated in the gas. If therefore we can collect all the ions produced by a charged particle as it is brought to rest, we have a measure of the original energy of the particle. To be effective, the ionisation chamber must collect all (or a fixed known percentage) of the ions produced and the amplifier must be linear or at least have a known variation of gain with signal-size.* The first requirement limits the system, for practical purposes, to the observation of α-particles. A β-particle of even ½ MeV will travel some hundred centimetres in air at ordinary pressures and would need a very large chamber to ensure its being brought to rest in the gas without striking one of the walls. Furthermore, this would lead to the use of very large voltages. Recombination of positive and negative ions occurs rapidly and this can be prevented only by the use of very strong electric fields to separate the ions quickly and to keep them moving fast. Fields of the order of 10,000 V/cm are needed to prevent appreciable recombination, and these are troublesome to maintain over a large volume.

The range of α-particles is however much less, even the fastest likely to be encountered from radioactive substances (say 9 MeV) having a range in air at S.T.P. of only 8·8 cm. This can be reduced by increasing the pressure, and by using a denser gas than air such as argon, krypton or even xenon. In argon at two atmospheres the range of a 9-MeV α-particle will be under 4 cm. The rise of pressure which makes possible a smaller chamber, while well worth while, has a disadvantage. This is that it requires a corresponding increase of electric field to prevent recombination of ions. It is better not to prevent such recombination entirely. A much lower field will reduce it to a small percentage. So long as this is known and constant it can be allowed for. To keep it constant the electric field in the working volume of the ion chamber must be everywhere the same, but this is much easier to achieve than an arbitrary distribution of

* We shall not discuss the construction of the electronic circuits used in particle detection as this is an extensive and specialised subject needing a book to itself. Circuit diagrams for some well-established types are given in Appendix 1. A small selection of references is given in the Bibliography at the end of the book.

field which would at its lowest point be large enough to prevent detectable recombination. The proportion of ions lost can be allowed for by calibration of the system with α-particles from a radioactive material giving a known release of energy. This method is in fact invariably used nowadays as it eliminates the need for accurate measurements of amplifier gain or of the response characteristics of the final recording system. Use of several known α-emitters for calibration enables us to check also the linearity of the entire equipment.

3.122 *Ion chamber for measurement of energy.* In a chamber suitable for the measurement of α-particle energies, the electrodes form a parallel-plate structure and the electric field is kept uniform by the use of a *guard-ring* round the collector which is kept at a potential very close to that of the latter. The radial extent of this guard-ring must be large enough to shield the working volume from the electrical effects of the surrounding earthed case. For this it is usually sufficient to make it 0·8 times the distance between the plates.

The radioactive source will usually be placed at the centre of the collector or of the opposing plate. It is essential that the working radius of these plates, and also the distance between them, should be greater than the range of any α-particles produced. If an α-particle strikes the metal of one of the plates before it comes to rest, some of its energy will be dissipated in the metal without producing any collectable ions so that the energy recorded will be too small. Similarly, if it passes outside the radius of the collector-plate, some of the ions produced will be lost to the guard ring.

The thickness of the radioactive source itself must be very small. Half of the particles emitted must inevitably enter the base on which the source is deposited and will then be undetectable. Of the remaining particles, even those which come off nearly parallel to the surface must lose only an unimportant part of their energy in the source material if their total energies are to be properly determined. A source deposit of 1 mg/cm² will absorb about 1 MeV of the energy of an α-particle even if this is passing perpendicularly through it so that it is desirable to limit the thickness to, at most, a few micrograms per cm².

The metal parts (usually brass) of the chamber shown in Fig. 8 must be well polished on the internal surfaces, and the insulation must be extremely good. Even glass is not altogether satisfactory owing to the readiness with which its surface deteriorates in moist conditions. Probably the best is P.T.F.E. (polytetrafluorethylene, also known as fluon, or teflon. Doubtless it will acquire many more names as more firms make it, since those firms, which cannot think of new materials to make, like to camouflage the fact by at least inventing their own dictionaries.) P.T.F.E. is strongly hydrophobic, is a first-class insulator, can stand soft-soldering temperatures without damage or distortion and is easily machined. It is expensive, but not expensive enough to deter anyone who can face without blanching the bill for the electronic equipment following the ionisation chamber. Equally satisfactory as insulators, although less satisfactory thermally or mechanically, are polyethylene, polypropylene, polystyrene or the old-fashioned sulphur. The brittleness of the latter makes it little used nowadays.

The gas filling the chamber is usually one of the inert gases, commonly argon, although where it is essential to keep the chamber as small as possible the more expensive inert gases krypton or xenon can be used. Pressures of two to ten atmospheres are usual. Some excess pressure is desirable to reduce the chance of contamination of the gas from outside as well as to reduce the size of chamber. Pressures above five to ten atmospheres lead to difficulties with gaskets.

Since it is troublesome and often difficult to clean the last traces of an old radioactive source from the interior of the chamber, each source is usually deposited on a small removable metal insert in the earthed plate. The insert can be held flush with the surrounding plate. Stainless steel discs 15 to 20 mm diameter are very convenient for this. Where the source is refractory, as many of the α-emitting materials are, it can be fired to burn off organic binders and many kinds of dirt and contamination.

The whole of the interior of the chamber must be kept scrupulously clean to avoid radioactive contamination, to avoid insulating layers which might affect the field-distribution or the collection-efficiency and to prevent the risk of contaminating the gas in the chamber with volatile impurities. For consistency of results the gas mixture must always be the same and it is necessary to be able to pump out the air from the chamber before filling it with the desired mixture. It may also be desirable to flush the chamber out two or three times with this before the final filling.

3.13 *Collection-times*

An important property of an ionisation chamber is the time of collection of ions. The mobilities of free electrons are far greater than those of positive or negative ions, which are about equal. Negative ions are not found to any appreciable extent in the inert gases so that the total collection-time in most ion chambers is determined by the time required by the positive ions. The mobilities of these are in the region of 1 cm/sec per volt/cm at S.T.P. in most gases except for hydrogen or helium, in which they are about five times as much. With 4000 volts/cm at a pressure of 4 atmospheres, the mean ion velocities are thus about 10 metres/sec, and the collection-time will be a few milliseconds. Hence the counting rate must be kept down to a few tens of particles a minute if the chance is to be kept small that the ions from two particles should be registered as from one.

The electron-collection-time in the same conditions would be two or three microseconds. At first sight, it might be thought that with a positive collector, which naturally collects only electrons, this electron-collection-time would alone be involved. If this were so, counting rates a thousand times greater would be achieved. Un-

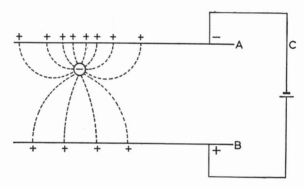

Fig. 9 Distribution of electric field and of induced charge due to a negative charge between two parallel plates. The much larger negative charge on A and positive charge on B, maintained by the battery potential, are not shown.

fortunately, the positive ions, although they travel to the opposite side of the chamber, do contribute to the collector current. Since this effect occurs in all cases in which electrons or ions are taken up by a conductor from an insulating region, it may be well to explain it. In general, it may be said that the electron-collector, the detecting instrument and the plate which takes up the positive ions are all part of one circuit. Current will continue to flow in this circuit until all positive ions, as well as all electrons, have landed.

In more detail, and more fundamentally, we can see the reason as follows.

In Fig. 9 is shown a negative charge $-Q$ in the space between two conducting plates A and B, connected by a wire C. The negative charge will induce a positive charge on each plate as indicated in the figure, the sum of these charges being equal to Q. The larger charge will be induced on the nearer plate. If $-Q$ starts from just outside plate A, almost the whole charge $+Q$ will be induced on A and practically none on B. When it moves across from A towards B, the positive charge induced on A will decrease and that on B will increase. By the time it has reached a point just outside B, almost the whole charge will be induced on B with practically none left on A. To permit this transfer of charge, a current must flow along the wire C during the whole passage of $-Q$ from A to B. The simple unsophisticated picture of an element of current entering the circuit at the instant an electron is collected is therefore quite wrong. At the instant our charge $-Q$ lands on plate B it simply neutralises the charge $+Q$ which it had induced there and the current in the wire C stops. Current flows in the external circuit only during the movement of the charge between the plates and stops at its instant of arrival. Even if we have only one ion or electron to collect, therefore, it has a finite "collection" time equal to its time of movement between the plates.

It may seem unrealistic to apply this picture to the collection of a single electron or positive ion, as this would mean that the charges induced on the plates would be less than one electronic charge, distributed over a considerable area. There is no real difficulty here. A metal surface consists of a network of positive ions permeated with a "gas" of free electrons. When we charge the surface of the metal positively, we establish an infinitesimal redistribution of the latter so as to leave the "centre of gravity" of the electrons a minutely greater distance from the surface than before. This mechanism can cope equally well with a fraction of an electronic charge and with a million million electronic charges per square centimetre.

The flow of a fraction of an electron through the wire may be regarded in a similar way.

Even when a large current flows through a wire we cannot properly think of millions of electrons tearing along it like motorcars on a road.

The electrons in a metal are wandering in all directions with velocities up to two or three thousand kilometres a second even when no current is flowing. When an electric field is applied, a general drift in the direction of the field is superimposed on the random velocities and it is this drift which constitutes the current. There are so many electrons that even at a current density of 100 amps/cm^2 the mean drift velocity along the wire would be only about 15 cm per hour. If one could see the electrons moving around in the wire one could not possibly tell whether a current were flowing or not. It would be enormously easier to detect the drift of population from the English countryside into the towns, which has been going on now for a couple of centuries, by watching the ebb and flow of traffic between town and country. The ratio of drift to "random" velocity in this case is a few ten-millionths, while in the case of electrons in a wire carrying 100 amps/sq.cm it is about $1/10^{11}$.

Returning to Fig. 9, then, the current in the wire C, whether it is

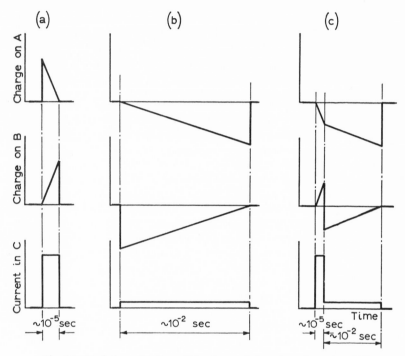

Fig. 10 Variation with time of the charges induced on plates A and B and of the current in C of Fig. 9 as a result of movement of charges between the plates.
(**10a**): Effect of transit of one electron from A to B.
(**10b**): Effect of transit of one positive ion from B to A.
(**10c**): Effect of transit of an ion-electron pair starting midway between A and B.

millions of millions of electrons per second or fractions of an electron per hour, can be equally well represented by a minute bias imposed on the pattern of random velocities. In Fig. 10a are shown the variations with time of charge on plates A and B, and of current in C, as an electron passes from the surface of A to that of B. In the gas of an ionisation chamber the electron will move with a constant mean velocity, giving a constant current in C. If a positive charge were to move from B to A, an exactly similar series of events takes place but a thousand times more slowly. This is illustrated in Fig. 10b. Negative charges are this time induced, and persist for a thousand times longer. The current is in the same direction as before. It also persists for a thousand times as long as in the electron case, but is a thousand times smaller since the total product of current and time, $\int_0^\infty idt$, must in each case be one electronic charge.

In Fig. 10c are shown the variations with time of charge on plates A and B, and of current in C, when one ion pair is produced midway between the plates. The electron then moves quickly to B while the positive ion moves slowly to A. The graphs of 10c are thus obtained simply by adding together the second halves of 10a and 10b. From the point of view of the detection equipment, the important quantity is the current flowing in C for this is what is actually observed. This clearly consists of a short, high, electron-portion and a long, low, positive-ion portion. In Fig. 10c these portions eventually make equal contributions to the charge transferred to the detecting equipment, since the ion pair was formed in the centre of the system. If it had been formed nearer to plate A than to plate B the electron portion would have contributed a larger charge than would the ion portion and vice versa. Comparison of the two contributions could therefore tell us where in the chamber an ion pair was formed. This is not a practical proposition for a single ion pair but is perfectly practicable when we have the thousands of ion pairs produced by an α-particle. If the path of the particle were not parallel to the plates we should of course find the position of the centre-of-gravity of the ion pairs produced. While practicable, the information gained by such a comparison is not usually worth the extra complexity of electronic equipment that would be needed.

3.14 Fast ion chambers

Of much greater interest is the use of an amplifier circuit* which will amplify only the short pulse of current due to the collection of electrons while neglecting the long pulse due to the positive ions. This enables us to count α-particles a thousand times faster for a given risk of accidental combination of the current from two different α-particles into a single pulse. Desirable as it is, this advantage was not taken for a considerable time, for a very good reason. This was that the whole point of using an ionisation chamber was usually to get an accurate measure of α-particle energy. If only the current pulse due to electron-collection was measured, then its size depended

* See Appendix I.

on the position of the path of the α-particle in the chamber as well
as on the energy dissipated. This position could not be known unless
the slow current pulse due to ion collection was also observed—in
which case we were back to the old slow counting rate.

3.15 *Gridded ion chambers*

The dilemma was resolved by the introduction of the gridded ionisation chamber. The principle of this is illustrated in Fig. 11. A grid

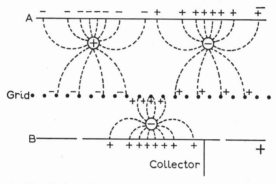

Fig. 11 Distribution of induced charge and of electric field in a gridded
ionisation chamber due to free charges between the electrodes. The
much larger charges on A, B and the grid, which are maintained by the
H.T. supply, are not shown.

of fine wires parallel to the plates is introduced, nearer to B, and
kept at an intermediate potential. The potentials with respect to B
might be −2000 volts for the grid and −4000 volts for plate A.

Then if the grid mesh is sufficiently fine, the electric fields of
charges between A and the grid are shielded from the collector.
Their movements do not induce currents in the collector circuit, but
only in the grid-to-A circuit. The source is placed on A and the
distance between A and the grid is sufficient to ensure that, at the
working pressure, none of the α-particles comes too close to the grid.
("Too close" in this connection means within a distance three or
four times the spacing between successive wires of the grid.) Then
the positive ions produced will all of them move off in their own time
to A, which is more negative than the grid, without making any contribution to the collector current. The electrons on the other hand
move towards the grid and though a few of them will land on the
wires, most will pass through to the collector. As soon as they pass
through the grid they begin to induce currents in the grid-to-collector

circuit, and since they all traverse the whole of the grid-to-B space, their contribution to the collector current is independent of their exact place of origin. The proportion lost to the wires is also constant and independent of the position of the path of the α-particle so that the electron current alone is now a good measure of the energy dissipated by the α-particle. The gridded ion chamber thus combines high speed with good energy-discrimination, is little extra

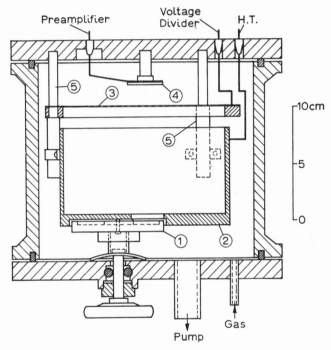

Fig. 12 Gridded ionisation chamber used at the Nobel Institute at Stockholm. 1. Five-position source holder, 2. Negative electrode, 3. Grid, 4. Electron-collector, 5. P.T.F.E. supporting insulators.

trouble to construct, and is used in most quantitative work with α-particles. In Fig. 12 is shown an example of a practical construction.

Having achieved the first object of using the fast electron collection times without interference from the slow-moving ions, a further gain is obtainable. This arises from the fact that, although the velocities of positive ions do not vary very much from gas to gas, the mean velocities of electrons vary a great deal. The mean velocity of an electron in a gas in the direction of an applied electric field is always very small compared to its random motions resulting from collisions with atoms of the gas. The ratio depends on the nature of the gas as well as upon the applied field and it is found that a considerable gain occurs when a gas is used the molecules of which are easily excited by electron impacts. This tends to suppress the build-up of random velocities. The mean-free-path of an electron is much greater at lower speed and the mean drift-velocity in the direction of the field gains more by the increase of free path than it loses by the greater frequency of energy loss. Quite a small proportion of such a gas added to argon will give the same advantage as the pure gas, while retaining the high density and ease of purification of the argon. A mixture of argon or a heavier inert gas with 5% or so of methane or carbon dioxide is therefore frequently used in place of the pure inert gas.

Using a chamber of the kind described, the energy of an individual α-particle can be determined to better than 1% in good conditions and the average energies of two separate mono-energetic groups can be compared to perhaps a part in a thousand. To achieve such results however requires stable voltages and very stable electronic circuits.

3.16 *Neutron chambers*

A great many more kinds of ionisation chamber have been designed for special purposes. We shall mention only two of them, designed for the detection of neutrons. Since neutrons produce no ionisation themselves they can be detected only indirectly. One way of doing this is to make use of the fact that, if a neutron is captured by the lighter isotope of boron, $_5B^{10}$, an α-particle is emitted, leaving behind the nuclide $_3Li^7$. The energy released in the reaction is 2·8 MeV. If, therefore, neutrons enter an ionisation chamber lined with a compound of boron or filled with a gas containing boron, α-particles will be produced which can be detected in the normal way. The usual procedure is to fill the chamber with boron trifluoride gas. This boils at $-101°C$ so that at room temperature it can be used at several atmospheres like the more usual fillings. The gas is inconveniently reactive chemically, but when really pure (which is difficult to achieve) the electron mobility is several times higher than for the normal argon mixtures, so that electron collection times are very small.

The chamber is suitable only for detection of slow neutrons; the probability of capture by the $_5B^{10}$ nucleus of neutrons with thermal velocities* is some thousands of times as great as it is for fast neutrons. The chamber is thus used just to detect the neutrons, not to measure their energy, which will be negligible compared to the energy released in the reaction. Hence there is no need to use a gridded chamber even if it is desired to use only the contribution by electron movement to the collector current. The size of a pulse varies according to the position of the path of the α-particle in the chamber, but so long as it is not unobservably small this does not matter.

Freed from the need to keep collector current proportional to energy released, there is also no need to stick to the parallel-plate design of chamber to maintain a uniform electric field. A

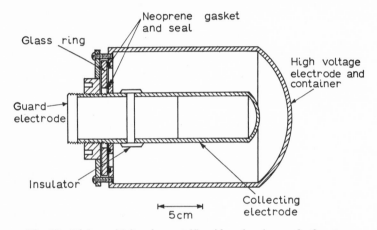

Fig. 13 High-sensitivity boron-trifluoride chamber, of the type described by Rossi and Staub.

coaxial system in which the central cylinder acts as collector enables us to obtain a much smaller capacitance between the collector and the surrounding electrodes for a given working volume of gas.

In Fig. 13 is shown a practical form of boron-trifluoride chamber. This will record some 25% of slow neutrons passing through the

* According to the principle of the equipartition of energy, neutrons which make a great many collisions in any material will finally acquire an energy-distribution identical with the normal thermal kinetic energy of translation of the atoms of the material. This energy-distribution thus depends only on the temperature of the material and not at all on the original energies of the neutrons. Such neutrons are described as having thermal velocities, or simply as *thermal neutrons*. Their mean energy thus clearly depends on the temperature of the material with which they are in equilibrium. When no temperature is specified, room temperature is assumed.

working volume, but is of little use for fast neutrons, of which it will record only about one in ten thousand. The slow-neutron efficiency can be improved either by increasing the size of the chamber —which for many purposes is already inconveniently large—by increasing the gas pressure or by increasing the proportion of boron 10 in the boron trifluoride. Ordinary boron contains 18·7% of boron 10 so that use of the pure separated isotope makes possible a reduction of five times in pressure or in the length of the chamber.

A detector for fast neutrons can be obtained by first slowing them down and then using the chamber just described. A better system, sensitive only to fast neutrons, is an ionisation chamber filled with hydrogen or methane. If fast neutrons pass through a gas, they

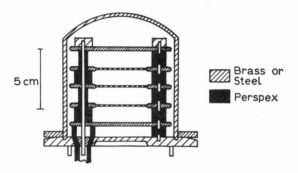

5 cm

▨ Brass or Steel
■ Perspex

Fig. 14 Proton-recoil chamber for neutron-detection of the type due to Barschall and Kanner. The first and third grids act as collectors, the rest of the electrodes being at high potential. It is filled with hydrogen or a hydrogen-argon mixture to 20 to 30 atmospheres.

will occasionally make close collisions with the nuclei of the gas atoms. If the nuclei are heavy, not much energy is transferred, but if they are protons with nearly the same mass as the neutrons, a large part of the energy of the latter can be handed over in a single collision. The proton then moves off with a high velocity and produces a series of ion pairs in the gas just as does an α-particle. The probability of such high-energy collisions is still very small so that large chambers at high pressure are needed for efficiencies of even a few per cent. Very high voltages may be avoided by dividing the chamber into a large number of sections. An example is given in Fig. 14, in which, however, only two collector-grids are shown. A more extensive discussion of neutron detectors may be found in *Experimental Nuclear Physics I*, edited by Segré, pp. 131 *seq.*

3.2 The Proportional Counter

3.21 *Principles*

In an ionisation chamber, the charge collected from the path of a single α-particle will increase with chamber voltage, due to the reduction of ionic recombination, until such recombination is reduced to negligible proportions. The charge collected then reaches a saturation value practically independent of voltage.

If the voltage rises further, the charge collected presently begins to increase again, more and more rapidly. This is due to the fact that some electrons can now gain enough energy between collisions to ionise the gas atoms which they strike. Each time this happens, a new ion pair is added to those formed by the original α-particle. The electron of the new pair may then join the primary electron in forming still further ion pairs. The total charge finally collected is then greater than the charge derived from the primary ionisation by a factor which depends on the field, pressure and dimensions of the chamber. It may be very large, of the order of hundreds of times. The process is often described as the development of an *avalanche*.

This clearly increases correspondingly the sensitivity of the ionisation chamber. This looks very valuable, but the value is practically limited in a number of ways. Firstly, the sensitivity is now extremely dependent on chamber voltage and on gas density. In an ordinary chamber with little ionic recombination, the sensitivity depends little on either of these. It is often easier to increase by a factor of a hundred the gain of the amplifier which is being used than it is to stabilise sufficiently well the voltage. Even where this stabilisation is carried out successfully, another trouble arises. This is that, even if we take our time and collect both ions and electrons, the charge collected from an α-track of given energy will depend on its position in the chamber. If it passed close to the collector (or grid in a gridded chamber), the amount of multiplication will be much less than if it passes close to the negative electrode, since the latter position gives the electrons much more space in which to ionise. (The positive ions produced are not themselves capable of forming any fresh ion pairs.) We thus lose one of the main advantages of the system, that of giving a good measure of the energy of the particles which it records.

The gridded chamber would avoid this difficulty, if the field were sufficient for secondary ionisation only between the grid and the collector. To get the full multiplication in this case, however, we cannot use only the electron-collection-time, as positive ions are now being formed in large numbers between grid and collector.

A double-gridded system with multiplication taking place only between the two grids would give us both a constant multiplication factor and short collection times, but this adds mechanical complication to the need for high voltage-stability between the grids.

3.22 *A practical counter*

A simple plan, which evades this last problem of mechanical complexity, is to use a coaxial system instead of a parallel-plate one. The positive collector then takes the form of a fine wire along the axis of a tubular negative electrode. The electric field in such a system is then inversely proportional to the distance from the central wire. It

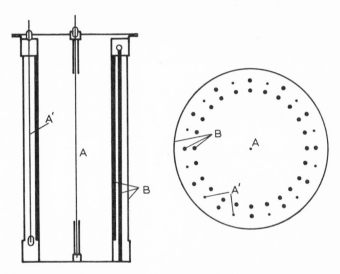

Fig. 15 "Wall-less" proportional counter of the type due to Curran *et al.* Only those particles are recorded from the main inner counter which do *not* produce counts in any of the ring of outside counters. Thus all β-particles counted are recorded with the correct energy.

is thus easy to ensure that this is high enough for electron multiplication only for a small distance from the wire. If the radius of the region in which multiplication takes place is 3 % of the radius of the outer tube, the volume of the region is less than 0·1 % of the whole. Thus little detection-volume is lost if primary particles are prevented from reaching this region and, even if they are not so prevented, very few will have any serious part of their paths inside it. If the gas pressure and voltage are high, there will be no difficulty in getting plenty of multiplication in the short distance available. Such a coaxial system using a constant gas-multiplication is known as a *proportional counter.* Multiplications of up to 1000 times can be used, but rather less than this is usual as the greater the multiplication the greater the stability of voltage required.

The composition of the gas, as well as its density, is of very great importance. Small amounts of impurity cause large changes of multiplication factor. It may even be worth while to keep a continuous supply of fresh gas flowing through the counter during a series of precise measurements.

The high sensitivity of the proportional counter enables it to record particles with energies down to a few keV. It is therefore useful for detection of β-particles or X-ray quanta of specified energy as well as of α-particles. In Fig. 15 is shown the design of counter used by Curran *et al.* to investigate the energy-spectrum of the β-particles from tritium. These have a maximum energy of 17 keV.

Counters may be used "in the proportional region" with a cheap non-linear amplifier simply for counting and may still be called "proportional counters" although the final output pulses may then be very far from true proportionality. The reason for such use will be discussed in section 3·34 below.

3.3 The Geiger Counter

3.31 *General principles*

This in its various forms is the most important of all the instruments used in applied nuclear physics. Structurally, it is similar to the proportional counter; indeed a given instrument can often be used either as a Geiger or as a proportional counter. The difference is produced by a further increase in the positive voltage which is applied to the central wire. This of course leads to further gas-multiplication and also brings into prominence a phenomenon which we have so far neglected. Besides the process of ionisation, the energetic electrons produce a good deal of atomic excitation. This will normally be followed by photon-emission. Most of the photons will strike the negative outer tube or cathode of the counter. Since they have usually an energy of only a few electron-volts, most of them are absorbed there without further effect. There is a small probability, however, that one of them will cause a photo-electron to be emitted from the cathode. This will be attracted to the central wire and will start a fresh secondary avalanche of its own. If this secondary avalanche produces enough photons to liberate a further electron there will be a tertiary avalanche and so on. As the voltage on the central wire is made more positive therefore, we reach a critical stage at which each charge-avalanche produces enough photons on the average to start another one. This process can go on indefinitely, and beyond the critical stage will not merely go on but will increase indefinitely. It will eventually be limited by some fresh factor. This may be internal, by the build up of positive space-charge round the central wire. This increases the effective diameter of the wire and reduces its accelerating field (*self-quenching*). Alternatively, a drop in the voltage of the wire may be produced by the external circuit (*external quenching*). In either case, the electric field will return to normal after a few tens or hundreds of microseconds. During this period the counter will be entirely or partially insensitive to further primary particles. The period is therefore known as the *dead-time* and is an important characteristic of the system.

The final charge collected does not depend at all on the number of ions first formed by the primary particle. Hence the Geiger counter is of no use for discrimination between particles of different energies. On the other hand, it can produce a large discharge as a result of the production of a single ion pair so that it is sensitive to the passage of even the most weakly ionising particles.

This high sensitivity, and the inability to distinguish different particle energies, makes it of no advantage to confine the whole path of a primary particle within the counter. The latter can therefore be made smaller without disadvantage. It will normally be filled with gas at only a few centimetres pressure, which absorbs less of the particle-energy available but which lengthens the mean-free-paths of electrons and makes efficient multiplication possible with much lower voltages on the central wire.

As was the case for the ionisation chamber and the proportional counter, the main filling of a Geiger counter is usually argon. An admixture of a few per cent. of a second gas is essential, however, and not merely desirable. This is because argon atoms, and the other inert gas atoms, can be excited into metastable states of considerable potential energy. The lives of these atoms in metastable states are long enough for some of them to reach the wall of the counter, without losing their energy by radiation, after the discharge in which they were formed has finished and the field round the central wire has been restored to normal. On striking the wall, the excitation-energy of the atom may be employed to eject a fresh electron, which promptly initiates an avalanche and a complete new discharge. Thus the presence of such metastable excited atoms could lead to an un-ending succession of separate discharges following a single initiating event; each of these discharges would be recorded by the rest of the apparatus as a new primary particle.

The excited argon atoms can be eliminated by collision with many kinds of polyatomic molecule, which use up the energy in dissociation of themselves. For a long time, alcohol vapour has generally been used but chlorine or bromine, which give the counter a longer life, are increasing in popularity.

Such *halogen-quenched* counters must be made of materials immune to attack by their chemically reactive contents. This makes them more expensive in the first instance. When the halogen molecule is dissociated, however, it re-associates later on, whereas dissociated alcohol molecules do not. An alcohol-quenched counter has thus a limited useful life, usually in the region of 10^8 discharges, at the end of which too little alcohol survives for effective quenching. The halogen-filled counters on the other hand have a useful life in the region of 10^{10} discharges. They have another major advantage; the production of photons capable of releasing electrons from the walls of the counter is much more efficient, so that the primary electron-avalanches can be much smaller at the critical stage of onset of the unlimited discharge or *Geiger condition*. Consequently a thicker and more stable central wire can be used and the tube can be operated at a lower voltage. Typical wire-voltages for the alcohol-quenched type would be 900 to 1500 and for the halogen-quenched type 450 to 600.

In contrast to the ionisation chamber, the collector of which experiences a voltage change of only about 10^{-4} volts per MeV dissipated in the gas, the collector of a Geiger counter may experience a change of several volts for a primary dissipation of 20 to 40 eV. Relatively little external voltage-amplification is therefore necessary to operate the most robust recording instruments.

Although the Geiger counter can detect any kind of ionising particle or radiation, it is normally used for the detection of β-particles. As indicated in Chapter 1, these have a continuous distribution of energy so that the measurement of the energy of an individual particle is rarely of importance.

3.32 *Detection efficiency*

The efficiency for detection of α- or β-particles is very high. For α-particles it is unity (100%) as nearly as can be measured. For fast β-particles it is not quite so good, for two reasons. Firstly, the ionisation by fast electrons is low. With the usual argon-alcohol filling at 10 cm pressure, such a fast β-particle (1 MeV or more) will produce on the average only about four primary ion pairs per centimetre. Then if the path in the counter is l centimetres, say, the chance that no ion pair should be found at all is $\exp(-4l)$. For

$l = 1$ cm, this is about 0·02. Since the β-particle cannot be detected if no ion pair is formed in the gas of the counter, the efficiency cannot be greater than $1 - 0·02 = 0·98$ (or 98%). For $l = 2$ cm, the reduction of efficiency from this cause will be only three in ten thousand. If l were a few millimetres only, as might happen if the particle crossed a corner of the counter, the efficiency could be quite low.

The second reason is that, while in the Geiger condition of operation every ion pair will lead to an avalanche, not every avalanche will build up to the limiting discharge. *On the average* an avalanche will produce enough photons to liberate more than one electron from the wall, thus giving rise to an increasing succession of further avalanches. If the primary α- or β-particle produces a lot of ion pairs with therefore a lot of primary avalanches, this process can be very reliable even if each avalanche gives rise on the average to only 1·1 or 1·2 fresh ones. When there are only one or two to begin with, however, there is quite a large chance that the discharge may die out by pure accident. A mere two or three avalanches would not be detected by the electronic system designed to record the full-blown discharge comprising many millions of individual avalanches.

The probability of such a failure after the successful production of a few primary ion pairs falls rapidly as the voltage on the centre wire increases. Hence, although the efficiency of a Geiger counter is well below 100% just above the critical or *threshold* voltage at which the Geiger condition is established, it rises rapidly as the voltage increases. By the time the voltage is 5% over the threshold value the efficiency of a good counter is likely to be over 90%. Unfortunately, it is inadvisable to raise the voltage to the point where *every* primary avalanche would lead to a full discharge. If this is done, numbers of additional discharges occur which are not initiated by fast particles. Probably this is because the elimination of metastable excited atoms, like all the other processes, is statistical in nature; if enough of them are produced there is a chance that one will escape destruction long enough to strike the wall and initiate a fresh, apparently independent, discharge.

The practical variations of the counting rate with voltage for a particular counter is shown in Fig. 16. The nearly flat region is known as the *plateau* and the operating condition should always be chosen near the middle of this. The fact that the plateau is not quite flat is mainly due to the occurrence of the secondary discharges mentioned above as the voltage is raised above the minimum value required to produce an avalanche. The best obtainable value for the "true" counting rate is therefore found by extrapolating the plateau back to the threshold voltage, i.e. to the point A of Fig. 16.

It rarely matters that the counter is missing or inventing a few counts so long as these remain a constant proportion of the true counts. To ensure this a stable high-tension supply is needed (the flatter the plateau the less important this will be) and in accurate work it is necessary at regular intervals to make a check. This is usually done by counting the particles from a constant source, such as a preparation of uranium, placed in a constant position with respect to the counter. Small day-to-day changes in the efficiency can then be allowed for.

The frequency with which such checks should be made depends

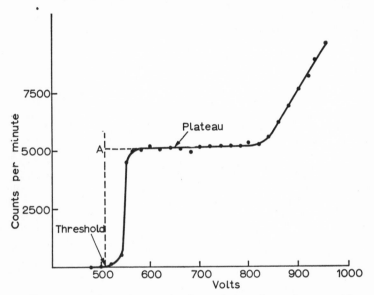

Fig. 16 Experimental curve showing variation of counting-rate with voltage for a Geiger counter (MX123). (P. Ellis.)

very much on the individual counter and on the constancy of its temperature and high-tension supply. A check before and after each important measurement may be regarded as ideal, but with most counters on most occasions a weekly or monthly check is quite adequate.

3.33 *Counter background*

Besides the efficiency, two important properties of any counter are the *background* and the *dead-time*. The background is the counting rate which occurs in the absence of any intentionally introduced source of radioactivity. It is due to a variety of causes; including cosmic rays, γ-rays from the floor or walls of the laboratory, radioactive contamination of the surrounding equipment or of the materials of the counter itself, and thermionic emission from the counter walls. The great sensitivity and entire lack of discrimination of Geiger counters lead to very much larger and hence more important backgrounds than occur in proportional counters or ionisation chambers. Background counts can be reduced by good screening of the counter with uncontaminated materials. They are naturally larger in larger counters with a bigger "target area". Some one to a hundred counts per minute may be involved.

3.34 *Counter dead-time*

The *dead-time* is the period following each count when the counter is insensitive to further incoming particles. As has been mentioned, this may be due to the accumulation of positive-ion space-charge which needs time to leak away or to a drop of the voltage on the centre wire, usually the former. The time for which the counter is insensitive for this reason may be some tens or hundreds of microseconds and is unfortunately far from constant. This makes it difficult to allow for and it is usual practice to introduce a constant, known dead-time which can be allowed for. This is done by a supply circuit which lowers the wire-voltage by 100 to 200 volts after each discharge for a defined period which must, of course, be longer than the longest possible natural dead-time of the counter concerned. Allowance for the particles which have been missed because they arrived during a dead-time is then easily made as follows. Suppose the fixed dead-time is t_1 minutes and the counting rate observed is n per minute. Then in each minute, the counter is insensitive for nt_1 minutes and the n particles observed must all have entered during the remaining $(1 - nt_1)$ minutes. Hence the true rate of entry of particles n_0 is given by

$$n_0 = \frac{n}{1 - nt_1} \qquad\qquad 3.1$$

If t_1 is known, therefore, n_0 can be calculated.

It is convenient to make the value of t_1 the same for all counters in the laboratory; then a correction table can be made up once and for all.

Equation 3·1 may be transformed to

$$n_0 = n + \frac{n^2 t_1}{1 - nt_1} . \qquad\qquad 3.2$$

If then we make a table of $\dfrac{n^2 t_1}{1 - n t_1}$ as a function of n, n_0 is obtained simply by adding the quantity read off from the table to the value of n. In Appendix 2 is given such a table for a dead-time of 600 μsec (10^{-5} minutes) used in the Physics Laboratory in the University of Birmingham, for which I am indebted to Dr. J. N. Kudahl. This dead-time was chosen to give a 1% correction at 1000 c.p.m. (counts per minute) and approximately m% at m thousand c.p.m. which makes rough mental calculations easy.

It has been pointed out by Westcott that an additional correction at high counting-rates is needed to allow for the drop in mean counter-voltage, which results from the appreciable mean current at such rates. This depends on the slope of the counter plateau as well as on the external circuit and even in favourable cases may amount to several percent of the dead-time corrections. This "sub-correction", however, is rarely greater than the practical uncertainty in the dead-time correction itself.

There is a considerable reduction in dead-time if a counter is used in the proportional region so that the discharge dies away after the first avalanche. The dead-time is then only a few microseconds and is nearly constant. In this case, too, there is a negligible number of secondary discharges, so that a really flat (though short) plateau can be obtained between the threshold and the onset of the "Geiger" region. The drawback is that a much larger degree of amplification is required to ensure registration of particles which have only a short path in the counter. This is expensive and, unless the amplifier is very carefully designed, leads to trouble due to overloading by the larger pulses. It does not matter if this means that the output pulses are no longer proportional to the original ionisation. In fact, pulses are likely to be limited electronically so that, whatever their original size, the pulses which reach the recording apparatus are all alike. Although the counter is still often referred to as a "proportional counter", this refers only to the original discharge, not to the final output.

The drawback of overloading the amplifier is that the latter can easily be set ringing, thus giving rise electronically to extra pulses which may easily be more numerous than those produced directly by the counter when it is working in the Geiger region. Nevertheless, the advantages are great and the use of counters working in the proportional region is likely to increase rapidly for accurate work at high counting-rates.

The Geiger counter can detect X-ray or γ-ray quanta only in-

c

directly, like neutrons in an ionisation chamber, by their production of secondary ionising particles. Either X-rays or γ-rays may eject electrons from the wall of the counter. These will then be efficiently detected but the efficiency of absorption of the primary quanta, just so as to eject electrons into the gas of the counter, is rarely more than one or two per cent. Where the strength of the source is adequate however, the cheapness and simplicity of the Geiger counter sometimes lead to its employment to detect such radiations. The

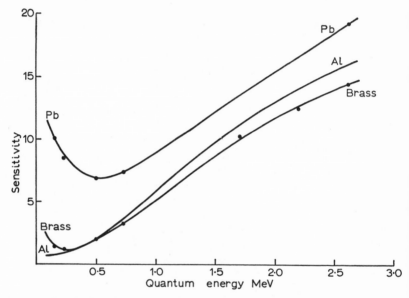

Fig. 17 Variation of γ-counting sensitivity of Geiger counters with wall-material and with quantum-energy. The ordinate is efficiency $\times 10^3$. (From data given Bradt *et al.*, *Helv. Phys. Acta*, **19**, 77 (1946).)

variation with quantum-energy of the total efficiency for detection of photons is shown in Fig. 17, for counters with cathodes made of three different materials.

3.35 *Types of Geiger counter*

Counters of many different forms are now made commercially.

3.351 *End-window counters.* When small active solid samples are to be investigated, the end-window counter is usually used. A diagram of the MX123 counter is shown in Fig. 18. This counter is halogen-quenched and has a "window" of mica, the thickness of which is about 2 mg/cm². * β-particles with energies of 30 to 40 keV and α-particles down to $3\frac{1}{2}$ MeV are able to pass through this.

When a thin flat source is brought close to this window, particles can be detected which are emitted over a solid angle of nearly 2π, so that over 40% of the disintegrations in the sample can be detected in favourable conditions. It is usually undesirable, however, to put the source as close to the fragile window as this requires, and 30%

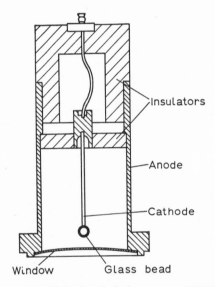

Fig. 18   Mullard end-window counter MX123.

of total disintegrations represents a very good average counting-efficiency.

This type of counter is easy to screen with lead or other absorber from extraneous sources of particles. A convenient form of lead "castle", designed by Dr. K. F. Chackett, is shown in Fig. 19. In this, not only is the counter itself well screened but a cavity approximately 10 cm cube below the counter window is also well screened. This cavity is lined with perspex, which is much less efficient than lead in emitting observable electrons when bombarded by casual γ-ray quanta from outside. It is equipped with a pair of grooved perspex

* The thickness of windows and absorbing foils is often given in terms of the weight per unit area rather than in terms of geometrical thickness. The differences between the loss of energy of charged particles in different substances—and even between gases and solids—is then relatively small.

sides into which shelves can be slid at various distances from the counter-window according to the strengths of the sources supported on them. If the shelf is not at the top of the set of grooves, sheets of absorber can conveniently be placed above the source. The front of the cavity is shielded during operation by a movable lead block. In this system the counter background will be reduced to 10 to 12 counts a minute. It is therefore very suitable for examination of weak sources of β-particles.

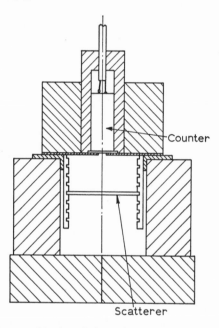

Counter

Scatterer

Fig. 19   Diagram of lead castle for end-window counter, with shielded shelves for sources.

For examination of very weak sources, end-window counters with extremely low backgrounds, of the order of one count per minute, have been designed and are available commercially.

A good plastic counter gives exceedingly little background due to radioactive contamination of its own materials. Ten centimetres of lead will stop most γ-rays from thorium decay-products in laboratory walls, etc. The main background left comes from cosmic rays which cannot be stopped by any reasonable quantity of local screening material. Their effect can be eliminated, however, as follows. If above the plastic counter is placed another larger counter or group of counters, any cosmic-ray particle (practically always a charged μ-meson at the bottom of the atmosphere) will necessarily produce a discharge in the upper counter before it passes through the lower one. The upper and lower counters can then be connected to an *anti-coincidence* circuit. This is one which will not record any count from the lower counter if it coincides with one in the upper counter or counters. All counts produced by cosmic rays are then removed from the background of the lower counter. Total background-counts of 0·3 or less per minute can then be attained. The advantage of this is so great that special units have been developed commercially consisting of one counter (the sample-counter) entirely within a larger "guard" counter except for the window. Both counters are of the continuous-flow type and the makers* will supply a complete counting system including counters, lead screening, anti-coincidence gear, scalers for recording the sample counts and a variety of equipment for automatic exchanging at predetermined times of up to 50 separate sources, the count from each being permanently recorded.

3.352 *Liquid counters.* When the activity to be observed is distributed through a liquid of appreciable volume, the end-window counter is unsuitable. Special *liquid counters* are made of the form of which an example is shown in Plate II. If active liquid fills the whole of the space around the thin-walled counting tube it will always have a definite geometrical position and the proportion of the emitted β-particles which penetrate the counter wall will be constant for any one β-particle energy. This proportion must be found by calibration with a known amount of radioactive material in the standard volume of liquid. A fresh calibration is needed for each type of β-emitter to be used and for each liquid in which it may be dispersed. This is so because the proportion of those disintegrations which take place in the liquid, from which β-particles may penetrate the inner counter wall, will depend upon the range of these β-particles in the liquid. For comparison of the activities of similar liquids containing the same active nuclide, no calibration is necessary.

In Plate III is shown a *flow counter* through which liquid may be passed continuously. It comprises a thin-walled internal helix of less than 1 ml capacity so that it can also be used with advantage in place of the counter of Plate II when small liquid samples are to be examined.

3.353 *Thin-walled counters.* For general monitoring purposes, a thin-walled glass counter may be used which β-particles can enter

* Tracerlab. Inc. 1601, Trapelo Road, Waltham 54, Massachusetts, U.S.A.

from any side. The cathode then consists of a thin film of aquadag or evaporated metal deposited on the inside of the glass wall.

3.354 *Gamma counters.* For γ-detection a thick-walled counter of similar form may be used or, more usually, one of the thin-walled type may be adapted to γ-detection merely by surrounding it with a sufficiently thick screen to stop the entry of β-particles but not the entry of the desired γ-quanta.

A photograph of a commercial model, with its portable supplies and meter is shown in Plate IV. The amplified output of the counter is fed to a loudspeaker giving an audible click for each discharge of the counter and also to a multiple-range meter which measures the mean discharge-current and hence the average number of counter-discharges per second. This is preferable to the loudspeaker for high counting-rates, when the sound output merges into a continuous buzz.

3.355 4π *counters.* None of the systems described so far detects all of the disintegrations in a sample and hence all need calibration if absolute rates of disintegration are required.

To measure these the 4π *counter* is used. This consists in essence of two separate Geiger counters with a very thin common dividing-wall on which the source is deposited. Then if the source itself is too thin to absorb any of the particles emitted, every one of these will enter one or other counter and every single disintegration will be recorded. There is a chance that particles may be recorded by both counters if they are ejected into one and are subsequently scattered back through the thin dividing-wall into the other. Hence the associated electronic equipment must take both counter outputs together so that a single count is recorded when either *or* both counters discharge. Since the source must be deposited on the dividing-wall, a 4π counter must be demountable. It is usually used with a continuous gas-flow. The thickness of the source-support must not be more than a few micrograms per cm² (0·1 to 0·01 microns) and is necessarily very fragile. It is usually impracticable to clean it and a new support should be made for each source. A form of 4π counter is shown in Fig. 20.

The shape of the interior of the counter appears to matter very little if it is used with really satisfactory electronic equipment. A "Siamese-twin" pair of ordinary straight-wire coaxial counters, side by side with a thin common wall, is just as effective as the double-loop-type counter shown in the figure. The former is also easier to make on a small scale, with correspondingly small background. The more asymmetrical is the form of the interior the greater is likely to be the range of initial pulse sizes, but if the amplifier is good enough to handle this range, no harm is done. It seems to be important, however, to have a good field-strength near the source, and to have a perfectly smooth anode wire without projections. The soldered or welded ends of this should preferably be screened by insulating sleeves. The dividing-wall must be electrically conducting to prevent the building-up of electrostatic charges and is usually made of an organic film of formvar or celluloid on to which is evaporated a few tens of Ångstroms of gold or aluminium. A small drop of a dilute

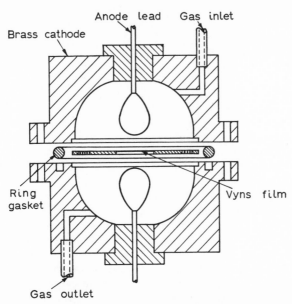

Fig. 20 4π counter. This is composed of two counters back-to-back, separated by a very thin conducting film on which the source is deposited. The example shown is due to Hawkins, Merritt and Craven of the National Physical Laboratory.

solution of the source-material can then be placed carefully on the centre of the film and dried before the counter is reassembled.

It is of great importance to see that the counter has a really good plateau before its results can be relied on. A poor plateau in a satisfactory design of counter usually indicates chemical contamination of the interior of the counter or of the gas supplied—in other words, dirt—which can be removed by appropriate cleaning.

Since a 4π counter is of interest only for accurate absolute work it is of particular value to operate both halves of the counter in the proportional region as discussed in section 3.34 above.

CHAPTER FOUR

DETECTION OF PARTICLES
II. SOLID AND LIQUID DETECTORS

4.1 Scintillation Counters

We have seen that gas-filled detectors have a low efficiency for the detection of hard (i.e. high-energy) X-rays or γ-rays. For a high efficiency, the first requirement is that the highly penetrating rays must be stopped—or at least must interact—in the detector. The thickness of argon at S.T.P. necessary to stop 99 % of 2-MeV γ-quanta would be about 600 metres, and the thickness for lighter gases would be even more. Gas as the active agent is therefore unpromising and it would be very desirable to use solids or liquids which have densities a thousand times larger. In some circumstances ionisation in solids or liquids can be observed, but the commonest solid and liquid detectors depend on the production of light by fast particles.

4.11 *Zinc sulphide detectors*

One of the earliest detectors was a layer of powdered zinc sulphide which emitted a just-visible flash of light when struck by a single α-particle. Particles were counted individually in a darkroom by an experimentalist with good sight and a strong lens. Counting of large numbers was then a very tedious job and, although physicists of the time went mad less often than the electronic equipment of today goes wrong, besides being a lot cheaper, their operational lives rarely exceeded a few million counts.

4.12 *Photomultiplier system*

Nowadays many other substances are used besides zinc sulphide and the light flashes are observed with the help of a *photomultiplier* rather than directly by eye.

In Fig. 21 is shown, in section, a typical arrangement. The light-producing material, known as a *phosphor*, is placed in contact with the flat glass end of the photomultiplier tube. A drop of immersion-oil to make optical contact is placed between the two to reduce the loss of light by reflection. The whole is surrounded by a light-proof screen. Either the phosphor itself or the inside of the screen is painted white to reflect light back to the face of the photomultiplier. Light emitted from the phosphor due to the transit of a charged

particle passes into the photomultiplier and strikes the photocathode
C. This consists of a very thin metallised film with a low work-
function which emits electrons when struck by light quanta. C is at
a negative potential with respect to the first *dynode*, D_1. This is an
electrode of beryllium-bronze or other material with a high secondary-
emission coefficient, i.e. a material which will emit several secondary
electrons for each primary electron from C which strikes it. While D_1
is some 150 to 200 volts positive with respect to C, it is a similar
amount negative with respect to D_2 and is so formed that secondary

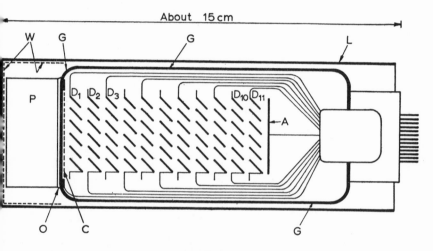

Fig. 21 Phosphor and photomultiplier. *P*—phosphor block; *G*—
glass envelope; *O*—oil to make optical contact between *P* and *G*; *C*—
light-sensitive cathode deposited on glass; $D_1 \ldots D_{11}$—secondary-emission dynodes; *A*—anode or collector; *W*—white paint; *L*—light-tight
aluminium case.

electrons will be attracted through to D_2. Thus each produces a
fresh group of secondaries which are attracted to D_3. The process is
illustrated in Fig. 22. A similar multiplication occurs at each dynode
until the final output is collected by the anode A. If the average
multiplication at each of the eleven dynodes is 4 times, then for each
electron emitted by the photocathode, 4^{11} or about four million,
electrons will arrive at the final collecting anode. This represents a
gain much larger than is obtainable from a proportional counter
although further amplification will still be needed to operate the
recording apparatus.

C*

4.121 *Counting efficiency.* The counting efficiency of a particular instrument depends on several factors. The first, as in all systems, is the solid angle subtended at the source by the sensitive volume of the counter. It is difficult to increase this much beyond a third of 4π with an external source but it can be brought to a value close to 4π if the source can be inserted in a suitable hole inside the phosphor. This is only rarely worth while.

The second factor is the efficiency of stopping γ-quanta. This depends mainly on the size and density of the phosphor block. γ-rays do not have a well-defined range in matter as do charged particles, but suffer a nearly exponential absorption. This is often erroneously supposed to be due to a fundamental qualitative differ-

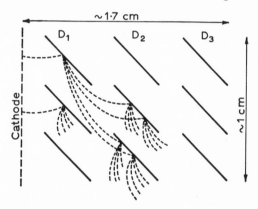

Fig. 22   Electron-paths in photomultiplier. *C*—cathode; $D_1 \ldots D_3$—dynodes.

ence between radiation and material particles. This is not so. If rabbits emerging from a cornfield during harvesting have to run the gauntlet of a uniform distribution of equally effective villagers armed with sticks, the number surviving to any specified distance can be shown to fall off exponentially with this distance without any need to invent a wave theory of rabbits. The important feature is that both rabbits and γ-quanta are stopped as the result of a single event, the probability of which is not affected by the length of path already traversed. The situation is thus mathematically analogous to that of radioactive decay. Here, too, the life of a nucleus ceases as the result of a single event, the radioactive disintegration, the probability of which does not depend on the previous life of the nucleus. The number of surviving nuclei falls off exponentially with time, just as our rabbits and quanta fall off exponentially with distance.

In contrast to this, the loss of energy by a fast charged particle takes place in a very large number of small steps. If 5 MeV α-particles pass through 2 cm of air, they each make nearly a hundred thousand collisions involving loss of energy. None is stopped, but none has the energy with which it started; the mean energy will now be about 2·8 MeV and they will be losing further energy at a considerably greater rate than before. After 3·5 cm of air and some 140,000 independent losses of energy, the average particle will be brought to rest. The probability that it should by chance have suffered even 10% fewer collisions is less than one in 10^{300}, which can be neglected for most purposes. In simpler—perhaps too much simpler—terms, the fact that α-particles have a well-defined range while γ-rays have an exponential absorption arises from statistics, not from physics.

To return to the efficiency of our scintillation counter, then, no practicable thickness of phosphor can stop *all* the γ-quanta striking it. If the absorption coefficient for γ-rays of a particular energy is μ, the proportion of quanta surviving to a distance $1/\mu$ cm, will be $1/e$ and the number absorbed will be $(1-1/e)$ or 63%. In a thickness $2/\mu$ cm, $(1-1/e^2)$ or 86% will be absorbed. To absorb 99% requires a distance of $4\cdot6/\mu$ cm. To give an idea of order of magnitude, this would mean about 18 cm of sodium iodide, or 53 cm of anthracene for 1 MeV γ-rays. Such dimensions would be inconveniently large and appallingly expensive. An absorption of 70% or less must usually be accepted. Even so, the efficiency is very high compared to the efficiency of a Geiger counter for detecting the same γ-rays. The efficiency improves still further as the energy of the γ-quanta decreases but, as we shall see, the system is unsatisfactory for the detection of quanta below about 20 keV owing to background difficulties.

4.122 *Response time.* Besides its efficiency for the detection of γ-quanta, the scintillation counter has another important advantage over gas counters. This is its quick response. The *response time* depends on several factors. Quanta absorbed in the phosphor give rise to fast electrons which are brought to rest in less than 10^{-10} sec. The useful part of their energy excites active regions of the phosphor. These excited regions then "decay" with the emission of light, much as radioactive atoms decay with the emission of particles. The excited regions have a well-defined half-life which may vary from a few nanoseconds (or milli-microseconds) for some liquid phosphors to several microseconds for some mineral crystals.

The light produced may suffer several internal reflections before finally reaching the photomultiplier, and may therefore travel some tens of centimetres in a phosphor of the size shown in Fig. 21. Most of the light, however, will be collected in less than one nanosecond. The transit time of the electrons down the photomultiplier may be thirty to sixty nanoseconds. If the time taken from photocathode to anode is constant, this is not important. The electron-pulses from several successive events in the phosphor could be following each other through the photomultiplier simultaneously and still be separable at the output end. In practice the time taken is not quite constant as the transit time from one dynode to the next depends on the exact path taken. The variation of current with time at the anode resulting from a single instantaneous pulse of light quanta reaching

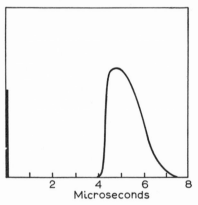

2   4   6   8
Microseconds

Fig. 23 Variation of collector-current with time following an instantaneous light-pulse at the photocathode at time zero.

the photocathode is shown in Fig. 23 for a system of the kind shown in Fig. 21 with 140 volts per dynode. The whole peak of current may take 30 nanoseconds to collect, but the main part of the rise time is less than one nanosecond. A fast amplifier which will detect only the rapid change which occurs during this rise time, can then be used to follow the photomultiplier. A possible circuit is given in Appendix I.

The longest time-constant in the system is the decay time of the phosphor and even here the time resolution can be improved if we look only at the fast initial rise of the rate of collection of light.

With a quickly decaying phosphor, the whole system can then be made to distinguish particles succeeding each other at intervals of less than one nanosecond. This is of the order of 10^5 times better than a typical Geiger counter.

Even where the very short discrimination time is not used directly as it is in coincidence work (see below), this may be valuable as it enables us to count particles 10^5 times faster than we could with a Geiger counter for the same percentage loss.

4.123 *Energy measurements.* The scintillation counter, like the ionisation chamber, gives an output which is proportional to the energy of the particle recorded.

The accuracy with which this energy can be determined is less in the scintillation counter for two main reasons. The first lies in the overall efficiency of conversion of energy.

Even an efficient phosphor, such as sodium iodide activated by a trace of thallium, turns only 20% of the energy of a particle into light. Since its main output is in the far violet region of the spectrum (4100Å) with a quantum-energy of 3 eV, emission of one quantum of light requires on the average a dissipation of 15 eV. This compares favourably with the 30 eV required per ion pair in an average ionisation chamber. The efficiency of collection of light-quanta by the photocathode of the multiplier is usually in the region of 30%, so 50 eV must usually be dissipated for each photon actually reaching the cathode. The efficiency of this cathode is about 10%, i.e. about 10 quanta must reach it for each electron liberated.

Altogether, therefore, the liberation of each electron at the photocathode will need on the average 500 eV of the energy of the originating particle or quantum.

Thus a γ-quantum with an energy of $\frac{1}{2}$ MeV, for example will on the average cause the liberation of 1000 electrons from the photocathode. In an ionisation-chamber, the dissipation of $\frac{1}{2}$ MeV would have given us some 15,000 ion pairs. The statistical fluctuations to be expected in either case are discussed in Chapter 6. Here we will simply quote the result that the percentage probable fluctuation varies inversely as the square root of the number of events involved. Hence the percentage probable error in the scintillation measurements is $\sqrt{15}$ times worse than in the gas-containing instruments.

It should be noted that this relative disadvantage applies only to measurement of the energy of an individual quantum or to discrimination between sets of quanta with very similar energies. Often we want to measure or compare the energies of fairly strong sources of γ-rays with well separated wavelengths. Statistical fluctuations can then be made as small as we like by combining the results from enough observations.

To measure the mean energy with a scintillation counter as accurately as with an ion chamber, we have to observe fifteen times as many with the former. Since we are likely with the scintillation counter to be able to count quanta a thousand times as fast, this presents no difficulty and indeed, since the counting efficiency is more than 15 times greater, the advantage lies with the scintillation counter.

The statistical reason for inaccurate individual energy-measurements is of course most important for the lower γ-energies. There is also a lower limit below which measurement of energy is impracticable. It depends on the background or *dark current* of the photomultiplier.

The removal of an electron from a metal surface absorbs energy in just the same way as does the removal of a molecule from a liquid (latent heat of evaporation). If the work done in removing one electron is ϕ *electron-volts*, the *work-function* is said to be ϕ *volts*.

The photocathode must have a low work-function to give a high efficiency of emission of electrons. Unfortunately this also leads to a considerable thermal emission of electrons even at room temperature. These electrons get multiplied just as effectively as do the officially authorised ones liberated by light from the phosphor. If they always occurred singly they would not be of much importance, but two or more may be emitted so closely together in time that the outputs resulting are indistinguishable. The frequency of such multiple emissions falls off rapidly with numbers, but in a standard instrument up to eight may contribute to a single pulse often enough to be important. Even with a sodium iodide phosphor this would correspond to the average result of absorption of a quantum of 4 keV energy. The dark-current pulses must be prevented from reaching the recording apparatus if very large background counting-rates are to be avoided. Inevitably therefore the average pulses due to 4 keV γ-rays must also be stopped. Statistical variations, of the kind already described, will lead to the loss of a proportion of γ-quanta of considerably higher energy.

The device which limits the size of pulses accepted is known as a *discriminator*. The adjustment of this cannot be perfect and in practice, if the dark-current background is effectively eliminated, γ-rays can be counted reliably only above some 15 to 20 keV.

The second main reason for inaccurate energy-measurements is of greatest importance when we wish to investigate γ-rays of one energy in the presence of a background of γ-rays of other, especially higher, energies.

To explain this we must consider briefly the mechanisms by which γ-rays transfer energy to the phosphor. There are three of these.

Photoelectric absorption. Here the whole of the energy of the γ-quantum is taken by a single electron, usually from the K-shell, which is thereby ejected from its atom and will be rapidly brought to rest in the phosphor. The ionised atom will emit a quantum of X-rays as the K-shell is refilled. If the whole of the γ-energy is to be available to the phosphor, this X-ray quantum must be absorbed in its turn with the liberation of a further electron. If the secondary quantum escapes, as may easily happen, the energy absorbed will be less than it should be by the amount of the X-ray quantum energy. The pulse of electrons from the multiplier will be correspondingly less also. In Fig. 24a is shown the distribution of number of pulses with size of pulse due to photoelectric absorption of a monoenergetic group of γ-rays in sodium iodide. The main peak represents capture of both the primary γ-quantum and all secondary radiations. The small subsidiary peak represents loss (i.e. emergence from the phosphor) of an iodine-K X-ray. This is known as an *escape peak*.* There are also escape peaks for L, M, etc., radiation loss but these are smaller owing to the less penetrating character of the radiation concerned and its consequent smaller chance of escaping from the phosphor. These smaller peaks are also much closer to the main peak and are not usually possible to resolve, though doubtless they bias slightly the position of the peak to lower energies. Escape peaks are important only at low energies; above $\frac{1}{2}$ MeV they are not normally resolved from the main peak.

Compton scattering. In this the γ-quantum "bounces off" an electron. The electron recoils with considerable energy, which is absorbed in the phosphor with production of light in the desired way, and the residual γ-quantum, with lower energy than before, travels on in a new direction. If it is also absorbed, all is well, but if it escapes, the energy of the recoiling electron is alone recorded. Since this may be any fraction of the primary energy, a continuous distribution of reduced-energy pulses may be recorded.

The pulse distribution which would be observed for a group of monoenergetic γ-quanta losing their energy by Compton scattering is shown in Fig. 24b.

The small diffuse peak *B* results from the loss of those secondary quanta which are scattered very nearly backwards. If we use a large crystal and a narrow beam of primary γ-rays, the number of secondary quanta of any kind escaping forwards or sideways can be made very small. There is however no way of avoiding the loss of the secondary quanta scattered backwards when the primary quantum interacts very near the surface through which it enters. The

* The sodium escape peak is usually too small to notice.

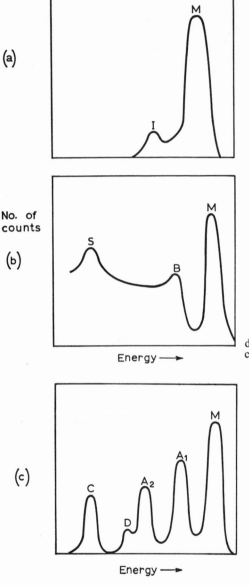

Fig. 24 Energy spectra produced by a sodium-iodide scintillation counter from monochromatic γ-ra
(a) Spectrum produced if quanta suffer photoelectric absorption. M—main peak corresponding to complete absorption of the quantum energy; I—iodine escape peak.
(b) Spectrum produced by Compton-scattering. M—main peak; B—backward escape peak; S—secondary peak.
(c) Spectra-produced by pair formation. M—main peak; A₁—peak corresponding to escape of an annihilation quantum; A₂ escape of both annihilation quanta; C—capture of one annihilation quantum; D—capture of two annihilation quanta.

peak S is due to secondary quanta Compton-scattered back into the phosphor from the shield around the counter.

Pair formation. A γ-ray with high quantum-energy may disappear in the field of an atomic nucleus, with the creation of a positron-electron pair. The rest-masses of these are each equivalent to 0·511 MeV so that pair-formation is impossible for quanta with energies less than 1·022 MeV. For higher energies than this, the excess appears as kinetic energy, shared between the positron and the electron. Since these will have ranges in the phosphor which are very small compared with the half-range of the original γ-ray, they will normally both be brought to rest in the phosphor and the whole of their kinetic energy will be recorded. As we saw in an earlier chapter, however, the positron will in due course be annihilated, together with an electron from the surrounding material, with the release in opposite directions of two fresh γ-quanta, each of energy 0·511 MeV. Either or both of these may escape, giving a pulse-height distribution of the form shown in Fig. 24c.

The peak at A_2, corresponding to loss of both secondary quanta, is usually small, since if one quantum is moving towards a neighbouring surface, the other will be moving away from it. Their combined paths must therefore be the full thickness of the phosphor along their line of flight. When a very short resolving time is used (requiring a more quickly decaying phosphor than sodium iodide), this peak may however be enhanced because there is an appreciable delay time between the stopping and the annihilation of the positron. The two annihilation quanta may therefore be emitted too late to be recorded as part of the same peak. Instead they will give yet two more peaks, C, D of 0·511 and 1·022 MeV, according to whether one or both quanta are absorbed. Each of these peaks will of course have its own entourage of X-ray escape-peaks and Compton-scatter continuum.

The complete spectrum of a single monoenergetic group of high-energy γ-rays may thus be highly complex. It is abundantly clear that measurement of the energy of a second monoenergetic set which is close to that of any of the secondary peaks will be difficult.

Similar escape-peaks are observed when proportional counters are used for measuring the energies of X-rays or soft γ-rays.

Apart from these peculiarities, the scintillation counter gives a straightforward background-count due to cosmic rays, surrounding contamination and so on. The high sensitivity to γ-rays makes this background much larger than it is for a Geiger counter and a standard 38 mm diameter by 25 mm deep sodium iodide crystal may give an unscreened background of a hundred or so per second, if all pulses above those due to photomultiplier thermal emission, or emission noise, are counted. A great deal of improvement is gained, in counting γ-rays of a known energy, if the electronic equipment includes a discriminator which will reject all pulses outside the main peak. For a high-energy γ-ray, the background may then be reduced to only a few counts a minute.

Having discussed the general properties of scintillation systems, it may be useful to tabulate the characteristics of some of the materials used. The requirements of high efficiency for absorption of γ-rays, high efficiency in transforming their energy to useful light and quick response are often conflicting and some compromise must be made. The characteristics of some of the more popular phosphors are shown in Table 4.1.

TABLE 4.1

Phosphor	Density	Refractive Index	Peak Wavelength	Decay Time	Intrinsic Efficiency	Linearity (α/β ratio)
Sodium Iodide (Tl)	3·7 gm/cc	1·775	4130 Å	0·25 μsec	20%	0·5
Caesium Iodide (Tl)*	4·51	1·788	4200-5700	0·7	10%	0·5
Anthracene	1·25	1·62	4470	0·03	10%	0·1
Stilbene	1·16	1·626	4100	0·004	6%	
0·5% p-Terphenyl in Toluene	0·86	1·50	3550	0·0022	3·5%	
NE 211	0·88	1·508	4250	0·0024	7·8%	
Lithium glass (Ce)†	2·6	1·57	3950	0·075	3%	0·23
Plastic scintillators	1·1					
e.g. Naton 136	1·1			0·0016	6%	
NE 102	1·03	1·581	4250	0·003	6·5%	
Zinc sulphide (Ag).	4·09	2·356	4500	0·2	30%	

* There is recent evidence that the efficiency of light production of caesium iodide is enhanced at low temperature. Efficiencies of up to 50% at liquid-nitrogen temperatures have been claimed.
† Lithium glasses may be obtained containing either natural lithium (7·5% Li⁶), or enriched to 96% in Li⁶ or to 99·999% in Li⁷.

The *peak wavelength* is the mean wave length of the band over which light energy is produced by the phosphor. It must match reasonably well the absorption band of the photocathode of the multiplier.

The *decay-time* represents the time over which the light output falls to $1/e$ of the initial value just after the absorption of a quantum.

The *intrinsic efficiency* represents the proportion of the total energy of the γ-rays absorbed which appears as useful light.

All phosphors give an output of light which is very nearly proportional to the energy absorbed where this is derived from γ-quanta or β-particles. Energy at very high density liberated by α-particles, however, is often less efficiently transformed to light. "Linearity" in the last column represents the relative efficiency of transformation of α- and β-energy.

The lithium glasses may be used to detect slow neutrons as well as γ-rays. Li^7 does not absorb slow neutrons but Li^6 does, according to the reaction Li^6 (n, a) $H^3 + 4.8$ MeV. The absorption-coefficients

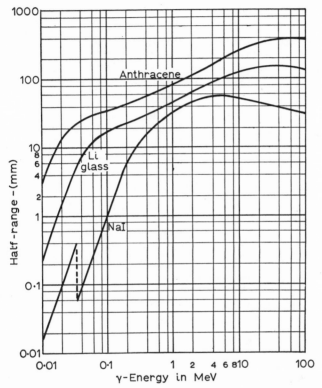

Fig. 25 Half-range of γ-rays in various phosphors. Calculated from the atomic absorption-coefficients given by Charlotte M. Davisson in *Beta- and Gamma-ray Spectroscopy*, ed. Siegbahn, 1955. The half-range in toluene is about 50% more than in anthracene; that in CsI is about 30% less than in NaI.

for slow neutrons of the natural- lithium- and Li^6-enriched glasses are 0.5 and 6 cm^{-1} respectively. Where slow neutrons must be counted through a large flux of high-energy γ-rays, the Li^7-glass can be used for this and compared with Li^6-glass.

In Fig. 25 is shown the *half-range*, or thickness required to stop 50% of quanta, in various phosphors, for γ-rays of different energies.

From this the approximate dimensions needed for a given efficiency of capture of γ-rays can be found. Over a considerable range of γ-energies, the absorption coefficients of different materials depend almost entirely on density. Fig. 25 can therefore be used to estimate such dimensions for other materials than those mentioned.

4.124 *Limitations of scintillation systems.* The practical achievement of the best possible performance is more difficult for scintillation systems than for any of the gas-filled instruments. The need for complete exclusion of light is a nuisance, the background counting rate is higher and the voltage-stability must be much better. This is because the secondary-emission coefficient of the dynodes varies rapidly with the energy of the electrons striking them and hence with the voltage between them. The overall gain of an eleven-stage multiplier varies as the eleventh power of this secondary-emission coefficient. In a practical system the gain may vary approximately as V^7. Hence a 0·1 % change of voltage may give a nearly 1 % change of output.

The mean current taken by the tube is not large, so that a high-current supply is not needed. A small output condenser of 0.01 μF or less is enough to handle the large pulsed currents.

Care must nevertheless be taken that the output-current of the multiplier is not too great. If the mean current is too large, the final dynode or the collector may be overheated. This might damage or even destroy the tube but would be easy to notice and to avoid. Even when the mean current is only a few microamperes, however, the individual output pulses must not be too big if accurate measurements of energy are needed. For this the output current of the multiplier must be proportional to the input and the multiplication at each stage must be independent of the number of electrons which pass through that stage. This will be true if the electric fields are unaffected. If the instantaneous currents to the later electrodes are large, however, space charges may be set up which will reduce the proportion of secondary electrons escaping from their dynode of origin. Furthermore, the potentials of the dynodes may change as a result of the resistance or inductance of their lead connections. This may seem very surprising when we are working with microamperes. Consider, however, the pulse due to a γ-quantum of 1 MeV absorbed in a fast phosphor such as anthracene. In a system similar to that already described, this could be expected to cause the liberation of about 1000 electrons from the photocathode of the multiplier. With eleven dynodes, this would give us an output pulse of 4.10^9 electrons from the final dynode to the collector. This represents quite a small charge; 6.10^{-10} coulombs. This charge should all be collected within some 3.10^{-8} sec, and half of it will pass in under 10^{-8} sec, corresponding to a current of 30 mA. Most of the *rise* of current may occur in 10^{-9} sec, giving a *rate of rise* of 30 million amperes per second.

Both the current itself and the rate of rise of current may cause trouble. With only 150 volts between electrodes some millimetres apart, 30 mA may well be greater than the space-charge-limited current. If so, no pulse so big as this could be observed, however large the original quantum-energy of the γ-ray, and at energies well below 1 MeV the system would no longer give a linear response. Some gain may be obtained by raising the voltage between the last pair, or last two pairs, of electrodes where the space-charge effects are most serious. This naturally makes correspondingly more severe the limitation of mean current set by consideration of heat dissipation.

The inductance of a centimetre of wire 0·2 mm diameter is roughly five microhenries. Then a rate of rise of current of $3·10^7$ amperes/sec would result in a drop in potential along a one-centimetre lead of the whole 150 volts available. Even at a tenth of this rate of rise of current, the space-charge problem would be materially exacerbated in the early part of the pulse. Again, an increase of voltage in the later stages would help but could hardly be large enough to maintain full linearity in the conditions considered.

Various alternatives may be used to solve the problem. The simplest is to reduce all the multiplier voltages, and hence the multiplication factor and the output current. If the peak current were reduced fifty times, both of the effects mentioned would be negligible up to an initial quantum-energy well above 1 MeV.

Another method, which may be more convenient when a standard stabilised voltage supply is used, is simply to take the output from the second or third dynode before the collector, where the currents are still small even under the normal working conditions. Both this and the previous method result in a smaller output pulse requiring larger subsequent amplification.

It will be clear that the effects of space charge and inductance are important only when the pulse derived from the phosphor is extremely short. If a ten-times slower phosphor such as sodium iodide were used, non-linear effects would arise only at ten-times larger multiplications. The obvious corollary is that one should never use a phosphor which is faster than is actually necessary for the experiment in hand. Where low counting-rates of high-energy particles or quanta are involved, it may even be desirable to use so slow a phosphor as potassium iodide (> 10 μsec) in spite of the serious extra background-count due to its own β-activity. The high speed of the organic phosphors is rarely of value except for coincidence measurements at high rates of counting.

In practice it is often impossible, for financial or other reasons, to choose the most appropriate equipment for each individual experiment. Furthermore, you may not have the detailed knowledge of the characteristics of each component which is needed to tell in advance whether the system will be sufficiently linear in response over the interesting range of energy. It is then necessary to calibrate the system using two or three sources giving γ-rays of accurately known energy. Even when the combination of phosphor and photomultiplier is known to be satisfactory for the job in hand this is in any case highly desirable as a final check on the overall linearity including all of the electronic equipment. The importance of the discussion above is not that it enables one to forecast a particular brand of trouble, but that it enables one to diagnose such trouble and to correct it, if it is too bad to allow for simply with the help of a calibration curve.

4.125 *Measurements on charged particles.* The strengths and weaknesses of scintillation counters have been discussed only in relation to the detection of γ-ray quanta. Scintillation counters, however, can very conveniently be used for measurements on α- or β-particles. The usable rates of counting may be tens of thousands of times greater than those of Geiger counters or gridded ionisation-chambers. The resolving times for coincidence work are correspondingly shorter. Even the least dense of the organic phosphors will readily stop the fastest β-particles which are at all likely to be observed from radioactive sources. They can therefore be used to measure energies above the limit that can be reached with a proportional counter (say 0.3 MeV in a large, high-pressure xenon-filled tube, or 60 keV in an ordinary argon tube). The size of phosphor needed is much less than for the absorption of γ-rays. A 5-MeV β-particle will be stopped by six millimeters of sodium iodide while 10-MeV α-particles would be stopped by less than 0·1 mm. Small blocks of phosphor have small background counting-rates and the thin sheets of phosphor suitable for α-particle counting may have backgrounds no larger than other types of α-counter since there is no room in so thin a layer for β-particles to dissipate more than a few tens of keV, which can easily be prevented from registering by a suitable discriminator. If a thin, light-proof window ($\frac{1}{2}$ to 1 μ of white-metal or aluminium) is placed over the phosphor, an external source can be investigated, while the ionisation chamber requires an internal source. The energy resolution will be perhaps five times worse than that of a good ionisation-chamber, but the greater convenience often more than compensates for this.

When simple detection and counting of α-particles is desired, rather than measurement of their energy, a screen of powdered silver-activated zinc sulphide is more sensitive, i.e. gives larger light-pulses per α-particle, than any other material. It may be obtained commercially already sprayed on to a perspex sheet. If desired, these sheets can be aluminised to a thickness of 2 to 3 μ so as to be entirely protected from bright lighting without the need for protection foils.

Alternatively, such a zinc-sulphide screen may be sprayed on to a thick piece of plastic phosphor suitable for the detection of β- or γ-rays. Then the light pulse from as little as 1 MeV dissipated by an α-particle in the thin zinc-sulphide layer will be much larger than that from a 10-MeV β-particle in the organic layer. Hence, with suitable electronic separation of large and small pulses, both α- and β-particles, or α-particles and radiation quanta, may be separately counted simultaneously from the same source.

Another useful method for measuring weak α-particle activities in appreciable quantities of material is to dissolve this material in a liquid phosphor. All α-particles will then be observable in spite of their short range in solids or liquids, while only a small proportion could emerge from the original material to be detected externally. This method can be used only where the α-emitting material is, or can be made into, a compound soluble in one of the phosphors available. A suspension of fine particles might be usable in thin layers but would usually absorb too much light to be useful in large volumes.

The method of dissolving the active material in a liquid phosphor may also be useful for soft β-emitters such as H^3 or C^{14}. Some liquid scintillators are based on organic phosphors dissolved in p-dioxane (diethylene dioxide). Since p-dioxane is miscible with water-soluble materials, many biological samples can be counted in this way.

4.126 Types of Scintillation Counter

The variety of types and shapes of phosphor is so great that it is impossible to give a comprehensive description of them. When planning to get any scintillation equipment it is advisable to write to the manufacturer who will on request send information concerning all phosphors currently available.

Here we shall give the sizes available only of cylinders, since these are mostly commonly needed. Rectangular shapes, or blocks with wells cut into them to take the source, can readily be made, within the same limits of size, if required.

At the time of writing, sodium iodide crystals may be obtained either unmounted or already sealed in moisture-proof cans, up to 9 in. (22·8 cm) diameter and 12 in. (30·5 cm) thick. It should be reiterated that the smallest crystal capable of recording adequately the desired activity is always the best, having both a better resolution and a lower background. It is often wise to accept a thickness which will give only a 60 to 70% efficiency in stopping γ-rays of the energy desired for this reason.

Caesium iodide crystals can be obtained up to 5 in. (12·7 cm) diameter and 3 in. (7·6 cm) thick.

Stilbene or anthracene can be obtained readily up to $2\frac{1}{2}$ in. (6·3 cm) diameter by 1 in. (2·5 cm) thick and may be obtainable in larger sizes by arrangement with the manufacturers. Plastic scintillators, which have extremely short decay-times but only moderate intrinsic efficiencies, can be made in effectively unlimited sizes. Standard blocks are available up to 24 in. (61 cm) in diameter and 24 in. long. They can also be obtained in thin, accurately controlled sheets and films down to 12 μ in thickness and up to 6 in. (15·2 cm) square, from which pieces can easily be cut to any desired shape. Plastic scintillators can also be made in flexible filaments of 1 to 2 mm diameter, or as thin-walled tubing 0·7 mm i.d., 0·4 mm wall. An assembly of parallel filaments in a hydrogenous matrix can be used for directional counting of very fast neutrons or as a fairly flexible probe for medical applications. A spiral or coil of the tubing can be used in a continuous-flow system to monitor the activity of liquids. Dual phosphors (an optimum thickness of zinc sulphide on a plastic scintillator) can be obtained in a similarly wide range of shapes and sizes.

Several liquid scintillators can be obtained and can naturally be used in vessels of any desired shape and size. Varieties capable of dissolving water-soluble active materials, of operating at low temperatures and of being much less inflammable than the somewhat hazardous toluene mixture may be mentioned.

Several special scintillators are made for the detection of slow neutrons. For example, boron-polyester mixtures can be obtained in a wide range of sizes. These detect the charged particles from the reaction B^{10} (n,α)Li^7. The cost is increased by only about 50% if the natural boron is replaced by boron enriched in B^{10}. The thickness of phosphor required for a given detection-efficiency is thus reduced five times, but is in any case so thin that very little γ-ray background will be observable.

The lithium-alkaline-earth-silicate glasses, activated with cerium, can be obtained as yet only in smaller sizes, as discs of up to 4·35 in. (11 cm) diameter and $\frac{1}{4}$ in. (6·4 mm) thick or as $2\frac{1}{2}$ in. (6·3 cm) diameter and 1 in. (2·5 cm) thick. Their detection efficiency is low for slow neutrons when natural lithium is used but they cost only one-third more if they are enriched in Li^6 when the half-range falls to 1·2 mm. These glasses are highly resistant to a variety of liquids and corrosive chemicals which would destroy most phosphors.

Any organic scintillator can detect fast neutrons with the help of "knocked-on" protons from the hydrogen contained. The thickness required for a given efficiency can be obtained as for γ-rays if the half-range of the neutrons is known. In anthracene, for example, this is about 3 cm for 1 MeV neutrons, increasing to 15 cm at 10 MeV. Both of these half-ranges are reduced to little more than half these values in the plastic or liquid scintillators.

Special "sandwich" phosphors are available, built up of thin co-axial cylindrical shells of clear plastic and of a zinc-sulphide suspension. These give both a shorter half-range than in anthracene, owing to the larger hydrogen concentration, a much larger light-output from the heavily ionising protons traversing the zinc sulphide, and little or no increase in the size of background-pulses due to γ-rays.

The range of photomultipliers which can be used with this immense range of scintillators is much less. Where equality of transit time for electrons from all parts of the photocathode is of extreme importance, a convex end and photocathode are used, so as to reduce appropriately the paths of the outermost electrons and to focus all electrons on to the central region of the first dynode. Most tubes, however, have an accurately flat glass end of $3\frac{1}{2}$ to 5 cm diameter, on the inside of which the photoemissive layer is deposited. This design makes it easy to apply the oiled surface of a flat phosphor block, to make good optical contact. When the phosphor crystal or block to be used is larger in diameter than the working face of the multiplier, a *light guide*, in the form of a short truncated cone of clear plastic, may be used to give a smooth taper from one size to the other with little loss of light.

For very large sheets or blocks of phosphor, photomultipliers are made with a much larger working face and photocathode, some 10 to 20 cm diameter. Electrons are collected from this to a multiplier composed of dynodes of the usual size and the system is not recommended where the wide variation of transit time of the photoelectrons to the first dynode would be important.

An extensive list of tubes is given in the Counting Handbook, issued by the Lawrence Radiation Laboratory, University of California. Some of the characteristics of three widely used tubes are given in Table 4.2 below and photographs of two are shown in Plate V.

TABLE 4.2

Tube	56 AVP	57 AVP	RCA 6810-A
No. of dynodes	14		14
Photocathode diameter	42 mm	200 mm	45 mm
Photocathode area	8 sq.cm	310 sq.cm	9 sq.cm
Gain at normal voltage	7×10^8	10^7	10^8
Current output	50 μA/lumen		
Dark current at full gain	5 μA		
Normal voltage	2·4 kV	2·7 kV	2·4 kV
Collector power max.	1 watt		
Mean collector current max.	2 mA		
Output peak max. (depending on voltage distribution)	100-300 mA	100 mA	120 mA
Pulse width (half height)	2×10^{-9} sec	57×10^{-9} sec	7×10^{-9} sec
Collector rise time	2×10^{-9} sec	25×10^{-9} sec	3×10^{-9} sec

4.2 Cherenkov Counters

A very fast charged particle, when travelling through matter, will radiate light by an entirely different mechanism, first discovered by P. A. Cherenkov.* The light is emitted only if the velocity of the charged particle is greater than the velocity of light in the medium. This does not contravene the principle of the theory of relativity. The velocity of the particle cannot be greater than the velocity c of light *in vacuo*, but the velocity of light in a medium whose refractive index is n is c/n. For emission of Cherenkov-radiation, therefore, the velocity of the charged particle must be between c/n and c. The refractive indexes of most glasses and plastics lie between 1·5 and 1·6. When the refractive index is 1·5, the velocity v of a particle must reach the critical value $\frac{2}{3}c$ before light will be emitted; for an electron this means an energy of 0·18 MeV and for a proton 320 MeV. In water, with a refractive index of 1·33 the corresponding energies are 0·26 MeV and 470 MeV respectively.

Cherenkov radiation can also be emitted by particles in travelling through gases, but in this case they must have such high energies that the effect is of value only in the study of cosmic rays. The

* In various books and papers the name is also given as Tserenkov, Cerenkov or Čerenkov owing to the somewhat individualistic approach of Western authors and editors to the transliteration from the Russian Черенков.

radiation does not depend on any property of the material other than the refractive index, but to be useful in detecting particles it is clearly necessary that the material should be satisfactorily transparent. If the special characteristics of Cherenkov radiation are to be displayed, the material should not produce light by any other mechanism when traversed by charged particles. Light from the material can be used to excite a photomultiplier just as can that from a phosphor. The combination of transparent medium and photomultiplier is known as a *Cherenkov counter*. It has the useful property that it is quite unable to detect any particles which have less than the critical energy however many should enter at the same instant.

A second important property of the Cherenkov radiation is that, like the bow wave of a ship or the shock-wave produced by a supersonic projectile, it is confined to a very narrow range of angles about the direction of movement of the particle. The radiation from any short length of track thus travels out along the surface of a cone whose axis is the path of the particle. The half-angle θ of the cone is given by

$$\cos \theta = c/nv.$$

When the particle is moving at a velocity only slightly greater than the critical value c/nv, when $\cos \theta \simeq 1$, the light moves in almost the same direction as the particle itself and the largest possible angle is $\cos^{-1} 1/n$. By observing the direction of emission of the light relative to the direction of movement of the particle, the velocity v can be found.

The properties of this type of counter are of great value in counting very fast protons from accelerators in the presence of a large flux of secondary particles, but the efficiency of the process is not large enough to be of much value for electrons from radioactive materials. An electron of a few MeV will give rise to only some 20 quanta per millimetre of path which have enough energy to eject electrons from the cathode of a photomultiplier, which is too little to give a pulse sufficiently above the background. The use of several photomultipliers in coincidence may help with particles which have very high energies, but there are few applications for which this seems worth while for counting individual electrons.

4.3 Semiconductor Detectors

Although the first of these was discovered over fifteen years ago, they are only now beginning to be important outside laboratories of nuclear physics. Development is taking place so fast that it is not advisable to go into much detail, which would almost at once be seriously out of date. Present performance (beginning of 1963), too, is barely up to that of the devices already described except in a few specialised fields. The potential performance is such, however, that in a decade's time one or other of the semiconductor devices may have made obsolete all of the common existing detectors except for the Geiger counter and the nuclear emulsion.

We will attempt to outline the principles and to indicate the strengths and weaknesses without discussing the more ephemeral constructional details which necessarily occupy much of the attention of semiconductor users at present.

4.31 *Semiconductor principles*

The semiconductor counter may be thought of as being very like a solid ionisation chamber. A gas is normally insulating, but if ions are produced in it by a fast particle, they can travel through it to collecting electrodes and can be recorded. A perfectly pure semiconductor, such as silicon or germanium, would also be an insulator at low temperatures, all electrons being bound in well-defined energy states or levels. A fast charged particle can produce ionisation, liberating some of these bound electrons, in a solid just as it can in a gas. In a similar way, the freed electrons can move through the solid to a positive collector. The current to this then indicates the passage of the primary particle. The detailed mechanism, however, differs considerably from that in a gas.

It is characteristic of an insulator or semiconductor that no empty energy-levels for electrons exist close to the highest filled levels. This is of course the normal situation for one individual atom in its ground state. Several electron-volts are usually needed to raise one of the electrons to the nearest empty level. In some ways a solid crystal may be likened to an enormous atom, with large numbers of electrons but a similar set of defined energy-levels. A considerable amount of energy is therefore needed to excite electrons in a semiconducting solid from any of their usual states into new levels. Once one has reached such a new level, however, the situation is unlike the case of an individual atom. In the more complex solid there are very large numbers of empty levels available with very similar energies, once we have surmounted the gap between them and the filled levels. The electron is therefore free to gain (or lose) small amounts of energy and, in particular, can gain energy from, and move under the influence of, an applied electric field. The excited electron is therefore said to have moved from the filled band (of energy-levels) to the *conduction band*. In a metal there is no such finite gap between filled and conduction bands so that there are always plenty of electrons able to move under the influence of an applied field.

In a true insulator the energy-gap between the filled and the conduction bands is so large that no ordinary influence of temperature or of minor impurities will change the situation. In a semiconductor, on the other hand, the gap is not so large. Hence thermal motions, even at room temperature, will occasionally displace an electron from the filled band to the conduction band. The number of electrons doing this, and the resulting electrical conductivity, will increase rapidly with temperature, in sharp contradistinction to the conductivity of a metal which falls as the temperature rises. Semiconductor behaviour is like that of a gas, which is normally an excellent insulator at low temperatures but which begins to suffer thermal ionisation and to become an increasingly better conductor above a few thousand degrees centigrade. Presumably solid insulators would also do this if made hot enough, but those which have been investigated have melted or even vaporised before a useful number of electrons have been able to cross the large energy-gap.

A pure semiconductor, at a temperature at which useful numbers of electrons have entered the conduction band, conducts not only as a result of these electrons moving towards a positive electrode but also as a result of the residual *positive holes* moving in the opposite direction. These arise as follows. The normal silicon crystal, for example, is electrically neutral, so that if an electron is moved away from one point of it, one unit positive charge remains behind. The positive silicon ion cannot move, but no net energy is required for transfer to it of an electron from the corresponding level of a neighbouring atom, making that, instead, the positive ion. Thus, though the atoms do not move, the positive charge can be handed from one to another. Since the charge has resulted from the extraction of a negative electron, it is known as a positive hole, or simply as a hole. When an electric field is applied, though properly speaking it is only electrons that move, towards the positive side, the effect produced is just the same as if the hole itself were moving towards the negative side. The process being more complex, the holes move more slowly in a given applied field than do the conduction-electrons. The difference of mobilities is only three to one, as compared to a difference of a thousand to one between the positive ions and the electrons in a gas, so that there is little need to separate the hole- and the electron-collection times. Furthermore, the mobilities of even the positive holes are exceedingly high, of the order of 10^3 to 10^4 cm/sec per volt/cm. This is greater than the mobility even of electrons in a gas at S.T.P. Since the solid devices may be hundreds or even thousands of times smaller than the gas ones, there is an enormous advantage in speed of action.

A pure semiconductor in which some electrons have been raised to the conduction band by thermal motions alone is known as an *intrinsic semiconductor*. Whilst easiest of all to talk about, it is extremely doubtful whether one has ever been made in practice. Electrons can be supplied to the conduction band by certain impurity atoms as well as by thermal agitation and even in the very purest zone-melted silicon or germanium (about one foreign atom in 10^{10}) the impurities are probably responsible for almost all of the observed conductivity.

Impurity atoms can produce conductivity in two ways. The silicon (or germanium) crystal lattice is formed of four-valent atoms, i.e. of atoms each of which has four electrons in its outer, valency, shell. All four are shared by the atom with its immediate neighbours. Each atom then reaches the chemically stable state with eight shared electrons in the outermost shell, four of its own and four from its neighbours. If now we introduce a five-valent atom such as phosphorus, it can fit well into the lattice if it shares four of its own electrons and four of the neighbours' but has then an odd electron over the stable eight. This electron is then very easily removed or, in other words, it is in an energy-level close to or even inside the band of conduction levels. This electron is then available for conduction. Such a five-valent atom as phosphorus is known as a *donor* impurity. When the electron moves off, it leaves a positively charged phosphorus-ion behind it but this does not now represent a mobile positive hole. This is because much energy would be required to move, to the vacant high-energy level, one of the stable set of eight electrons in the normal low-energy levels of any of the silicon neighbours. Since the charged atom of phosphorus cannot hand on its charge and since of course it cannot move itself any more than can any of the surrounding silicon atoms, it can contribute nothing to the conduction.

Not only do such five-valent atoms contribute no positive holes to the conduction-mechanism themselves, but they may capture those derived from elsewhere. Thus suppose we have such an atom whose odd electron is in a level a little *below* the conduction band. It is therefore not free to move without some extra energy, though at a level well above that of any of the filled-band electrons in the surrounding silicon. If, however, a positive hole, being passed along freely from one silicon atom to the rest, reaches the intruder, the latter will at once accept the hole by handing over its high-energy electron. This, as it drops to the lower empty level in the silicon,

liberates the excess energy as a photon or as kinetic energy of vibra-
tion of the crystal lattice. None of the neighbours has then an
electron with energy enough to raise it to the now-vacant level in the
impurity atom so that, while its positive charge remains as great as
ever, the contribution of that particular hole to the conductivity of
the crystal has finished.

If a large number of five-valent atoms are present in the silicon
lattice, therefore, large numbers of conduction-electrons will be available but, for all practical purposes, no positive holes. Silicon with
such an impurity is known as a negative—or *n-type*-semiconductor.

The second way in which we can make an *impurity* or *extrinsic
semiconductor* is by introducing trivalent atoms such as boron into
the silicon lattice. This results in an exactly complementary behaviour. The boron atom has only three electrons in its valency
shell and hence, when it has shared these with its neighbours in the
lattice and received a return contribution from them, it has still not
obtained the stable configuration of eight. It has in fact exactly the
electron configuration, though not the net positive charge, of a
silicon atom elsewhere in the lattice which has lost an electron to the
conduction band and which is therefore momentarily the site of a
positive hole. If a wandering conduction-electron approaches, it can
at once be captured into a low-energy state near that of the normal
filled band of the silicon, with liberation of energy to its surroundings. Such a trivalent atom as boron is accordingly known as an
acceptor impurity. Once the electron has been accepted by the
boron, it cannot emerge again without the return of an equivalent
amount of energy, which is highly improbable. Once again then, we
have converted a highly mobile, but this time negative, charge into a
static charge incapable of taking part in conduction. The positive
hole from which the captured electron came, however, has not been
interfered with at all by the boron and in fact has been given a
permanent existence by the removal of the electron with which it
could eventually have recombined. If many boron atoms are
present, every time a hole-and-electron pair is formed from silicon
by the lifting of the electron into the conduction band, by whatever
mechanism, the electron will quickly be captured and the hole left
free. Furthermore, little thermal energy will be needed to transfer
an electron to a boron atom direct from one of its silicon neighbours
thus creating a free hole in one step. Whichever way operates most
quickly, the presence of many trivalent atoms in the silicon lattice
will make the silicon into an impurity semiconductor just as did the
pentavalent atoms, but this time we have a positive-hole *p-type*
semiconductor.

4.32 *Bulk detectors*

For efficient use as a particle detector, the ideal system would be a block of material conducting at both ends but having no charge-carriers of either sign in the body of the block apart from those produced by the particles to be detected. The obvious method of achieving this would be to obtain a block of a true insulator, making contact with the faces by evaporating a gold film on each. Then electrons liberated by an α- or β-particle into the conduction band, and the holes produced at the same time, could be pulled out of the block by a field between the faces and the corresponding current could be recorded.

Unfortunately, the mobility of electrons or holes, in the insulators tried up to the present, is far too small. This may well be a matter of lack of purity, but meanwhile semiconductors, in which mobilities are high, must be used.

There is no difficulty in making a block of silicon conductive over a shallow region near each of two opposite surfaces. It is not yet practical, however, to eliminate the naturally occurring carriers of one or other sign in the main body of the block. These result in a non-stop flow of current which swamps the small extra pulse of current due to passage of a fast charged particle. We would not be concerned with the natural background-current if it were perfectly steady. A direct-current component will have no influence on the pulse amplifiers used to observe the particle currents. Unfortunately, the current in a semiconductor—or in any other conductor—is not constant but suffers from statistical fluctuations the amplitude of which is proportional to the square root of the current times the band-width of the amplifier. The latter must be large if short pulses are to be amplified correctly so the current must be kept down as low as possible.

Cooling the block to −196°C (liquid nitrogen) reduces the background-current to a value at which pulses due to 1·3 MeV γ-quanta have been observed with an energy resolution of 10%. This is two or three times worse than can be obtained with a sodium iodide crystal and photomultiplier. Background-currents can also be much reduced by a process known as *gold doping*. This consists in contaminating the silicon with a trace of gold. Gold atoms give rise to a complex series of energy-levels which will capture and hold either electrons or positive holes. At −130°C this can reduce the number of conduction carriers by some 10^5 times, again making it possible to observe pulses. This, however, leads to the loss by capture also of some of the electrons and holes produced by the charged particle that it is desired to observe, the holes in particular having a half-range of only about 1 cm before being trapped.

Neither system is yet of practical value, but the great simplicity of a single crystal block with a contact each end makes it so attractive compared to the phosphor-photomultiplier system that development to a usable state is likely to be rapid.

4.33 *Junction detectors*

The presently practical devices produce a useful carrier-free region in a different way, by using the properties of an *n-p junction*. Suppose in adjacent layers of a single sheet of silicon there are in one an excess of p-type impurities and in the other an excess of n-type impurities. Then on one side of the boundary there are at first free holes and on the other free electrons. If both of these diffuse across the boundary they will be captured very quickly by the opposite type of impurity to that from which they came. The region of the boundary will then be rapidly depleted of carriers of either sign, but a negative space-charge will build up on the p-side of the boundary and a positive one on the n-side, until there is a sufficient electric field to prevent any further carriers from reaching the boundary at all. A shallow layer will then be kept almost entirely insulating, and no charge will be transferred between the two sides. If an external voltage of 100 volts or so is applied in the same direction, such a layer can be increased in thickness to as much as a millimetre in good conditions.

Electrons and holes liberated in the region by the passage of an α- or β-particle will now be swept in opposite directions very rapidly by the strong electric field and the pulse of current which they comprise, and which may have a peak value of the order of 1 μA for a 1 MeV particle, is easily detected in the absence of any important irrelevant background.

4.331 *Analysis of properties of junction-detectors.* Some of the quantitative properties of the system can be understood better with the help of a short piece of analysis. Consider a coordinate system with the axis of x perpendicular to the plane of the n-p junction, with $x=0$ in this plane, and x positive in the n-type material.

Suppose that the density of positively charged donors in the n-region is N_n per cm³ and that the density of negatively charged acceptors in the p-region is N_p per cm³. Then the space charges in the two regions are $e \cdot N_n$ and $e \cdot N_p$ respectively, where e is the electronic charge. Consider first the n-region.

D

Here, by Poisson's relation, if we assume that the potential does not change in the y or z directions,

$$\frac{d^2V}{dx^2} = +4\pi e N_n/k$$

where k is the dielectric constant of the material. Then

$$\frac{dV}{dx} = 4\pi e N_n(x_n - x)/k,$$

where x_n is the distance from the boundary at which the electric field becomes zero.

Then $V = 4\pi e N_n(x_n x - \frac{1}{2}x^2)/k$ if we take $V=0$ where $x=0$, i.e. measure potential from the boundary.

Where $x = x_n$ the field has dropped to zero and this must therefore be the limit of the depleted region within which we have charged donor atoms. If outside this region the voltage is approximately constant, the total voltage V_n on the n—or donor—side of the junction is the value of V for $x = x_n$, i.e.

(a) $$V_n = 2\pi e N_n x^2_n/k \qquad 4.1$$

Similarly, the voltage on the acceptor or p-side, V_p, is given by

(b) $$V_p = -2\pi e N_p x^2_p/k. \qquad 4.1$$

Now, for overall neutrality, the total number of charges on each side of the junction must be the same.

$$\therefore \qquad N_p x_p = N_n x_n.$$

It is usual to make one layer very thin to facilitate the entry of α- or β-particles. If this is the acceptor layer, of p-type silicon say, x_p will be very small compared to x_n. Hence $N_p \gg N_n$, and to make the p-layer thin we must produce in it a much larger concentration of acceptors than there is of donors in the n-type region.

Now from equation 4.1,

$$V_n/V_p = N_n x^2_n/N_p x^2_p$$
$$= x_n/x_p$$

But $\qquad x_n \gg x_p,$

hence $\qquad V_n \gg V_p.$

For practical purposes, therefore, the total voltage across the whole system will be little more than V_n, and the thickness of the depletion layer will be little more than x_n.

Equation 4.1a thus tells us the dimensions of, and voltage across, the depletion layer. We are most interested in the thickness since the depletion layer is also the sensitive layer, and this thickness determines the maximum energy of particle which can be stopped within it. For a given voltage, $x_n \propto 1/\sqrt{N_n}$. For a thick layer, therefore, we want a low density of donor impurities. This is equivalent to saying that we want the n-layer to have a low electrical conductivity. We cannot change e or k so the remaining factor to notice is that x_n is proportional to $\sqrt{V_n}$. This is highly important as it gives us at once a facility available in no other detector; the ability to vary the sensitive volume electrically to suit the particles we wish to count. Thus if we want to count α-particles with a range in silicon of 20 microns, we can use a low voltage and the system will be nearly insensitive to fast β-particles. On the other hand, by raising the voltage we can extend the sensitive region to the point where it can absorb the whole energy of a 1 MeV β-particle with much larger range. In the units built so far a change in the thickness of the sensitive layer of some four or five times is as much as can be achieved, but even so the advantages are very considerable.

One more quantity can be obtained from the thickness x_n of the carrier-free region. This is the capacitance per unit area of the counter. In effect we have two conducting plates separated by x_n so that this capacitance C_n is $k/4\pi x_n$ per unit area or

$$C_n = (kN_n e/8\pi V_n)^{\frac{1}{2}},$$

which varies inversely as the square root of the voltage applied.

The resistivity of the silicon employed is inversely proportional to the number of donors, so that for practical use we can substitute numerical values for all of the other known properties of the silicon, giving us approximately as design figures,

$$x_n \simeq 0.5\sqrt{\rho V_n} \text{ microns} \qquad\qquad 4.2$$
and
$$C_n \simeq 1.8 \times 10^4 (\rho V_n)^{-\frac{1}{2}} pF/\text{cm}^{-2},$$

where ρ is the resistivity in the absence of the depleting field.

If we construct our junction counter from a very thin layer of n-type silicon on a base of p-type silicon, the effective extent and voltage of the counter will be x_p and V_p respectively, and the capacitance will be $C_p = (kN_p e/8\pi V_p)^{\frac{1}{2}}$ per unit area giving us in practical units

$$x_p = 0.3\sqrt{\rho V_p} \text{ microns} \qquad\qquad 4.3$$
and
$$C_p = 3 \times 10^4 (\rho V_p)^{-\frac{1}{2}} pF/\text{cm}^{-2}$$

The numerical constants differ slightly from those of equation 4.2 because the mobility of the holes in the p-type silicon is less than that of the electrons in the n-type, giving a somewhat higher resistivity in the p-type for the same number of impurity atoms and carriers.

4.34 *Methods for production of semiconducting detectors*

The layer construction of the detector implies that the particles to be detected must enter through one of the layers. If the whole energy is to be recorded, the sensitive region must come as close as possible to the surface. This necessitates a high concentration of impurity atoms over a very small distance. This is at present achieved in two ways. The first is by diffusion. A sheet of high-resistance p-type silicon (i.e. with a low concentration of the acceptor impurity) is heated to 800° to 900°C in phosphorus vapour or in contact with a phosphorus compound for a few minutes. Phosphorus then diffuses in for about 0·1 microns, swamping the p-type impurity and replacing it with a much higher concentration of n-type. Particles can thus reach the actual junction by traversing only 0·1 μ of material. In effect therefore the counter has a "window" which is less than 0·1 μ thick. Even for α-particles this is negligible and the counter can fairly be regarded as "windowless".

The second method, which produces what is also known as a *surface-barrier* detector, is simply to allow oxidation of the initially clean surface of n-type silicon or germanium. A fully effective p-type surface layer is then produced, even thinner than the surface layer in the diffusion detector.

In either case contact is made to the two faces of the composite block by the evaporation on to them of a very thin gold film (0·04 mg/cm² or so) to which leads may be cemented.

The surfaces usually need protection from the air, and when in use from light, to which they are also somewhat sensitive. The final structure is simply a tiny wafer, from a few millimetres to a centimetre or so in diameter, on a pair of fine leads. The simplicity compared to the ionisation-chamber is most striking.

As has been indicated, these detectors are only in their infancy and any statements of practical performance should be improved upon by the time this is published. Apart from the limitations of existing counters themselves, they need wide-band amplifiers with high gain to exploit their potentially short resolving time. The use of photomultipliers following scintillation counters has made a very high gain unnecessary for the amplifiers used with these. Ionisation chambers need amplifiers with a high gain, but the comparatively long pulses produced by such chambers have needed only small band-widths. Rapid development of amplifiers is therefore also taking place. In spite of these difficulties, junction detectors have been made which will detect β-particles down to about 20 keV, and give measurements of their energies, linear up to 350 keV, with an error of only $\pm 7\frac{1}{2}$ keV. α-particle groups have been resolved to 15 keV at 6 MeV.* This is already as good as the resolution of the best ion-chamber, and this should be by no means the limit. In fact, the high possible resolving power may represent the strongest of all of the many advantages of semiconductor devices. The reason for this is quite fundamental.†

We have seen that the liberation of each electron from the photo-cathode of a scintillation system involves on the average the absorption of 500 eV of energy from the primary particle in the phosphor. Production of an ion pair in the ionisation chamber requires some 35 eV on the average. Production of a free electron and a positive hole in a semiconductor needs on the average only about $3\frac{1}{2}$ eV. Hence in the junction detector a given primary particle will produce ten times as many charges for collection as it could in the ionisation chamber. The random statistical fluctuations in the number produced are then reduced, as a proportion of the total, by $\sqrt{10}$ times and the energy-resolving power ultimately attainable should be better than that of any detector yet constructed.

Numerically this means that a probable error of only 0·13% at 1 MeV or $0·13/\sqrt{E}\%$ at E MeV may be obtainable.

Such an accuracy will take some time to achieve and it is worth pointing out that in some cases it may never be obtainable. In the preceding paragraphs it has been assumed that the whole of the energy of the primary particle is dissipated in the detector. If this is true, it does not matter whether the primary particle produces the free holes and electrons directly or through intermediary particles. Suppose, however, that we wish to detect the passage of a fast primary particle which passes right through the detector without much loss of energy. We might, for example, use two well-separated counters as a sort of telescope. Only those particles would be recorded which passed through both counters and if these were small and far apart the direction of movement of the primary particles could be found with great accuracy.

Now to reduce background, it would be valuable also to record only those particles which dissipated the expected energy in each counter, the energy-range accepted being made as narrow as possible. On the arguments above, it might be supposed that if the first counter were a perfect solid-state one and if say 1 MeV on the average were dissipated there, an energy-range of $\pm\frac{1}{4}\%$ of 1 MeV, or 5 keV,

* Commercial detectors with a performance at least as good as this are now available.
† If the reader is unfamiliar with the calculation of errors, the following paragraphs should be left until after section 6.2 has been read.

should accept most of the desired particles. This would not be so. Primary particles of several MeV, in traversing matter, produce secondary electrons or δ-rays with energies up to several keV. The number of these in a short length of primary track will be small and will therefore fluctuate largely. Hence the energy dissipated in the thin counter will show unavoidable variations corresponding to a probable error of perhaps 5% rather than to the 0·13% which we could expect for a 1 MeV particle coming to rest in the counter.

The only consolation for this undesirable and quite inescapable result is that the increase of error in such cases due to the present imperfections of solid-state, or even scintillation, counters is relatively unimportant so that the foremost counter in such a telescope can be freely chosen for its convenience rather than for its energy-resolving power.

4.35 Lithium–drifted Detectors
See Appendix VII, p. 338.

CHAPTER FIVE

DETECTION OF PARTICLES
III. DETECTORS ALLOWING DIRECT DISCRIMINATION

The detecting systems described so far will tell us whether a particle has passed through them, when it passed and often how much energy it dissipated in the process. To use them satisfactorily, however, we need a good deal of independent information as to what kind of event is occurring. If we observe a particular large pulse from a scintillation counter we have no direct way of telling whether it is due to one of the γ-quanta we may be looking for, to an α-particle from some contaminating natural radioactive material, to a cosmic-ray meson or to an accidental coincidence of two or more β- or γ-rays of lower energy. When we have a sufficiently strong source whose nature is known, as is usually the case, this is not important. We shall have to allow for a background but, so long as this is steady and small, its origin is unimportant. When, however, we are faced with a novel or complex situation it may well be important to investigate events qualitatively before settling down to count them.

Several instruments are available for this. We shall consider briefly the cloud chamber and the bubble chamber and rather more fully the photographic emulsion. All these make possible the direct visual discrimination of different kinds of particles.

5.1 The Cloud Chamber

If air saturated with water vapour is suddenly cooled, it will become supersaturated. If the degree of supersaturation is great, much of the water vapour will at once condense into a dense cloud of fine droplets. At moderate degrees of supersaturation, however, droplets will be formed only if there are suitable "nuclei" on which condensation may start. Fine dust particles, positive or negative ions or perhaps even individual hygroscopic molecules may all act as condensation nuclei. If moist air is cleaned of all other nuclei and then suddenly cooled to a moderate extent, droplets of water will indicate the presence of ions. If a fast charged particle passes through just before or just after the cooling process, the ion pairs along its path will be quickly surrounded by a corresponding series of water droplets. Observation of the latter then shows exactly where the particle has passed and, since the density of ionisation depends on the speed and charge of the primary particle, the frequency of the water droplets along the track gives a good idea of this speed and charge. The row of droplets, or *track*, left by the primary particle can be seen directly in a good light against a dark background. It can also be photographed for a permanent record.

There are two main ways of cooling the air suitably.

5.11 *The expansion chamber*

In this, the moist air is allowed to expand suddenly by opening a valve which connects the main chamber to a second volume at lower pressure. The moist air cools on expansion and is in a sensitive state for a few hundredths of a second. After this it is warmed by conduction from the walls and must be recompressed and resaturated before the next expansion. The source of water is usually a wet black velvet sheet on the floor of the chamber. The black velvet forms a good background for observation through the strong glass top of the chamber. For photographic work it is usual to provide a brief, high-power electronic flash immediately after the expansion. This avoids the waste of power and risk of non-uniform heating of the chamber that would result from the close proximity of a continuously burning high-power lamp.

The expansion chamber has the advantage of giving a uniform degree of supersaturation over a large proportion of the volume of the chamber. The supersaturation varies with time after the expansion, but if, as is usual in serious work, a light flash lasting only a few tens of microseconds is used, the effects of this will be negligible. A serious disadvantage is that the sensitive time is short and the *recycling time* needed before a fresh expansion is possible may be several minutes. By special techniques the recycling time may be brought down to ten seconds or less, but this is difficult and even so the instrument will be sensitive for less than 1 % of the time, with a corresponding inefficiency in its detection of randomly occurring particles. Photographs are usually taken simultaneously by two cameras so placed as to take a convenient stereo pair of photographs. In Plate VI is shown a single photograph of the track of a fast carbon ion. The tracks of several secondary electrons knocked from the atoms of the gas by the carbon ion can also be seen. These are immediately distinguishable not only by their lower drop-density but also by their frequent and erratic changes of direction. The heavy carbon ion must pass exceedingly close to an atomic nucleus to be deflected appreciably, but the electrons, over twenty thousand times lighter, can be seriously deflected at much greater distances and hence suffer such deflections a great deal more frequently. Their tracks can therefore be distinguished at sight without need for special equipment or measurements.

As in the case of the ionisation chamber, sources of particles must be put inside the cloud chamber or close to suitably thin windows in its walls.

5.12 *The diffusion chamber*

The second method of producing supersaturation by cooling is embodied in what is known as the *diffusion cloud chamber*. In this a large temperature gradient is set up between a warm top, where also the source of water or other liquid is placed, and the bottom which is cooled to well below the freezing point of the liquid, usually by solid carbon dioxide. Liquid evaporates at the warm top of the chamber and diffuses downwards into cooler air. At a certain level in the chamber, the saturation point is reached, and below this the diffusion maintains a steady supersaturation which will cause the condensation of droplets on ions appearing there at any time. The sensitive region is usually a good deal less than half of the total depth of the chamber. This is not serious as in any case the cameras used will give sharp photographs only of a track lying fairly near the plane on which they are focussed. More important still, the size of drops will vary with depth so that tracks at different levels will look different even when they are perfectly focussed and due to similar particles. Photographs of tracks in the diffusion chamber have always a serious background, partly due to the presence of old, diffuse, tracks and partly to the steady rain of droplets produced in the part of the chamber below the sensitive region. The fact that the pressure is constant makes it easier to make very large chambers, however, for observing the complete tracks of particles with large energies. For visual observations the poorer quality of tracks is unimportant and the gain of some ten thousand times in detection efficiency over a period of time is extremely valuable.

Water is not the best liquid to use in a diffusion chamber. Ethyl or methyl alcohol is usually used, the latter being better, but less pleasant to handle because if you breathe too much of it you may become blind as well as merely drunk.

An important advantage of either kind of cloud chamber is that it can be used in conjunction with a magnetic field of reasonable magnitude to determine the momentum of the primary particles when the charge of these is known. Where the mass is also known, the energy is calculable. The sign of each particle is indicated by the direction in which it is deflected and the momentum mv can be determined from the radius of curvature r of the track, using the relation

$$mv = Ber$$

where B is the magnetic induction in gauss and e is the charge on the particle in e.m.u.

5.2 The Bubble Chamber

This is the exact inverse of the cloud chamber. Instead of looking for droplets of liquid in a supercooled vapour, we look for bubbles of gas in a superheated liquid. In the absence of any nuclei for bubble formation, a liquid can be superheated considerably without boiling. Ions of either sign can act as nuclei, so that a fast charged particle moving through such a superheated liquid will leave behind it a track formed by a row of bubbles of vapour. Once formed, these bubbles will grow rapidly, or even explosively, and must be photographed very quickly if accurate measurements are to be made. Like a supersaturated vapour, the superheated liquid can be kept in a sensitive state for only a very short time. It is usually heated under pressure to a steady temperature a little below the boiling point at the pressure applied. It is then brought into a sensitive state by lowering the external pressure suddenly to below the vapour pressure. If the chamber is initially completely full of liquid, this involves only a very small change of volume, owing to the low compressibilities of liquids. The light-flash and camera-shutter for observation will be timed to operate a millisecond or less after the expansion. Direct naked-eye observation is not usual and could be useful only for highly penetrating particles as in most liquids the lengths of the tracks will be a thousand times less than in the gas of a cloud chamber.

The first liquid used for a bubble chamber is said to have been beer, but liquids of more easily determined constitution are more usually employed. Organic compounds such as propane will operate at convenient pressures and will give good tracks at a temperature near to room temperature. When it is desired to study interaction between the entering particles and nuclides in the liquid itself, it is much better to use a pure element. Liquid argon, liquid hydrogen (particularly for research on fundamental particles) and even liquid helium have been used. The chambers using organic liquids, and still more those using liquid hydrogen, represent serious fire hazards and all bubble chambers are capable of explosions which, though not in the megaton range, can do a lot of damage in a laboratory. It has therefore been suggested that one should never engage a designer of bubble chambers who has *not* produced one explosion or who *has* produced more than one. If he hasn't produced an explosion yet, he won't take the danger seriously until he has done so; if he has produced more than one, he is careless.

In Plate VII is shown a photograph of tracks in a hydrogen bubble chamber.

5.3 Photographic Emulsions

Both the cloud chamber and the bubble chamber, though they give a lot of useful information, require complex and expensive equipment. Much qualitative and some quantitative information can be obtained far more cheaply using a photographic emulsion. Sometimes an ordinary process plate may be used, but for most purposes X-ray film or one of the special emulsions designed for experimental work in nuclear physics will be necessary.

A photographic emulsion consists of a large number of fine grains of silver bromide dispersed in gelatine. The emulsion is usually spread in a thin layer on a plastic film or a glass plate. When a fast charged particle passes through a grain it may produce a few free uncharged atoms of silver, just as does light when it falls on a grain These atoms can act as starting points for the chemical reduction of the whole grain to metallic silver by a mild reducing agent or *developer*. As a charged particle traverses the emulsion, it will leave behind it a row of developable grains. After development, these grains become a row of specks of silver which show the path traversed. Undeveloped grains of silver bromide can be removed by *fixation* or solution in sodium thiosulphate ("hypo") which does not dissolve the developed specks of metallic silver. An emulsion is dense and the range of particles in it is small—less than one two-thousandth of their range in air. The individual grains are also small and a microscope is needed to see the track left by an individual particle.

The simplest use of an emulsion is like the use of a simple ionisation chamber—to indicate the passage of a large number of particles, without distinguishing the effects of individual particles. The first observation of the radioactivity of uranium by Becquerel was made with photographic emulsions and these are still the most widely used detectors in use today. An example is the film badge used by people working with radioactive materials. It consists simply of a small piece of X-ray film which is wrapped in black paper. The radiation to which it has been exposed is measured by comparison with a series of similar films exposed to known amounts of radiation and developed at the same time. This device is described more fully in section 14.12.

Emulsions may be used in a similar way to compare the strengths of radioactive sources. A series of sources at a standard distance from a corresponding series of films can be compared, using a photometer to measure the fogging produced on the film in a standard time. This method is probably used less often than it should be when large numbers of weak or short-lived sources are to be examined. No more work is required for a month's exposure than for ten minutes so long as everything can be left undisturbed in the interval. Thirty or forty samples can be dealt with simultaneously without appreciably increased cost, while thirty or forty Geiger counters with scalers, or even simple ionisation chambers, would cost a great deal more than one. When very weak specimens with large areas are concerned, a Geiger counter with a low background is more sensitive than a film, but when a very small (say, 2 mm diameter or less) source is involved, a film placed directly against it and left for a month or two could detect considerably smaller activities than could the counter.

This leads us directly to the next use of emulsions. If an extended source of α- or β-particles is placed in contact with an emulsion, the developed emulsion will show directly not only the total activity but its distribution over the source. The picture resulting, produced by the direct action on the emulsion of the radiations from the source itself, is known as an autoradiograph. This technique is often important, for example in finding the distribution of radioactive elements in minerals (see Plate VIII) or of radioactive compounds in biological structures (see Plates XVI to XVIII). In good conditions the resolution may be much better than 1/10 mm, but is always limited by the range of the particles being observed. If a β-emitter gives particles with a range in emulsion up to a millimetre, even a point source will give rise to a grey disc, whose density falls off steadily from the centre out to a millimetre radius. A point source of α-particles with a range of 20 microns, on the other hand, will give a correspondingly smaller disc of confusion and may allow location of the point source to a few microns. This method will be described in more detail in later chapters. Here it will suffice to say that for best sensitivity a film coated with emulsion on both sides, such as Kodirex or Ilford industrial G X-ray film may be used, while for the best definition a single-sided film such as Flurodak (Kodak) is preferable. When great sensitivity is not the main consideration and when the pattern of distribution is more important than the absolute activity, an ordinary process plate is better than any of the X-ray films. All of the latter give a good deal of background fog, i.e. when developed they are distinctly grey without any exposure at all. The process plates are entirely free of such background fog and hence the image due to the source shows up with much clearer contrast. The advantage of this is so great that it wipes out a large part of the advantage in sensitivity of the X-ray films.

The methods described so far are probably used in over 98% of the work done with photographic emulsions in applied nuclear physics. Their discriminating power is however very limited.

5.32 Nuclear emulsions

A more exacting, but much more powerful, technique has been developed since the war, particularly by Powell and the Ilford laboratories, using emulsions specially designed to detect the tracks of individual particles. These emulsions enable us to get information which could be obtained in no other way so, although their use is less frequent, it is extremely important. They are made in several grades, of differing sensitivity. The least sensitive ones can detect, of the particles in which we shall be interested, only fission products, α-particles and the slowest protons. The most widely used grades will detect α-particles of all energies and protons of energies up to 80 MeV or a little more. The most sensitive will detect β-particles, even when these are moving so fast that their ionisation is at a minimum. In Tables 5.1 and 5.2 are shown some characteristics of

TABLE 5.1

Characteristics of Ilford Nuclear Emulsions

Mean Grain Diameter Emulsion Group	$0.27\,\mu$ (G)	$0.20\,\mu$ (K)	$0.14\,\mu$ (L)
Particles detectable by the emulsion	*Code name of the emulsion*		
All charged particles, any energy	G5	K5	L4
Protons up to 80 MeV Slow electrons just detectable		K2	
Protons up to 7 MeV		K1	
Protons up to 5 MeV α particles		K0	

G5, K2 and K5 emulsions are available diluted with gelatine to two to four times the normal gelatine/silver ratio.

K1 and K2 emulsions are available loaded with 16 mg of lithium or 23 mg of boron per ml.

10 μ and 50 μ plates are available and either plates or *pellicles* (sheets of emulsion without support) may be obtained in thicknesses 100 μ, 200 μ, 400 μ, 600 μ, 1000 μ and 1200 μ (to ±10%).

All unloaded emulsions are also available in gel form.

TABLE 5.2

Composition at Room Temperature, 58% Relative Humidity

Ag	.	.	.	.	1·82 gm/ml	10·2 × 10²¹ atoms/ml
Br	.	.	.	.	1·34 ,,	10·1 ,, ,,
I	.	.	.	.	0·0120 ,,	0·056 ,, ,,
C	.	.	.	.	0·277 ,,	13·9 ,, ,,
H	.	.	.	.	0·0534 ,,	31·6 ,, ,,
O	.	.	.	.	0·249 ,,	9·4 ,, ,,
N	.	.	.	.	0·074 ,,	3·2 ,, ,,
S	.	.	.	Total .	0·0072 ,,	0·135 ,, ,,
					3·83 ,,	

All to 1% standard deviation.

Over a small range on either side of 58%, an increase of 1% in relative humidity adds about 2 mg of water to each ml of emulsion.

several of the Ilford series of emulsions, which are at present the ones most widely used.

In Plates IX to XII are shown tracks of several kinds of particles magnified a thousand or so times.

The β-particle tracks (Plate XII) are more scattered than in the cloud and bubble chamber photographs. This is due to the larger atomic number of the nuclei in the emulsion responsible for the scattering. (The probability of scattering at any given angle varies as Z^2, where Z is the atomic number of the nucleus responsible.)

While the difference between the track of a β-particle and that of any of the heavier particles is obvious to the merest beginner, it needs a little experience to distinguish between α-tracks and proton tracks. The α-tracks are denser and less scattered and with some practice can be distinguished very well so long as they are of reasonable length—say 20 microns or more.

5.321 *Energy measurement.* Once the particle responsible has been identified, the range, or length of its track in the emulsion, can be used to find its energy. Naturally, for this the particle must be brought to rest, i.e. the track must end, in the emulsion. High-energy particles may well pass right through to the glass, where their tracks will stop, however much energy the particles may still possess. Tracks which reach the glass can easily be identified by focussing up and down and, though they may be useful to show the number of particles which have struck the emulsion, must not be used to determine energy. Careful measurements of range as a function of energy have been made by several research teams and their results can be used for calibration purposes.

In Appendix III are given the results of these in a convenient form. Different nuclear emulsions differ only slightly in density and the error made in using the same table for all of them is less than the error of measurement of a single track. Where many similar tracks are measured to find an accurate mean energy, it may be worth while to check with the manufacturers the characteristics of the particular emulsion used.

The measurements are best done with a calibrated eye-piece scale in a good microscope. A high magnification is needed; at least a $\times 45$ ($\frac{1}{8}$) objective should be used, with $\times 10$ or $\times 15$ eyepieces. Only the projected length in the focal plane of the objective can be measured in this way; the distance moved by the particle in the third dimension must be found by focussing the microscope carefully first on one end of the track and then on the other, using the highest-powered objective available, and recording the distance moved by means of the fine-focussing control. Pythagoras's theorem applied to the horizontal and vertical measurements will then give the true range. The vertical measurement cannot be used directly but must first be corrected for shrinkage of the emulsion during processing. If, as is usual, the emulsion is on a glass plate, it cannot shrink side-ways, but will shrink about 2·3 times in thickness.* The shrinkage varies somewhat with the humidity of the air in which the plate is stored after processing but except in very accurate work the effect on the correction factor is less than the experimental error.

The ranges of individual β-particles are difficult to measure and are not usually important. Owing to their extensive scattering it is necessary to estimate the total length of path following all the meanderings of the track. This will be called *along-the-track* measurement. It is a laborious process and is usually carried out only for one or two sample tracks to give an idea of the mean energy. A range-energy table up to 250 keV is given in Appendix III.

A new method which is a good deal easier when the mean energy of a number of tracks is needed, has been investigated by Levi and Rogers. The starting point of the track in the emulsion is taken as origin and the distance from this is measured to the furthest point of the track. This distance is evidently the radius of the smallest sphere, with centre at the origin, that can be drawn to enclose the whole of the track. It is therefore called the *radius* of the particle track. The main practical difficulty is to see that two or more β-tracks do not get mistaken for one. If a hundred or so tracks are measured and the number with radius greater than r is plotted on semi-logarithmic paper against r, a kind of absorption curve is obtained from which the energy may be found (see Appendix III).

* If the final soak, before drying, is in a 5% solution of glycerol, this shrinkage factor is reduced to 2·0.

Where the strength of source is adequate, it is quicker and better to measure the absorption in aluminium as described in section 6.22.

The ability of an emulsion to detect individual particles makes it an exceedingly sensitive detector of radioactivity. The limits are fixed by the length of time for which it can be exposed to the source and, like any other detector, by the background of tracks due to particles other than those from the source.

5.322 *Fading and background.* The α-sensitive emulsions can be exposed for several months without special precautions. For longer periods they must be kept cool, at a low humidity (not completely dried in vacuum or by phosphorus pentoxide, as this will cause the emulsion to crack and even to tear pieces of glass away from the surface of the plate) and in an atmosphere free of oxygen. In ordinary air the silver atoms forming the latent image are slowly oxidised so that after a few months the α-particle tracks formed at the beginning of the period are rather thin and weak and after six months to a year may have faded away altogether.

The background of α-tracks is usually very small, perhaps one track per sq.mm per month in an emulsion 100 μ thick. It must be remembered, however, that an emulsion is an integrating detector that cannot be switched off and that it will have been picking up background steadily since manufacture was completed. New plates should always be used when a low background is needed. Individually wrapped plates should be avoided if possible, as, though convenient, the black wrapping paper is usually appreciably α-active, very probably with plutonium from atomic bomb tests.

Unless the source is less than 20 microns or so in thickness, no α-particle will come in from outside during the exposure and, apart from impurities in the source, the background will originate only from a trace of thorium in the emulsion and a rather larger trace in the glass. Th^{232} itself will give rise to a single α-particle and the lives of the daughter-nuclei are such that a second α-particle from the same nucleus is unlikely in the period of the exposure. Later members of the thorium decay-series, however, have short-lived descendants and any α-emission will be accompanied by at least two more. Hence the tracks they produce will occur in groups of three or more originating from a point. These groups are the well-known "thorium stars" (Plate IX) which will soon be recognised by anyone working with nuclear emulsions. Sometimes the tracks will appear to diverge not from one point but from two points a micron or so apart. This is due to the fact that one of the members of the decay chain is a gas, emanation 220. This has a half-life of 52 seconds, during which time, if it is not trapped by recoil in a gas-tight grain of silver bromide, it may diffuse a little way through the gelatine before decaying again.

Since there is much more thorium in glass than in the emulsion itself, most of the tracks seen will appear to emerge from the glass. Often one sees a group of tracks, as in Plate XIII which clearly originated at one nucleus just outside the emulsion. These tracks can be recognised and neglected. They are eliminated if the emulsion is poured on to perspex instead of glass. Under the best conditions the emulsion is so good that the real limitation for long-lived sources is neither fading of tracks nor background but the patience of the experimenter.

The limitations for β-detection are similar but more severe. Fading of old tracks often sets in after a few weeks in unprotected plates, but storage in dry nitrogen prevents this just as it does in the case of α-particles. The integrated background between manufacture and use, even if plates are specially delivered, may easily be hundreds of tracks per square millimetre if one does not live in Ilford. It is possible to eradicate these tracks by keeping the plates for 24 hours in moist air at 60°C but this may well reduce sensitivity. If background is really important it is much better to pour fresh emulsion oneself, as described later in this section and in Appendix IV, just before starting the experiment.

The rate of addition of fresh background tracks in an unscreened plate is likely to be some 200 to 300 per mm^2 per day, mainly due to γ-rays from material surrounding the laboratory and from cosmic rays. In a well-screened chamber this can be reduced several times and in a screened chamber deep underground can be reduced nearly a hundred times. The main source of background is then from K^{40} in the glass. If glass is replaced by Perspex, we can get down to one or two tracks per mm^2 per day. This is approaching the point at which the β-tracks due to decay of C^{14} (half life 5700 years) become important. C^{14} forms rather more than one part in 10^{12} of ordinary carbon in living matter and is therefore an ineradicable part of the gelatine of the emulsion. Gelatine cannot yet be synthesised and is derived from calves' feet. Its radioactivity could doubtless be avoided if the calves' mothers and the calves themselves were fed all their lives on C^{14}-free food. This could be produced in an airtight greenhouse by supplying the growing food-plants only with carbon dioxide obtained by burning coal or from limestone, in both of which all the C^{14} has long since decayed.

This looks to be a promising new industry for Guernsey, which has plenty of both cows and greenhouses, but so far as is known to the author has not yet been attempted.

If only those tracks are counted which can be seen to pass through the surface of the emulsion which faced the source, the background is reduced yet further to a few per mm² per month. This is very much better than for any other instrument. To attain such low values, however, involves a degree of inconvenience—particularly the requirement for deep underground storage, which must be at 100 metres or more—that gravely reduces the advantage in simplicity of emulsions when compared with other detectors.

5.323 *Neutrons and γ-rays.* γ-rays and neutrons cannot be detected directly. The former, however, will give rise to Compton- or photo-electrons which will appear to start within the emulsion rather than to come from the surface and hence the presence of the primary quanta can be inferred. Similarly, fast neutrons will make occasional collisions with hydrogen nuclei in the emulsion. The track of the resulting "knock-on" proton can be seen and measured. If we know the direction from which the neutron came, and note the direction as well as the length of the proton track, the original energy of the neutron can be calculated.

Slow or thermal neutrons cannot be detected in this way, although if very large numbers of them are present, enough will be absorbed by silver and bromine nuclei to produce radioactive isotopes of these elements. When these decay, by β-emission, the β-tracks would be visible in a G5 or similar emulsion and the neutrons thus detected indirectly. This is not a very satisfactory method, as it does not permit an easy distinction between slow neutrons and γ-rays. To detect slow neutrons unambiguously it is usual to use one of the α-sensitive emulsions, such as K2, into which is incorporated a salt of boron. When the nuclide B^{10} absorbs a neutron it breaks up into an α-particle and a lithium nucleus according to the scheme

$$B^{10} (n, \alpha) Li^7.$$

The energy liberated in this process is 2·8 MeV which drives the two product-nuclei apart fast enough to give a very characteristic short track in the emulsion. A boron-loaded emulsion 100μ thick will stop nearly 30% of thermal neutrons but some of these will be absorbed by silver or bromine so that the efficiency of detection by such an emulsion is nearer 10%. This compares quite favourably with a boron-trifluoride counter.

Apart from its ability to distinguish the character and energy of different particles, the nuclear emulsion makes possible a far better determination of the point of origin of the particle than does any other method.

The point at which even a β-particle enters the surface of an emulsion can be found to a couple of microns, while the point of entry of an α-particle can be found to less than a micron.

To make use of this it is usually necessary to record the exact original location of the emulsion with respect to the source before it is removed for processing. Since this is not easy, it may be better to deposit the emulsion direct on the source and to develop it *in situ*. Then source and tracks can be examined together under the microscope, and the exact region of the source from which tracks emerge can be identified. Naturally this can be done only when the source is not changed beyond recognition by the processing solutions and where loss of radioactivity from it to the solutions either could not occur or would not matter.

Messrs. Ilford supply G5 emulsion in gel form. Brief instructions for melting, pouring and maturing this, which must of course be done in a dark room using a suitable safelight, are given in Appendix 4.

Nuclear emulsions must be handled and processed carefully if good results are to be obtained. They lack the protective coating of most ordinary negative material and are very susceptible to scratches and pressure marks which develop to thick and unsightly marks and lines which may easily obscure the lower parts of the emulsion and can on occasion be mistaken for particle tracks by inexperienced observers. Some examples are shown in Plate XIV. Another result of the lack of protection is that nuclear emulsions must never be allowed to come into contact with foils of reactive metals such as aluminium or iron. Either of these metals reacts with silver bromide to give finely divided metallic silver, together with aluminium or iron bromide. The latter salts are hygroscopic and, since the reaction goes much more quickly when the materials are damp, quickly erodes deep holes into the metal and leaves the emulsion covered by a sticky black mess in which all information is entirely lost. Foils of stainless steel, copper, silver, gold and platinum may safely be used in contact with the emulsion, but most other metals are dangerous.

Processing is usually slow, as nuclear emulsions are used in a much thicker layer than are those of the common negative materials. A process plate may often be developed in one minute, fixed in three minutes and washed in ten more. A thick nuclear emulsion may take hours to develop, a day or so to fix and a day to wash. Details of one method of processing are given in Appendix IV.

Once the emulsion is processed and properly dried, it is very hard and is not easily damaged. If surface fog or shallow pressure marks occur, these may be effectively removed by rubbing the surface briskly with a small wad of lens tissue soaked in alcohol, without damage to the main emulsion.

A serious drawback of all work with nuclear emulsions is the time taken to collect data. Where observation of only a few events or of small areas of emulsion are sufficient, the method is of enormous value. Where large numbers of tracks are to be counted, much time must be spent in looking down a microscope. This is not tiring, or trying to the eyes, if a properly adjusted binocular instrument is used by a practised observer, but can be extremely tedious. Where much work is to be done it is usual to follow the example of the industrialist faced with a lot of semi-skilled repetitive work, and hire one or two girls as full-time scanners. Although scanners are not paid very much, this destroys the advantage in cost of the method over electronic counting. However, there are some experiments for which this method is still the best or even the only possible one.

CHAPTER SIX

STATISTICAL ERRORS, ABSORPTION AND SELF-ABSORPTION

Quantitative Measurements

6.1 General Considerations

In the application of nuclear physics we are unlikely to discover a new isotope or even unknown properties of a known isotope. The basic questions that we want our measurements to answer are: how much activity do we have present, what active nuclide is responsible and where is it? The last question is usually met in the form of "what is the exact distribution?" and, being largely qualitative, will not be dealt with here.

As in all scientific work, we try to design experiments so that we do not have to answer more than one question at a time. Thus, if we are investigating the quantity or character of one active nuclide, we shall start by defining accurately its position with respect to the detector. When we are interested in spatial distribution we shall take care to use sufficiently well-known amounts of a single active nuclide at a time. Naturally we may find cases when all questions must be solved at once, as when we are confronted with a mildly active lump of rock or when a part of a cyclotron laboratory is unexpectedly found to be highly radioactive, but neither of these situations can be described as a designed experiment.

Suppose first that we want to measure the activity of a small sample of a known β-emitter. This is placed in front of an end-window Geiger counter in a "castle" of the kind shown in Fig. 19, and the count recorded in a suitable interval. The only problem is to decide on what interval would be suitable. This depends on the accuracy required. The emission of β-particles, or any other radioactive decay in the conditions in which we normally observe it, is a *stochastic process*, i.e. a process in which each event is quite independent of all other events. The β-particles will not follow each other at regular intervals, like bullets from a machine-gun, but will come at irregular intervals and the "decay rate" discussed in Chapter 2 is only a statistical average, not an accurately defined quantity. We can talk precisely about the decay rate or half-life only of an infinite number of radioactive atoms. Neither of these terms has any meaning when applied to a single atom. For moderate numbers of atoms the terms are meaningful but not perfectly precise. Correspondingly, the number of particles counted in any finite time interval will fluctuate about the value to be expected from the number of radioactive atoms present and their geometrical position with respect to the counter.

6.2 Counting Statistics

We shall not give the proof here, but it can be shown, if N is the average number of counts to be expected, that the measured number will be likely to differ from this, with a *standard deviation* of $\sqrt{N}$, or a *probable error* $0.67 \sqrt{N}$. The definitions of these quantities are as follows.

If the average of n quantities, N_1, N_2, N_3 . . . N_n is $\overline{N}$, the *standard deviation* S is defined as the root-mean-square of $(\overline{N}-N_1)$, $(\overline{N}-N_2)$ . . . $(\overline{N}-N_n)$. The standard deviation is a reliable guide only where considerable numbers of measurements are made; it is a good guide for values of n over 10 and may be useful down to $n=4$ or 5. For such small numbers, however, the simple mean of the magnitudes $|\overline{N}-N_1|$, $|\overline{N}-N_2|$, etc., is slightly less unreliable, because less liable to exaggerate the effect of one unusual reading, and is much less trouble to calculate. In counting work there will inevitably be a standard deviation of $\sqrt{N}$ on a count whose mean should have been N, even when experimental errors of the usual kind (inaccuracy of timing, uncertainty of source position, etc.) are negligible. In practice this statistical source of error is usually the main source and $\sqrt{N}$ is nearly enough the total standard deviation from all causes. It should never be forgotten, however, that even in counting work it *is* possible to make other errors than statistical ones and, especially in accurate work, they too must be considered.

If a large number of values of N are measured, a quantity r can be defined so that half of the observations lie between $(N+r)$ and $(N-r)$ and half outside these limits. r is then defined as the *probable error*. In all kinds of measurements this is a much more useful concept than the term "possible error", which at first sight appears more attractive. An imaginative experimentalist can think of a lot of things that could *possibly* have gone wrong with an experiment. He may give a large numerical value while his more down-to-earth colleague who considers only the obvious sources of error may give a small one from the same data. It may be a little displeasing to find that an accepted value is outside the range of your probable error, but it isn't reprehensible, as it would be if you had given the possible error.

For counting processes, or any others obeying the same simple statistical laws, the probable error is simply calculable from the standard deviation, being 0·67 times this, i.e.

$$r = 0·675 \, S.$$

Where a result must be acted on, naturally it is often necessary to have greater confidence in the result than the mere even chances given for the expectation that it should lie between $N \pm r$. A range giving any desired degree of confidence can be calculated from the probable error, so long as the error distribution is *normal*.

For a detailed explanation of this term the reader should consult a book on elementary statistics; one or two are mentioned in the bibliography at the end. Here we will say only that the error distribution is normal in counting or other stochastic processes and approximately so in cases where the total error is the resultant of a large number of small independent errors. It is then highly unlikely that all of these will accumulate in one direction.

In Table 6.1 below is given the probability that a count should

TABLE 6.1

No. of Probable Errors, m	No. of Standard Deviations	Proportion of Results lying in the range $N \pm mr$
		%
0	0	0
1	0·67	50
1·48	1	68
2	1·35	82
2·97	2	95·4
3	2·02	95·7
4	2·70	99·3
4·45	3	99·73
5	3·37	99·926
5·93	4	99·9935
6	4·04	99·9946
7	4·72	99·99977
7·42	5	99·999943

differ from the expected value by a series of numbers of probable errors and standard deviations.

It is not worth considering any higher values of m as the probability of such deviations is already less than one in a million and the chance that the electronic equipment has a brainstorm during the count or that the experimentalist misreads the figures on the scaler is a good deal larger than this. It is just such events that make most real-life distributions of error "abnormal". Perhaps the main value of the table is that it can be used as a reliable indication of trouble other than the expected statistical ones. Any reading which is more than three standard deviations from the mean should always be looked at with suspicion, although suspicion, as in other fields, is not proof of guilt. In the course of a year's experimental work, many thousands of readings or counts may be taken, and about three per thousand would be *expected* to be three standard deviations from the mean. My own experience is that of readings giving an appearance of such a deviation, two-thirds are due to a slip in arithmetic or in plotting on a graph, which is easily rectified from the original data. Of the rest, enough are due to changes in apparatus or conditions to be worth a brief search. A difference of five standard deviations, when both experimenter and apparatus are working correctly, will rarely occur more than once in a fairly strenuous lifetime.

We have so far discussed errors entirely in terms of deviations from the mean of several similar observations. Often, when working with a rapidly decaying substance, one count cannot be expected to be the same as the next, and we cannot therefore take the mean of several. This does not matter. Suppose that N_1 is the value we have observed and N is the mean we would like to have observed if we had been able to take plenty of readings. Then the standard deviation we want is $\sqrt{N}$. But N_1 is not far from N if N is fairly large; for example there is a 95% chance that it lies between $N \pm 2\sqrt{N}$.

Thus if we take the standard deviation to be $\sqrt{N_1}$ we are modifying the strict definition in the ratio $\sqrt{N \pm 2\sqrt{N}} : \sqrt{N}$. If N is large, this reduces to $(1 \pm 1/\sqrt{N})$.

Even for as few as a hundred counts, this makes only a 10% difference to the standard deviation, itself only 10% of the actual count we want.

It is normal practice in counting work to forget about these hair-splittings and simply to say that the standard deviation of a count N_1 is $\sqrt{N_1}$ and the probable error is $0 \cdot 67 \sqrt{N_1}$. Whether or not the practice appeals to the purist, it is harmless if N_1 is large enough, i.e. whenever the statistical methods may be relied on at all. It may be noted that it does not matter in principle whether we make a total count all at once or in parts. Suppose we make a series $N_1, N_2, N_3 \ldots N_n$ of the same constant source. Then the standard deviations will be $\sqrt{N_1}$, $\sqrt{N_2}$, etc. Now, if we add two quantities with errors, we do not simply add the probable errors or standard deviations as this would imply that the errors in each were necessarily in the same direction. As is shown in any book on elementary statistics, one treats independent errors in the same way as vectors at right angles, the resultant being the square root of the sum of their squares.

Thus the standard deviation of $N_1 + N_2$ is

$$\sqrt{\left(\sqrt{N_1}\right)^2 + \left(\sqrt{N_2}\right)^2} = \sqrt{N_1 + N_2}$$

and similarly the standard deviation of $N_1 + N_2 + \ldots + N_n$, or ΣN, is simply $\sqrt{\Sigma N}$, which is exactly what it would have been if the whole ΣN had been recorded in one count.

No increase of accuracy is obtained therefore, and extra work is entailed, if a lot of short counts are taken and averaged rather than a single long count. It is a good practice where possible, however, to take two counts in the time available, and to check that they are the same within the expected statistical limits, before going on to the next source. Any misreading of the scaler will then be noticed, and a third count taken if necessary.

We are now in a position to plan an experiment to attain any desired accuracy. Some examples may be helpful. If we want to have a standard deviation of 1% in our result, we must record at least 10,000 counts. If we want to be 99% certain of being within 0·1% of the true answer, about $2\frac{1}{2}$ standard deviations (see Table 6.1) must be 0·1%. The standard deviation must therefore be 0·04%, 1/2500 of the actual count. The latter must then be 2500^2 or over six million. If we are counting β-particles on a Geiger counter at 5000 a minute it would take over 20 hours and a Geiger counter cannot always be relied on to give a constant efficiency to 0·1% over such a period. A check of its constancy, to be any use, would have to be at least twice as accurate as the main count, taking nearly four days. But the counter can't be trusted over such a period to this accuracy. Hence we should have to alternate for shorter periods, say four hours on the uranium standard and one hour on the sample for the better part of a week, day and night.

It is well to recognise that radioactive measurements are inherently imprecise and to avoid, as far as possible, experiments which require anything better than 1%.

6.3 Geometrical Errors

Having dealt with the basic problem of counting-statistics, we shall consider the positioning and form of the source.

6.31 Beta-particles

Errors in position with respect to the counter affect the result both by varying the air absorption and by varying the effective solid angle subtended at the source by the counter. Except for very soft β-emitters at large distances, control of the solid angle is much the more important so it will be considered first.

6.311 *Displacements along the counter axis.* When the distance h between an end-window counter and a source on its axis is large compared to the effective diameter of the counter window, D, the solid angle is simply $\pi D^2/4h^2$. Hence a 1 % error in counting rate will be produced by $\frac{1}{2}$ % change in distance which at 5 cm say will be only $\frac{1}{4}$ mm. At closer distances the solid angle varies less rapidly

Source—window distance (in window diameters)

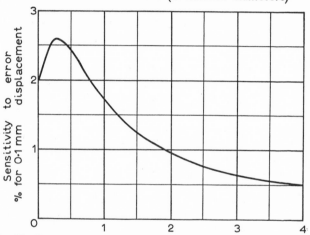

Fig. 26 Percentage error in counting rate produced by 0.1 mm shift of a small source along the axis of a 1 cm diameter end-window counter. For a y-cm window the assumed shift would be 0.1 y mm.

than $1/h^2$ so that the sensitivity to displacement changes more slowly, as shown in Fig. 26. There is maximum of sensitivity to errors in distance when the source is about a quarter of the window diameter away, but for practical purposes we can say that there is little change in susceptibility to error for distances along the axis between O and D. Over this range an error of 1 % in count would be produced by a change of about 0·005D. Thus for a standard end-window counter with a window 2 cm in diameter, 1 % error in count would be given by 0·1 mm movement of source. Few source arrangements make it possible to achieve this accuracy reliably and none that are known to the author could achieve anything like 0·1 %, for which h would have to be fixed to 10 microns.

Although, as in much of this section, it is not important to remember detailed figures, it is well to remember that many competent experimenters have failed to realise the sensitivity of result to the position of the source, and have puzzled over a 5% difference between nominally similar sources, whose only fault was to be not quite flat and so to lie at slightly different distances from the counter.

6.312 *Displacements transverse to the counter axis.* Displacement of a small source sideways from the axis of an end-window counter is much less important. Even at the most critical distance, a 2-mm

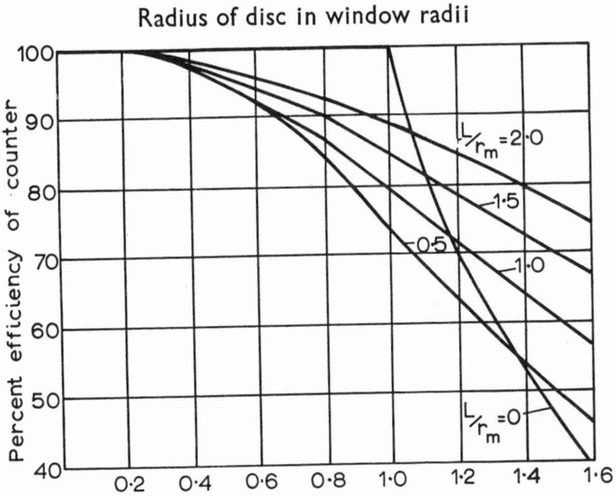

Fig. 27 Efficiency of counting particles from a uniformly active symmetrically placed disc source as compared to an axial point source at the same distance L from the window (radius r_m) for five different values of L/r_m (cf. Fig. 29).

displacement from the axis of a 25-mm window will be needed to produce a 1% effect. The source can often, therefore be placed centrally well enough by eye.

The effects of sideways displacement may become important if we cannot use a small source. Particles from the outer parts of a plane source of large area have less chance of entering the counter than do those from the centre. If the area of the source is comparable with or greater than that of the counter window, this may be very important.

In Fig. 27 is shown the change in counting efficiency for uniformly active discs of various diameters as compared to a point source on the axis. The efficiency has been calculated for five different distances from the window, neglecting the effects of absorption or scattering of particles by the air. The curves can be used for end-window counters with any diameter since both the disc diameter and distance from the window are given in terms of the window diameter.

If the source is non-uniform in activity per unit area, or irregular in outline, it is very difficult to allow for. The issue is further complicated for the softer β-emitters, such as Ni^{63} (0·063 MeV, β^-) or C^{14} (0·16 MeV, β^-), by the fact that particles from the outer regions suffer greater loss by absorption in the air. Calculated sets of correction curves are not reliable in such cases, and it is best to calibrate the actual counter that it is proposed to use by making counts of a small source of the same kind at different points along a line perpendicular to the axis of the counter, at the desired distance from the window. The calibration curve can then be used to correct the figures for a large source. The autoradiographic method is best for large non-uniform sources, a photometer being used to compare the activity of different areas with that of a standard source exposed on the same film.

For sources of area comparable to that of the window, an alternative plan is to deposit them not on a plane but on a concave dish. The profile of this can be so formed that the counter window subtends the same angle at every point of the surface of the dish. The shape, even for a given counter, will vary with distance from the face of the counter and it is worth while to design and make such dishes only when many repetitive measurements of a similar kind are to be made. An example would be any radiochemical investigation in which many successive samples in solution were deposited on a tray and dried before counting. It is difficult to control very closely the distribution of the sample as it dries, but this would not matter if all points of the tray were equally effective. For sources close to the window, it has been shown by Denis Taylor using P^{32} (1·69 MeV, β^-) that the correct profile is very close to spherical with a radius of curvature comparable with the diameter of the counter window. In Fig. 28 are reproduced Taylor's curves for a particular counter, the G.M.4. The shape of dish suitable for a particular distance from the window should be the same for any β-emitter since there is very little difference in the amount of air traversed by the β-particles from different parts of the dish.

It may be noticed in Fig. 28 that at very close distances there is a slight *drop* in detection efficiency exactly on the axis of the G.M.4 counter. This is due to the existence of a small *dead space* or insensitive region just opposite the end of the counter wire, and is not of practical importance; at most only a few tenths of a per cent of counts are lost.

6.32 *Absorption by air*

For the very soft β-emitters absorption in the air and in the window of the counter may be more important than the geometry. Because of their distribution of energies, β-particles show a nearly exponential absorption as is shown later in this chapter. The half-range in air for C^{14}, for example, is about 2·3 cm and for Ni^{63} less than 5 mm. In this case arguments from solid angle have little

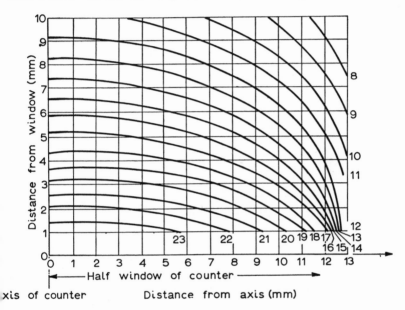

Fig. 28 Distribution of counter-efficiency for a G.M.4 counter with an end window 23 mm diameter using a small source of P^{32} (β^-, 1·7 MeV). The figures on the heavy curves show absolute counter efficiency %. (Adapted from *The Measurement of Radio Isotopes* by Denis Taylor.)

relevance and the sensitivity to source position may be great. For C^{14} a displacement of 0.1 mm along the axis would produce a change of absorption of 0·4%, which is still less than half of the geometrical effect for sources nearer to the counter than about 4 cm, but for Ni^{63} a displacement of 0·1 mm would cause a change of nearly 2%, which would be *more* serious than the geometrical effect.

Paradoxically, the effect of moving a small Ni^{63} source a little away from the counter axis will be less than for a β-emitter with higher energy, because far fewer particles will reach and penetrate the outer parts of the window.

In Fig. 29 is shown the approximate proportion of all the β-particles emitted which reach some part of the window of an end-window counter, for β-particles of high energy from a small, axial, thin source.   As has already been said, such curves and other figures given in this chapter must be taken as useful guides, not as Tablets of the Law.   Where accurate absolute figures are required there is no way of avoiding a lot of work, usually based on a suitable 4π-counter.   The making of accurate absolute measurements generally is difficult and highly specialised and no attempt will be made here

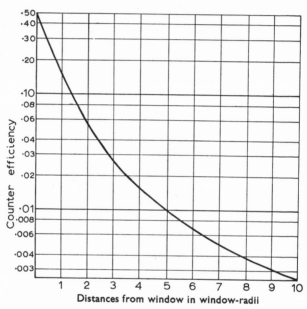

Fig. 29   Counter efficiency as a function of distance from the counter window of small thin axial sources of high energy.

to discuss it.   Many useful techniques are given in *Metrology of Radionuclides* and this, or some other similar work, should be consulted before making a practical start.

To use the curve of Fig. 29 we must allow for the absorption of the air and of the window itself, the weight-per-unit-area of which is given by the manufacturers.   Data for this are given in Appendix 3.

6.33 *Back-scattering of beta-particles*

One final trouble afflicting the would-be counter of β-particles from thin sources must be mentioned.   This is the effect of *back-scattering* from the base on which the source is deposited. The effect is negligible for α-particles but may be large for β-particles. It is some 15 to 20% greater for negative electrons than for positrons. β-particles are scattered so much that by the time they have penetrated a small part of their total range into the base they will be completely isotropic and just as likely to return towards the surface as to go further in. Those that emerge from the surface will never come back while those which travel on may well do so and hence the total reflected, especially from heavy elements, may be a large fraction of those entering the base. A *back-scattering factor* may be defined for any given counter and source as the ratio of the count observed to that which would be observed if the same source were placed in the same position on an infinitely thin support. If a small source is very close to the window of a counter, and the back-scattering factor is 1·5, as it might easily be, we could effectively collect all the particles emitted into a solid angle of 3π rather than the 2π which is all we have a right to expect.

This may be an advantage, increasing the efficiency of the counter by a useful amount, but to find the absolute strength of the source the effect must be eliminated or allowed for. It can be eliminated by depositing the source on a thin layer of material with small atomic number. Cellophane or Melinex are convenient materials, and there is not much point in making them thinner than a few microns. If serious back-scattering is produced by 5 μ of Cellophane, it will also be produced by a few millimetres of air behind 0·01 μ of Cellophane, and be much less easy to allow for into the bargain since the geometrical conditions are less well defined. It is often best to put sources on a standard thickness of plastic or aluminium and to allow for the back-scattering where necessary.

The back-scattering factor increases with the thickness of the backing until the thickness reaches about $\frac{1}{3}$ of the maximum range of the particles, after which the change with further increase of thickness is negligible. For thick backings, the back-scattering factor varies little with energy for β-emitters with maximum energies above 0·5 MeV, but depends on the solid angle subtended by the counter at the source since the back-scattered particles are far from isotropic. This anisotropy changes rapidly with the atomic number of the backing, favouring particles which leave normal to the backing for high atomic number Z but disfavouring them in the lightest elements. The back-scattering factor therefore increases more rapidly with Z when a counter subtends only a moderate solid angle at the source than it does for a 2π counter.

In Fig. 30 is shown the effective back-scattering from a thin source on thick layers of various elements for two different counters; one subtending 2π at the source and the other subtending one steradian. The results show that thick layers of the lightest elements which are conveniently available give serious back-scattering.

One more important source of error remains to be discussed. This is self-absorption in the source, which is of great importance for the

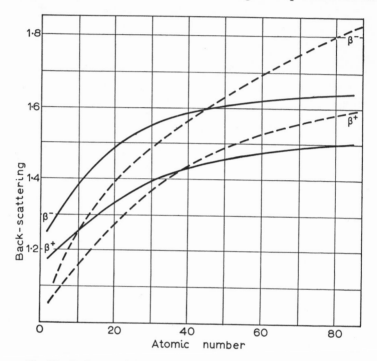

Fig. 30 Back-scattering coefficients for β^+ and β^- plotted against the atomic number of the scatterer.
Solid curves: solid angle 2π.
Broken curves: solid angle one steradian.
From H. H. Seliger, *Phys. Rev.*, **88**, 408, 1952.

softer β-emitters. We will leave this, however, until after a discussion of the methods of measuring absorption.

6.4 Gamma-ray Counting Efficiency

We shall now consider the counting of γ-rays. Here absorption effects by the air and by the window of the counter are normally negligible. Allowance for the solid angle subtended at the source by a scintillation counter is complicated however by the fact that the effective window area depends on both quantum-energy and on the distance from the source. Quanta with high energies from a nearby source may pass through the corner of a phosphor without absorption while the rays from a distant source, travelling nearly parallel to the axis of the phosphor, have the full thickness to traverse. It must also be remembered that secondary X-ray or Compton-scattered quanta are much more likely to be lost if the primary quantum interacts near a surface of the phosphor, so that calculation of the effective solid angle for absorption of the whole energy is very hard.

The best method of determining the effective total efficiency is by calibration. This may always be done with the help of a known standard source, but if the equipment is available the need for this can be avoided by a simple and ingenious method.

Many neutron-deficient nuclides are available which give a γ-quantum, γ_1 say, which follows emission of a positron so quickly that they may in practice be regarded as emitted simultaneously. When the positron comes to rest, it will be annihilated, together with an ordinary electron, with the liberation of a pair of 0·511 MeV γ-quanta. These are emitted in opposite directions. If the positron is absorbed close to the source, the two secondary γ-quanta will appear to emerge from almost the same place and at almost the same time, as the direct quantum γ_1. If then we have two detectors, one of which detects γ_1 while the other detects one of the annihilation quanta γ_2, the detectors will respond practically simultaneously. (The slowing down of the positron and the annihilation process in dense material will take less than 10^{-7} sec.) The source should not be directly between the two detectors. If it is, then if one annihilation quantum passes through one detector, the other annihilation quantum will necessarily pass through the other, complicating the detection of γ_1. The detectors can be tilted so that the source still lies on the axis of each.

We can put both outputs into a *coincidence unit* which will not respond to a signal from either detector alone but will respond if it gets a signal from both simultaneously. (More accurately, if it gets both signals within a short, defined, interval; the *resolving time*.) This is not difficult to arrange if the two detectors are made to give electrical pulses of reasonably similar size. One way in which it can be done is shown in Appendix 1. Commercial units can also readily be obtained.

We shall assume for simplicity that γ_1 is of sufficiently different energy from γ_2 to be distinguished by the detecting systems used.

E

This is not essential but simplifies the analysis of events. We suppose that in each case a discriminator is used, following the photo-multiplier and amplifier, which will reject pulses larger than those desired as well as those smaller. Let the proportion of γ_1-quanta recorded by detector 1 be ϵ_1 and of γ_2-quanta by detector 2 ϵ_2. Then if the absolute strength of the source is n disintegrations per minute, detector 1 will record $\epsilon_1 n$ quanta per minute and detector 2 will record $2\epsilon_2 n$ quanta per minute (since two annihilation quanta arise from each disintegration). Each of these counting rates is re-corded on a separate scaler. Each pulse is also passed to the co-incidence unit, the output of which is recorded on a third scaler. Now the probability that any particular γ_1-quantum should be absorbed in detector 1 is ϵ_1 and for each quantum that *is* absorbed in detector 1, there is a probability $2\epsilon_2$ that *one* of the annihilation quanta arising from the same disintegration will be absorbed in detector 2. Then the probability of *both* detectors responding is $2\epsilon_1\epsilon_2$. This is the probability for a recorded coincidence from the decay of any one nucleus. Hence the total coincidence rate is $2\epsilon_1\epsilon_2 n$ per minute. Then we can find n simply by dividing the product of the two separate scaler rates, $\epsilon_1 n \times 2\epsilon_2 n$ by the coincidence rate $2\epsilon_1\epsilon_2 n$. We have thus obtained the absolute disintegration rate of the source and from the two separate detector rates can obtain ϵ_1 and ϵ_2. Since the effective window-area of each detector varies with distance, a complete cali-bration for the energies of γ_1, γ_2 requires measurement of ϵ_1 and ϵ_2 for a series of distances. The backgrounds recorded by the three scalers must of course all be allowed for. The background coincid-ence rate due to accidental coincidences between independent decays is usually small if the source is not too strong and the resolution time of the coincidence unit is short. The background coincidence rate can be found by allowing detector 1 and detector 2 to record the outputs of two independent sources giving similar counting rates. If each detector is protected from the source which is being recorded by the other, there will be no true coincidences. The background of the detector of the higher-energy γ-quantum, γ_1, say, can be measured in the usual way, but that of the other detector may be more difficult. This is because, while it is set to record γ_2 only, the escape of some Compton-scattered γ_1-quanta may permit the dissipation by such γ_1-quanta in detector 2 of energy near enough to that of γ_2 to be recorded. If γ_2 happens to lie close to an important escape-peak of γ_1, the situation will be worse. If the discriminators used are good, the error produced will not be large, and a fair estimate of it can be made by observing the coincidence rate after setting *both* detectors to record γ_2; when in the absence of the effect there should be only background coincidences (remembering that the source does not lie directly between the detectors).

The same method can usually be used for β-γ or α-γ coincidences. There is then no objection to the two detectors lying on opposite sides of the source. In this case any small counting rate in the particle counter due to γ-rays can easily be found by using an absorber thick enough to block the particles but too thin to have any important effect on the γ-rays.

There are no cases in which two successive β-decays occur quickly enough for β-β coincidences to be useful for finding simultaneously the efficiencies of two β-counters and only one or two β-decays followed by quick α-decays.

In our analysis we have assumed that the probability that a given γ_2 quantum should be recorded by detector 2 was independent of whether the corresponding γ_1 had entered detector 1. This is not usually the case where two successive quanta are emitted by the same nucleus. To take an extreme, if the pair of γ-quanta were normally emitted in the same direction as each other we should never find any coincidences. The pair of annihilation quanta produced when a positron is captured by an electron are always emitted in exactly opposite directions. If we allow positrons to be annihilated symmetrically between two identical scintillation counters, therefore, every time we have a quantum passing through detector 1 there will be an identical one passing through detector 2. The experiment may still be useful, but does not give the strength of the source or the counter efficiency. It tells us only what proportion of those γ-quanta, which do actually enter the phosphors of the two counters, dissipate enough energy there to be recorded.

In most of the many cases when two γ-quanta are emitted in cascade from the same nucleus, there will be a close angular correlation between the two directions in which the emissions take place. The coincidence method can then be used to learn something about this angular correlation but we cannot simultaneously find the absolute source strength from one set of counts from the three scalers. Where β-emission takes place without a spin change, the emission is spherically symmetrical about the emitting nucleus, so that β-γ measurements may be used, but where a change of spin is involved there may be a correlation of both β and γ directions with the nuclear spin which will again prevent us from using the method.

6.5 Absorption Measurements

The first requirement of any absorption measurement is that the geometrical arrangement of source and detector should be well defined. The second is that the response characteristics of the detector should be very well understood.

6.51 *Alpha-particles*

Both of these requirements may be illustrated by the different absorption curves which can be obtained from the same source of,

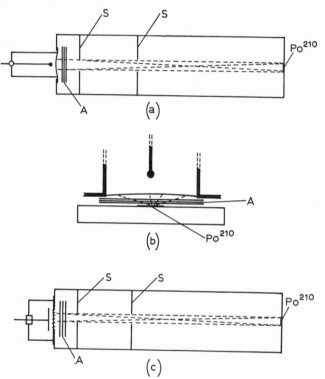

Fig. 31 Different arrangements of counter and absorbers to investigate the absorption characteristics of mono-energic α-particles.
A-A absorber foils, S-S, stops.
(*a*) Geiger counter and collimated beam in an evacuated chamber.
(*b*) Geiger counter and uncollimated source.
(*c*) Shallow ionisation chamber and collimated beam in an evacuated chamber.

for example, α-particles from Po[210], by three different methods. α-particles from Po[210] are rather simple in behaviour. Apart from a group of short-range particles forming only a few per million of the whole, all of the particles emerge with the same identical energy of 5·30 MeV, giving each one a range of 3·6 cm in dry air at S.T.P. or

14·8 μ (4·0 mg/cm²) in aluminium. The reasons for this well-defined range, uniform to about 2%, have already been discussed. Suppose we examine a parallel beam of such particles, with a detector such as a Geiger counter which can record any charged particle whatever its energy. Then if we put aluminium foils with successively increasing thicknesses in front of the counter, as in Fig. 31a, the recorded counting-rate will give us a curve such as (a) of Fig. 32. With a foil thicker than 16 μ none of the particles reaches the counter and with a foil thinner than 14 μ all of them do. Only with foils very

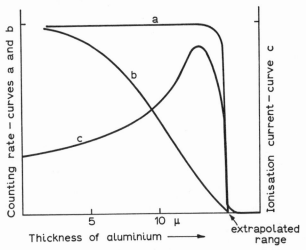

Fig. 32 Absorption curves for α-particles from Po²¹⁰.
(a) Collimated beam and Geiger counter.
(b) Uncollimated beam and Geiger counter.
(c) Collimated beam and shallow ionisation chamber.

close to the mean range do we find a change of number due to the 2% or so of *straggling*, which arises from the small statistical variations in energy lost in the first 95% or so of the range.

For α-particles the collimator must be evacuated to 0·01 mm pressure or less to avoid loss of energy in the air.

If we have the same source, uncollimated and close to the counter, as in Fig. 31b, many of the α-particles will pass through the foils at large angles of incidence and hence will have longer tracks in each foil. At extreme angles they may thus have their entire range within the thickness of a single foil. Each extra foil will cut off particles within a fresh range of angles of incidence and we finally get a curve of the form of Fig. 32b.

Finally, we show the result of using a collimated source as in Fig. 31a but replacing the Geiger counter by a shallow ionisation chamber (as in Fig. 31c) connected to an electrometer. We then observe, not the individual particles, but the mean ionisation within the chamber which results from all the particles entering over a period depending on the relaxation-time of the electrometer. Before any absorbers are introduced, the particles traverse the chamber at high speed, having lost only the energy taken up by the thin window of the chamber. Most of that energy will be dissipated inside the metal of the collector plate without producing collectable ions. The ionisation produced in the gas of the chamber by a very fast charged particle is low because the electric field of the charge has little time in which to act on the atoms near which it passes. When some absorber foils are introduced, each particle enters the chamber more slowly and ionises more intensely. As in the arrangement of Fig. 31a, the actual number of α-particles entering is unaffected.

The measured ionisation current therefore increases, as shown in Fig. 32c, as the thickness of aluminium foil increases. When the residual energy falls below 100 keV or so, and the particles are moving with velocities no greater than those corresponding to the energies of outer electrons of the gas-atoms, the ionisation by each particle begins to drop again and falls rapidly to zero. Some particles, too, will ionise over only a part of the width of the chamber before being brought to rest. The combined result is the steeply falling part of the curve of Fig. 32c.

We thus see three entirely different absorption curves for the same identical source. Each, properly interpreted, gives us useful information concerning the particles responsible. The most useful piece of information is common to all three; the *extrapolated range*. This is the range figure obtained by extrapolating the steeply falling part of the curve to the range axis. This eliminates the effects of straggling and, since it gives us a figure nearly independent of the kind of apparatus used, is of considerable practical value. From it, using the known range-energy relation for the material of the absorbing foils, the original energy of the particles can be found. Most absorption measurements are made with the object of finding the value of this energy.*

Apart from this, the information supplied by each piece of equipment is different. The Geiger counter arrangement of Fig. 31a shows us that we have a single group of particles, essentially all of which have the same energy, to perhaps 2%. It would be difficult to detect

* Note that the extrapolated range, though a useful quantity, has no simple relationship to the mean range which has a far simpler physical significance.

1% or fewer of the particles which had a lower energy, but if sufficiently strong sources were available we might be able to detect one in ten thousand which have appreciably larger energy.

The arrangement of Fig. 31b would probably fail to detect even a 20% sub-group with lower energy, and certainly would not reliably indicate a 10% energy-spread in the main group. It would however be much more suitable for looking for high-energy groups since its large acceptance angle would make it possible to use sources thousands of times smaller than could be used as in Fig. 31a, and would not involve the trouble of setting up a vacuum system.*

The ionisation chamber arrangement will be slightly more sensitive than either counter arrangement to low-energy groups but less so to high-energy groups. Its main useful features are its insensitivity to a background of β-particles and the information it gives concerning the ionisation rate due to particles with different residual ranges. Nowadays it would be more usual to use a scintillation counter with a very thin phosphor. This would have both the same advantages combined with the ability to record individual particles. It would have greater sensitivity for very weak sources and additionally the capacity to detect low-energy groups, by means of a discriminator, a good deal more easily than either of the other systems. For occasional use it is however a lot more expensive.

Systems like those of Figs. 31a and c which employ only a small solid angle in order to get a nearly parallel beam of particles are sometimes said to have "good geometry" while 31b represents "bad geometry". These are very unsatisfactory terms since, as we have seen, the arrangement of 31b is actually better for certain purposes. An even móre repulsive pair of terms are "high geometry" and "low geometry". Having a definite conscientious objection to their use, I do not propose even to say which of these is which. It is best simply to say whether a large (wide) or a small (narrow) solid angle is used.

6.52 *Absorption of beta-particles*

The general remarks made in section 6.51 concerning the importance of the experimental arrangements apply here, but the properties of β-particles make the situation more difficult. Firstly, β-particles are very easily scattered through large angles (see Plate XII). This means that even a monoenergetic beam of particles, which is well collimated

---

* If very rare high-energy groups were in question it would be advisable to use gold foils rather than aluminium. Aluminium will occasionally produce a long-range proton, by the reaction Al^{27} (α, p) Si^{30}, which could not be distinguished from an α-particle by the Geiger counter. The potential barrier of gold, however, is high enough to make this reaction impossible for all practical purposes.

when it reaches the absorbing foils, will be by no means collimated by the time it is half-way through them. Apparatus of the form shown in Fig. 31a will produce a curve much more like that shown in Fig. 32b than that of 32a. Secondly, apart from one or two rare nuclear isomers such as Se^{77m} and Nb^{93m} which emit monoenergetic sets of conversion electrons, β-particles do not have all the same energy in the first place. As shown in Chapter 2, the normal β-emitter gives a continuous spectrum of electrons with energies anywhere between zero and a well-defined maximum which is characteristic of the nuclide concerned.

For both of these reasons an ionisation chamber arranged as in Fig. 32c gives no information which cannot be obtained with a Geiger counter.

In practice, then, nearly all absorption measurements on β-particles are done with an arrangement such as Fig. 31b. Unfortunately, the fact that other arrangements are no better does not make this one good. A typical absorption curve will look like Fig. 33a. The high-energy end of the curve no longer meets the range axis at a finite angle as does curve 32b. The curve is asymptotic to the range axis and the last point clearly above the background of the counter will depend on the strength of the source and on the counting time. To find the limiting β-energy we must use a β-ray spectrometer. Fortunately, it is only rarely in applied nuclear physics that we need to measure such an energy accurately and if we are prepared to accept an approximate value we can get it from the absorption methods quite well, using the simplest arrangement of Fig. 31b. The way in which this is done depends on the fact that, for a normal β-spectrum, the curve of Fig. 33a is very nearly exponential over more than a 100 times change in counting rate. If therefore we draw a semi-log plot, we shall get a nearly straight line over a useful range. Two typical cases are shown in Fig. 33b. In curve 1, either there are no associated X-ray or γ-ray quanta, or the counter does not detect them, while in curve 2 there is some penetrating radiation observable. One cannot always trust even a "pure" β-emitter such as P^{32} to give a curve such as (1) since *bremsstrahlung** radiation will be produced in the absorber by the fast electrons themselves and will penetrate further than they do. Much depends, therefore, on the sensitivity of the detector to such radiation.

The great majority of β-emitters in any case produce γ-emission as well, often at the rate of more than one γ-quantum per β-particle.

* *Bremsstrahlung*, or *braking radiation* consists of X-rays produced by the normal process of stopping electrons in matter and is emitted during the violent accelerations of the charged electrons which occur when these pass close to an atomic nucleus.

In this case we get a straight initial part of the curve only because of the low detection-efficiency of the counter for γ-rays. A scintillation counter with a sizeable block of phosphor would give no useful result for the β-particles at all.

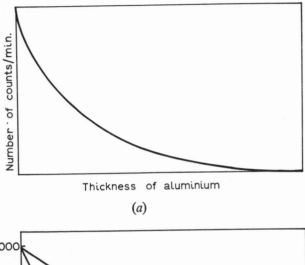

(a)

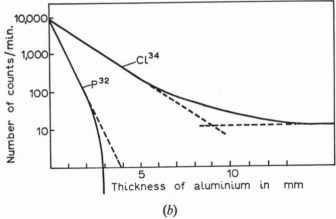

(b)

Fig. 33 *Absorption of β-particles.*
(a) On linear scale.
(b) On semilogarithmic scale. P^{32} is a pure β^--emitter; Cl^{34} emits β^+- and γ-rays.

A curve of the form (1) can easily be derived from one of form (2) by subtraction from all the earlier readings of the part of the count due to γ-rays, as indicated by the flat part of the curve. This is not altogether easy in practice because the "flat" part is never quite flat

E^*

and the shape of the bottom of the derived curve is naturally extremely sensitive to the exact quantity subtracted. Nevertheless, it is possible to produce a curve with fair experimental repeatability which is pretty steep at the end and which therefore gives a fair figure for the limiting range of the β-particles. With the help of a range-energy curve, this gives the upper limit of energy of the β-spectrum, though much less accurately than was possible for α-particles.

We can get the same result without very much increase in error and with a good deal less trouble, simply from the slope of the initial straight-line part of the curve. A different range-energy relation is then used, showing the energy simply as a function of the absorber thickness required to lower the counting rate by a factor of 10. This is particularly convenient to the practical man because he need not concern himself with the thickness of either the window of the counter or of the source itself so long as these are not great enough to take us off the straight part of the semi-log plot. Both must be properly taken into account to use the limiting range, and in many practical cases the source thickness especially may be very imperfectly known. It is often unnecessary to subtract the γ-ray count unless this is over 1 % of the initial counting rate, as its subtraction will have only a very small effect on the initial slope of the line.

Both types of range-energy relation are shown in Appendix III. The reader has been sufficiently warned not to expect too much of their accuracy but they often give us much the most convenient method of determining or confirming which of a small number of possible nuclides is responsible for an observed activity.

Absorbing materials

At this point it is worth while to discuss briefly the best materials to use as absorbers. The penetrating powers of α-particles are so low that it is difficult to obtain any solid material in thin enough foils to plot an absorption curve in detail. The whole of the steep part of the curve in Fig. 31a may take place within 0·2 μ of aluminium and it is difficult to obtain—or handle—foils less than 1 μ in thickness.

For heavier metals the situation is even worse, although gold can be obtained in such thin foils as to be usable in some cases. A practical method of getting round this difficulty is as follows. The earlier part of the curve can be obtained using a small number of, say, 3 μ foils. When the steep part has nearly been reached (as shown by a pilot experiment) no further foils are added but the whole stack is rotated through a measured angle about an axis perpendicular to the direction of the beam of particles. The particles then traverse the foils along a line which is no longer normal to the foil surface and which gives them a calculably longer path in the material of each foil. By this means the increase of path in the solid between each pair of measurements may be made as small as desired. A further advantage is that arrangements can more easily be made to rotate the stack than to add to it without disturbing the vacuum.

An alternative system, often used for α-particles, is not to use localised solid absorbers at all but to use a system such as that shown in Fig. 31a with a variable measured pressure of gas instead of a vacuum. The weight per unit area of gas between source and counter is easily calculated from the pressure and the length of the tube and can of course be varied as much or as little as desired.

β-particles have considerably longer ranges up to some millimetres of aluminium, and hence metal foils can very conveniently be used. Aluminium, which is cheap and readily obtained sufficiently pure, is employed. As we shall see, however, it is sometimes desirable to use other materials, so it is useful to indicate the way in which they differ from aluminium. The loss of energy by charged particles depends mainly on their interaction with the electrons along their path. Hence the *stopping power* of a material depends mainly on its electron density. For β-particles especially, however, it depends also on the probability of scattering by nuclei, which is greater in the heavy elements. Consequently the relative stopping power of different materials is not quite the same for α-particles as it is for β-particles, and is not even quite the same for fast β-particles as it is for slow ones.

Figures for α-particles for a number of materials are given below.

TABLE 6.2

Comparative Stopping Powers for α-particles averaged over the range 0 to 6 MeV

Material	Atomic Number	Weight of Material equivalent to 1 mg of Aluminium	Thickness of Material equivalent to 1 μ of Aluminium
Hydrogen	1	0·26 mg	7900 μ (Gas, S.T.P.)
Carbon	6	0·7(5)	0·8(8) (Graphite)
Air	—	0·8(1)	1700 (Gas, S.T.P.)
Copper	29	1·4	0·4(1)
Silver	47	1·8	0·4(6)
Tantalum	73	2·3	0·3(8)
Platinum	78	2·4	0·3(0)
Gold	79	2·4	0·3(4)
Lead	82	2·5	0·5(5)
Emulsion	—	1·5	1·0

Rough values for other elements may be found from the empirical fact that over moderate ranges of atomic weight A, the third column varies approximately as $\sqrt{A}$ and the fourth as $\rho/\sqrt{A}$ where ρ is the density. When greater accuracy is needed for heavy particles of a particular energy, reference may be made to more extensive works such as *Experimental Nuclear Physics*, ed. Segré (Vol. I), or one of the larger handbooks, in which range-energy relations are plotted for a number of substances (and on an extraordinary variety of scales). Where greater accuracy is needed for work with β-particles, more detailed methods of analysis have been proposed by, among others, Feather* and by Bleuler and Zunti.†

6.53 *Absorption of radiation*

Equipment for γ-ray measurements is more expensive than that for β-particles. If any large amount of work is to be done, however, a scintillation-counter will work out cheaper in the long run than the time of even the lowest-paid technician. (Naturally, having some experience as an experimentalist, the author realises that this is not the *sensible* comparison to make. In *all* cases the cost per useful result of the good, well-paid technician is less than that of the lowest-paid one. It is however a required qualification for any administrator controlling extensive funds that he should be unaware of this fact. The theoretical reason for this is obscure but the observational evidence is overwhelming. Hence the comparison actually made is the one which should always be cited when applying for costly apparatus.)

6.531 *Measurement of absorption.* As for α- and β-particles, the results of measurements of γ-ray absorption depend a good deal on the method used. The theory has been worked out with considerable accuracy for the three possible modes of interaction between γ-rays and matter: Compton scattering, photo-electric absorption and pair-formation (see section 4.123). At high energies (above 10 MeV) nuclear absorption also begins to be important, owing to the onset of such reactions as the photo-emission of neutrons. Most γ-ray work is done at lower energies than this; γ-rays from a given source, unlike the β-particles, are usually *monochromatic*, i.e. all having the same quantum-energy, or distributed between a small number of monochromatic groups. A parallel beam of rays, will then show a truly exponential fall in number with distance. The exponential coefficient will depend only on the energy of the γ-quanta and on the material of the absorber. Careful measurements have been pub-

---

* N. Feather, *Proc. Camb. Phil. Soc.*, **34**, 597 (1938).
† E. Bleuler and W. Zunti, *Helv. Phys. Acta*, **19**, 375; **20**, 195 (1946 and 1947).

lished for a considerable number of materials, so that if the coefficient can be measured the energy can be found. For the higher energies, lead is almost universally used. Curves showing the *half-range** of γ-rays in lead as a function of energy are given in Appendix III.

A beam from a nearby source, entering the absorber over a wide range of angles, will show a more complicated law of absorption. This is due to the appearance in the beam, as it travels through the absorber, of various secondary radiations: X-rays, Compton-scattered γ-rays etc. Some of these may be derived from primaries which would not themselves have reached the counter in the absence of the absorber. Measurements are usually made, therefore, with collimated beams.

A serious difficulty in measuring accurately a large degree of attenuation in an absorber, even when using only a narrow-angle beam, is the scattering of γ-rays from surrounding objects. Soft γ-rays can be prevented from reaching the detector by unofficial routes by appropriate screening around source and detector and by careful collimation. This is unsatisfactory for hard γ-rays the main beam of which becomes mixed with a lot of softer secondaries from the inside of the screening itself. It is therefore better to suspend source and counter at large distances from any reflecting surface, with the stack of absorbing sheets midway between them. Even then a trace of scattering by the air will occur and must be taken into account as part of the background. The difficulty of the problem is illustrated by the efforts which have been made for its solution. Fig. 34 shows the arrangement used by Cowan. The counter was connected through 100 ft (30 m) of coaxial lead to the scaler and supplies, and a winch was used to lower and raise the whole working assembly when the lead absorbers were to be changed.

The absorption of γ-rays of a given energy is usually given in terms of an *absorption coefficient*, which we will call a. This is defined so that the initial intensity I_0 drops to $I = I_0 \exp(-ax)$ after penetrating a thickness x of absorber. This relation can be used to calculate the total cross-section σ for absorption or scattering of the γ-rays by a single atom, since $a = N\sigma$, where N is the number of atoms/cm³. The absorption may also be given in terms of the half-range, which is the distance in which the intensity drops to half. From the relation above, $r_{\frac{1}{2}} = (\ln 2)/a$ cm. The *mean range* is $1/a$.

The absorption may also be given in terms of a *mass absorption coefficient* μ. This is defined so that the initial intensity I_0 drops to $I = I_0 \exp(-\mu w)$ after penetrating w gm cm⁻² of the material. Then $w = \rho x$ where ρ is the density so that $a = \rho \mu$. The mean range can

* I.e. the thickness of lead reducing the γ-ray intensity by a half.

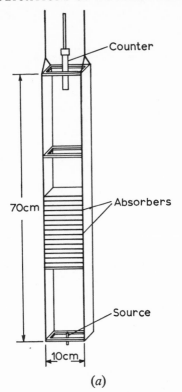

(a)

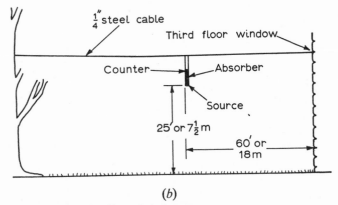

(b)

Fig. 34 *Absorption of γ-rays.*
(*a*) Carrier for source, absorber and counter.
(*b*) Apparatus in working position.
From C. L. Cowan, *Phys. Rev.*, **74**, 1842 (1948).

then be written $1/\mu$ gm cm^{-2} and the mass-half-range is $\ln 2/\mu$ gm cm^{-2}.

Over an important range of γ-ray energies, from 1 to 4 MeV, there is little difference in the mass-half-range between different elements, so that the weight per unit area required to produce any given attenuation is much the same for all materials. In Fig. 35 is shown the variation of half-range with γ-energy between 0·1 and 20 MeV.

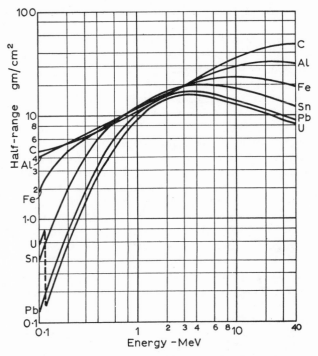

Fig. 35 Half-range of γ-rays between 0·1 and 40 MeV for several elements. From Gladys W. Grodstein, National Bureau of Standards Circular (1957).

Some fresh factors appear at low energies, so we shall leave discussion of these until after we have discussed the range of energies displayed in Fig. 35. Since there is a maximum half-range for each element there are usually two possible γ-ray energies for each half-range. The ambiguity can be resolved by comparing the half-range in two different elements, one preferably being carbon or aluminium. Carbon is the best absorber of those shown in Fig. 35 to use for discrimination in the 2 to 5 MeV range. The conventional lead, which shows only about 10% change in half-range between 2 and $6\frac{1}{2}$ MeV, is no use at all for this purpose.

Nowadays absorption methods are rarely used to find γ-ray energies in the region above 0·1 MeV. A good sodium-iodide or caesium-iodide scintillation counter with suitable electronics can determine the energy far more quickly and accurately as well as being able to deal with mixtures of γ-rays of different energies. Unless the energies are very well separated, a mixture is difficult to analyse by absorption methods. Below 0·1 MeV, a proportional counter can be used.

6.532 *Radiation shielding.* The importance of a knowledge of γ-ray absorption in applied work lies rather in the need to know on the one hand through how thick a layer of material they can be detected, and on the other how thick a screen must be used to attenuate them to a safe or negligible intensity. This level may be set by safety requirements or by the need to reduce γ-ray background in counting-experiments.

Since screening of people and of apparatus from γ-rays may be quite an expensive part of any project, it is worth while to discuss briefly the choice of materials. As can be seen from Fig. 35, if we can stop γ-rays of 2 to 3 MeV, we shall be able to stop all other γ-rays as well. In this region there is really very little to choose between different elements and it might be thought that screens should therefore always be made of whatever material has the lowest cost per unit weight. Delivered to the laboratory, the cheapest material per unit of weight is indubitably water. This, however, like sand which is probably the next cheapest, is messy to handle. Concrete is probably the cheapest material obtainable in units convenient to handle and of any desired shape. Why then is it so usual to use lead, which is perhaps a hundred times more expensive? The most obvious reason is that it takes less space. A lead wall which will attenuate 3 MeV γ-rays by a factor of 100 would be about 10 cm thick, while a concrete wall would be nearer 50 cm thick. For a large installation, when space is no object, concrete should certainly be used.

When only a small source (i.e. small in volume, not necessarily weak) is to be screened, lead is far better for a geometrical reason. This is most easily explained for the case of a point source to be screened in all directions, i.e. by a spherical absorber. Then, since all materials have much the same mass-absorption coefficient, the product of radius R and density ρ of the sphere must be much the same for all of them, so that ρR represents an adequate number of half-ranges for the attenuation which is desired. Now the total mass of the sphere will be $4\pi R^3 \rho / 3$. But ρR is a constant independent of the material, K say. Hence the mass of the sphere, substituting K/ρ for R, is $4\pi K^3/3\rho^2$. Hence the total weight of screening *increases* as the density gets less because the volume of the sphere goes up as the cube of its radius, and the mass per unit area in the path of α ray from the centre increases only linearly with density or radius. Paradoxically, therefore, lead is used instead of cheaper materials for screening small sources largely because the total screening then weighs much less and is more easily carried around. If a standard 20-lb lead isotope-can were to be replaced by one of concrete which gave the same attenuation, the latter would weight nearly a ton (which would not even be cheaper than the lead) although the mass-absorption coefficient of concrete is not much less than that of lead.

For really strong sources, lead itself has an uneconomically low density and uranium metal is used, in spite of the greater cost, giving a screen $2\frac{1}{2}$ times lighter and $1\cdot6$ times smaller in linear dimensions. For such sources a composite shield, uranium inside and lead outside, could actually be cheaper than lead alone, since very little uranium near the centre can replace a great deal of lead on the outside. At present prices, however, the optimum radius for the uranium is so small a fraction (less than a tenth) of that of the lead that the small gain is not worth the extra trouble just to reduce cost of material. As a working rule one would conclude that sources of small dimensions are usually best screened by lead and sources of large dimensions, such as accelerators, by concrete.

The screening of detectors depends on much the same considerations, with the extra requirement that the screening material should itself be free of radioactive contamination. This rules out concrete straight away as it usually contains a significant contamination of thorium in the granite chips of the aggregate. Lead is also liable to a (smaller) degree of contamination with heavy radioactive elements and it is better to use iron, which usually carries little contamination, when the lowest backgrounds are needed. When a strong γ-ray source is to be used near to a sensitive detector, it is most economical to put roughly half of the screening round each rather than all of it round the source—so long as this does not permit a radiation-level hazardous to health in the laboratory.

6.54 *Low-energy gamma-rays*

We shall now return to the absorption of soft γ-rays and X-rays. In this region, between say 1 and 100 keV-quantum energy, the half-range in general increases rapidly with increase of the energy of the quanta and decreases with increase of atomic number of the absorber. There are, however, some very striking anomalies, as shown in

Fig. 36.* In each element, at a particular energy, there is a sudden fall in the half-range by a factor of some six or seven times. The point at which this occurs represents the energy at which the incident γ-rays just become capable of ejecting one of the two tightly bound K-electrons in the atoms of the absorber. The loosely bound outer electrons are much less effective for γ-absorption by photo-electric emission owing to the fact that in interacting with them it is less easy

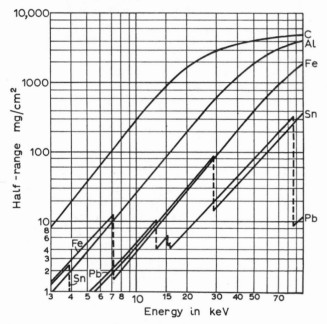

Fig. 36 Half-range of X- or γ-rays between 3 and 100 keV for several elements.
From *Handbook of Chemistry and Physics*, 44th Edition (1962-3).

to transfer to the rest of the atom any spare momentum and angular momentum, which must go somewhere if both are to be conserved.

The energy at which the jump in absorption occurs is known as the *K-absorption-edge* of the element concerned and, being the ionisation energy of the K-shell, varies from element to element very closely as Z^2. In the heaviest elements one can see also the L-absorption

---

* Some divergence at 0·1 MeV may be observed between Fig. 35 and Fig. 36. The compilers have presumably weighted differently the experimental data. Much bigger differences are given by quite small changes in geometrical arrangement.

edges but these are smaller in magnitude and multiple, since the eight electrons of the L-shell do not all have the same identical energy.

The existence of these absorption edges makes it easy to make a close estimate of the energy of an X-ray or soft γ-ray by an absorption method. For example, suppose that, by measurement of the absorption in aluminium, we decide that we have an X-ray in the region of 9 keV. Then we repeat the measurement of the absorption in nickel, copper and zinc. If we find that the half-ranges in copper and nickel are both low, in the region of 30 mg/cm^{-2}, while that in zinc is in the region of five times greater, we know at once, from very rough measurements, that our X-ray energy lies between 9·0 and 9·65 keV (the absorption edges of copper and zinc) which is only a few times worse than one could get in this region with a proportional counter and good electronic equipment.

This measurement also enables us to distinguish between X-rays and β-particles of somewhat greater energy, which is difficult to do by means of a single absorption measurement.

This method can be used indirectly to identify with certainty the element giving rise to characteristic X-rays without recourse to chemistry. Decay by electron capture or the internal conversion of γ-rays is always followed by X-ray emission. The characteristic K-X-ray of the product atom rather than of the parent is emitted because emission takes place long after all the nuclear events are finished. ("Long after" in this connection means 10^{-10} sec or less, but the nuclear rearrangement following K-capture or the internal conversion of a γ-ray takes less time than this by a factor of many millions.)

Thus if the X-ray which we identified above as being between 9 and 9·65 keV was emitted from a K-capturing nucleus, we can say at once (from the tables of X-ray emission lines) that the daughter-nucleus must have been germanium and hence the K-capturing parent-nucleus must have been arsenic.

In Appendix V is a table of the K-absorption edges and the predominant K-X-ray emission wavelength of the elements.

6.6 Self-absorption

In all our discussion of absorption measurements so far, whether of α-, β- or γ-rays and whether from point sources or extended sources, we have neglected any effects due to the thickness of the source itself. These effects are of great practical importance in finding the true total activity of practical sources, and are somewhat different for the three different types of radiation. We shall suppose in the whole of the following discussion of thick sources that the radioactive atoms concerned are uniformly distributed through the homogeneous material of a source of uniform thickness. This is not usually difficult to arrange.

6.61 *Alpha-particle sources*

The situation in the case of a mono-energetic α-particle emitter is fairly simple, since the particles travel in straight lines and have all the same range. If we are using a narrow-angle beam, as in Fig. 31*a* above, the thickness of the source will not affect the proportion of particles reaching the counter (neglecting the thickness of the counter window) until the whole of the energy of the α-particles coming from the source is absorbed in the source material, less of course the small energy needed to penetrate the window of the counter. Beyond this thickness, further addition of active source material will not affect the counting rate.

For any source thickness other than zero, the absorption curve will no longer be of the simple form shown in Fig. 32*a*, but will begin to drop as soon as the sum of the thicknesses of absorber and source reaches the available range of the particles from the back of the source. From there the curve will drop linearly till it finally reaches the axis at the same point as before, when only the α-particles from the top surface of the source remain. The experimental determination of an absorption curve thus gives quite a good method of estimating the effective thickness of the source.

If the uncollimated α-source of Fig. 31*b* is thick, the situation is somewhat different. Here the counting rate, in the absence of additional absorbers, will be reduced as soon as the thickness of the source is sufficient to stop the most obliquely moving α-particle from its rearmost layer from reaching the counter. Hence particles will be lost with quite thin sources and, in the limit where the source is right against the counter window, any thickness of source will involve loss of some particles. The absorption curve will remain of the general form shown in Fig. 32*b* but cannot be used directly to deduce the thickness of the source. An absorption curve should not be measured for α-particles in this way, except perhaps to find a quick, rough estimate of the extrapolated range.

If the thickness of the source and the solid angle subtended by the counter are both accurately known, allowance can be made for loss of particles in the thickness of the source. The calculations are simple enough if the source is close to the window of the counter, as an α-particle source will usually be. The result is particularly simple in the important special case in which the source is so thick that no further addition to it will affect the counting rate.

Suppose that there are *n* breakdowns per second per unit volume in the material of the source and that the range of each α-particle in this material is r. Consider a layer of thickness dx at a distance x below the surface (see Fig. 37). Then only those particles will reach the surface which are emitted within an angle θ to the normal to the surface, where $\cos \theta = x/r$. This limits them to a cone of semivertical angle θ, which contains a solid angle $2\pi(1 - \cos \theta)$.

The number of disintegrations per second which occur within the layer is ndx per unit area, and since the α-particles from these will be

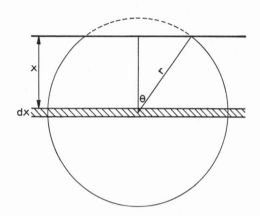

Fig. 37 Calculation of α-particle emission from a thick source.

distributed over a solid angle of 4π, the number from this layer which reach the surface will be

$$2\pi(1 - \cos \theta)/4\pi . \ ndx \text{ per unit area per second}$$
$$= \tfrac{1}{2}(1 - x/r)ndx.$$

Then the total number reaching the surface per second, n_t, from a thickness t of source, will be given by

$$n_t = \int_0^t \tfrac{1}{2}(1 - x/r)ndx \qquad 6.1$$

or
$$n_t = \tfrac{1}{2}n(x - x^2/2r)_0^t \qquad 6.2$$

∴ $n_t = \tfrac{1}{2}n(t - t^2/2r)$ per second per unit area.   6.3

For very small values of t this reduces correctly to $n_t = \tfrac{1}{2}nt$ since nt is the total number of disintegrations per second per unit area in the whole thickness of the source, and the particles from half of these disintegrations emerge from the source.

When t reaches the value r,

$$n_{max} = \tfrac{1}{4} nr \text{ counts/sec/unit area} \qquad 6.4$$

No particles reach the surface from layers below $t=r$. Thus we have the very simple result that the number of α-particles which reach the surface of a thick source is just a quarter of the number of disintegrations which take place within a layer of the source one α-range thick. (Compare this with a *thin* source in which *half* of the total of particles emitted emerge from the surface.)

Thus a window-less counter, directly against the source so as to detect the whole of the particles which leave the surface, can very simply and directly tell us the strength of the source in disintegrations/sec/unit volume. If it is not already known, the range of α-particles in the material of which the source is made can easily be found in a subsidiary experiment. When the strengths of two thick sources made of similar material are to be compared, and not measured absolutely, this will not be necessary.

If the counter has a window of appreciable thickness, as will usually be the case, the proportion counted is equally easy to calculate, though the resulting expression is not quite so delightfully simple. All that we have to do is to allow for the effect of an extra absorbing layer on top of the source. It is convenient to give this thickness as a fraction of the total range of α-particles in the window material. Suppose that the thickness of the window is a times this range. This means that, in Fig. 37, we suppose our radioactive atoms to be uniformly distributed from a depth ar downwards so that the integration limits in equation 6.1 are taken between ar and t instead of 0 and t. To save space we will give here only the result for the special—but common—case where the thickness of the source is r or greater. The general case is easily written down if required. The number reaching the new surface, i.e. penetrating also the window of the counter, is then

$$n'_{max} = \tfrac{1}{2}n[(r-ar) - \tfrac{1}{2}(r - a^2r)]$$
$$= \tfrac{1}{4}nr(1 - 2a + a^2)$$
$$\therefore \qquad n'_{max} = \tfrac{1}{4}nr(1-a)^2 \qquad 6.5$$

In the case of a thin source, the proportion getting through the window would be $(1-a)$ of all disintegrations as can be seen at once by considering the permitted range of solid angle. In other words, the efficiency of the counter is $(1-a)$ for the thin source but $(1-a)^2$ for the thick one. Again, if we need to know the absolute disintegration rate we must know ar as well as r, but for comparison of sources with the same counter this is unnecessary.

For both thin and thick sources it is worth noting a small advantage to be obtained from the existence of the counter window. This is that, since only those particles which reach the window within a limited angle of incidence can penetrate it, there is no need to have the source inconveniently close. The maximum angle of incidence which will be accepted is given by $\cos \theta = a/r$; a/r is the fraction of the total α-range that the window represents. This is likely to be $\frac{1}{10}$ to $\frac{1}{4}$ in practice. Even at $\frac{1}{10}$, θ is reduced from $90°$ to $84°$ and a small source may be placed anywhere within the range 0 to almost $\frac{1}{10}R$ from the window of the counter, where R is the radius of the counter window without at all affecting the counting rate, though we must remember the addition of a small extra layer of air to the window thickness.

One final property of a thick α-particle source may be noted. This is that the emerging particles are partially collimated. The original emission is of course isotropic, but those starting off normal to the surface can reach it from greater depths than can those starting at angles of incidence θ other than zero. When the thickness of source is greater than the particle range, the numbers of particles emerging from the surface per second per unit solid angle will be proportional to $\cos \theta$, with a maximum normal to the surface and falling to zero tangential to it. The collimation is too little to be of much use but can cause errors if it is not remembered.

Although the numbers vary with direction, the energy distribution does not; at each angle there are equal numbers of particles in equal intervals of residual range. This is quite different from a thin source, from which all particles at all angles have the same energy.

6.62 *Self-absorption in beta-particle sources*

The effect of source thickness on β-emitters is less simple. β-particles do not travel in straight lines but, if their energy is high, will often travel through an absorber for some distance before they are scattered very far from their original direction. Our approach to the problem is in general the same; we consider the proportion of β-particles emitted in a thin layer at a depth x in the source (Fig. 37) and integrate the numbers from all layers down to a depth equal to the maximum range of the particles concerned. We cannot follow individual particles at a definite solid angle, but must consider the complete emission of each layer of the source.

To the accuracy to which we can regard the absorption of β-particles as exponential, we can solve the problem analytically. The method used, although founded on a law which is only approximately true, has the advantage that it applies to sources of any area, even if this is larger than the window of the counter.

Suppose again that we have n disintegrations per unit volume per second. Suppose that the counter efficiency for detecting β-particles from atoms at the top of the source is ϵ. We shall suppose in the following argument that the source, whether thick or thin, is placed on a thick support of inactive material of similar atomic number. The effects of back-scattering of particles from below the source will then be independent of the depth in the source from which they come. As has been shown above, this back-scattering may be a considerable percentage of the total count observed. For a thin source close to the counter, ϵ may therefore be more than $\frac{1}{2}$ (see Fig. 30 above). Then the efficiency of the counter for detecting particles from atoms at depth x will be $\epsilon \exp(-\alpha x)$ since the effect of the overlying layer of thickness x reduces the counting rate to an extent $\exp(-\alpha x)$. (This is not true for α-particles but is true for any particles showing an exponential absorption.)

Then the number of particles counted per second due to disintegrations in a layer of thickness dx at depth x is $n\epsilon \exp(-\alpha x)dx$ per unit area and the total number emerging from the surface of a layer of thickness t will be, by integration,

$$n_t = n\epsilon(-1/\alpha)(\exp-\alpha x)_0^t \text{ per second per unit area.}$$

$$n_t = n\epsilon[1-\exp(-\alpha t)]/\alpha \text{ per second per unit area.}$$

$$\therefore \quad n_t = nt\epsilon[(1-\exp(-\alpha t)]/\alpha t \text{ per second per unit area.} \qquad 6.6$$

This expression will be valid to an accuracy dependent on that of the law of exponential attenuation; i.e. to a few per cent. It will as usual be better for comparing similar sources than for absolute determination of activity. When the thickness is small, this expression reduces simply to $n = \epsilon nt$, as would be expected, since nt is the total number of disintegrations per second per unit area in the whole thickness of the source. A graph of the expression $[1-\exp(-\alpha t)]/\alpha t$ is given in Fig. 38 for a usable range of αt. For any thickness such that $\exp(-\alpha t)$ is less than say 0·01, when αt is a little over 4, this term may be neglected, giving us

$$n_t = n_{\max} = n\epsilon/\alpha \text{ counts/second/cm}^2. \qquad 6.7'$$

The distance $1/\alpha$ is the mean range of the β-particles so that in practical terms a thick source behaves just as would a thin source, coinciding in position with that of the top surface of the actual thick source, and containing the same number of disintegrating atoms as exist in one mean range of β-particles in the thick source. The mean range is $1/\ln 2$ or 1·443 times the half-range. The mean range is thus 0·434 times the thickness that reduces the intensity by a factor of 10, which is given for a range of β-particle energies in Appendix III.

Owing to the fact that β-particles of quite moderate energies have a range of a millimetre or so in most materials, we do not usually have to deal with sources for which the "large-thickness" formula 6.7, is valid. It is of importance mainly for liquid counters and for very soft β-emitters such as C^{14} (0·156 MeV). When comparing two different β-emitters, even with the same counter, it must be remembered that the efficiency ϵ includes the effects of absorption in the window of the counter. It will therefore not be the same for different β-energies although, as shown earlier in the chapter, the difference is important only for the softer emitters.

The full relation 6.6 is most often of use to determine whether or

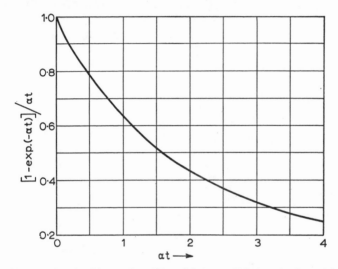

Fig. 38 Graph of $[1-\exp(-\alpha t)]/\alpha t$ against αt. This function is useful in the determination of β-particle emission from a moderately thick source. (See equation 6.10.)

not a source can be regarded as sufficiently close to perfectly thin. Where application of 6.6 results in a correction of only a factor of two or so from the thin-source value, it is probably reasonably reliable —so long as the back-scattering conditions are constant. If they are not, i.e. if the source is on a thin backing, an increase from zero of the source thickness may at first actually *increase* the proportion of the total number of emitted particles which are counted, owing to scattering within the thickness of the source. This occurs only where the counter angle is well below 2π, and the size of the effect is not easily calculated.

Unlike α-particles, the β-particles emitted from a thick source will remain closely isotropic except for particles of the highest energies. For these last the situation will be far too complex for useful calculation and rarely of practical importance.

The whole of this section may perhaps best be summarised in three simple instructions to those who wish to find the strengths of β-sources: (1) don't use thick sources, (2) don't use thick supports, (3) if you can't obey both of these, don't expect much absolute accuracy.

6.63 *Self-absorption in gamma-ray sources*

Owing to the much greater penetrating power of γ-quanta this is of importance only in special cases. Such cases would include strong sources of very soft γ-rays such as those from Fe^{57m} (14 keV) which are used in Mössbauer experiments, and the investigation of the activity of large specimens of rock or of screening material for counters.

In conditions in which a thin source would give an exponential absorption curve (i.e. with small solid angles and no detection of secondary radiations) the formulae 6.6 and 6.7 can still be used. The absorption curves for β-particles, however, are roughly exponential for any geometrical arrangement since whatever may be the original directional pattern of the particles is wiped out by scattering at a quite shallow depth in the absorber. This is not true for γ-rays, the primary quanta of which travel along straight lines just as do the α-particles. The absorption characteristics are quite different from α-particles, however, so that in no circumstances can the equations 6.3 or 6.4 be applied to γ-rays.

It is not useful to recalculate the case considered for α-particles; that of a source which was thick, but small in area and close to the window of the counter. Even if the front of a thick γ-source is close to the window, the rear of it will not be.

The only case which is worth while to consider analytically is that of a source which is large in area compared to the window of the counter, such as one will have when investigating screening material as mentioned above or when searching for uraniferous ores.

Suppose again that we have n disintegrations per second per unit volume of the material. As before (see Fig. 37), the number of γ-rays emitted into a cone of semi-angle θ will be $2\pi n(1 - \cos \theta)/4\pi$ per second. The attenuation of these before they reach the surface will vary with angle, so we must consider only an element of solid angle. The number emitted at angles between θ and $\theta + d\theta$ will be $\frac{1}{2}n \sin \theta d\theta$ per sec per unit volume. Those emitted from a depth x in a layer of thickness dx into this range of angle will be

$$\frac{1}{2}n \sin \theta d\theta dx \text{ per sec per unit area.}$$

The distance they have to go to reach the surface will be $r = x \sec \theta$. Hence they will suffer an attenuation $\exp(-ax \sec \theta)$, where a is the absorption coefficient.

Hence the number which actually emerge from the surface into the range of solid angle $2\pi \sin \theta d\theta$ from the layer dx will be

$$dn_{x,\,\theta} = \tfrac{1}{2}n \sin \theta \exp(-ax \sec \theta)d\theta dx \ \text{cm}^{-2} \ \text{sec}^{-1} \qquad 6.8$$

and the total flux emerging from the surface will be

$$n_t = \tfrac{1}{2}n \int_0^t \int_0^{\pi/2} \sin \theta \exp(-ax \sec \theta)d\theta dx \ \text{cm}^{-2} \ \text{sec}^{-1} \qquad 6.9$$

where t is the total thickness of the active material. This is not a very attractive expression but is easy enough to handle in special cases.

Integrate first with respect to x. Then we have

$$n_t = \tfrac{1}{2}n \int_0^{\pi/2} \sin \theta (-1/a \sec \theta)[\exp(-ax \sec \theta)]_0^t \, d\theta \qquad 6.10$$

$$= \tfrac{1}{2}n \int_0^{\pi/2} \sin \theta / a \sec \theta [1 - \exp(-at \sec \theta)]d\theta \qquad 6.11$$

$$\left. \begin{array}{l} = \dfrac{n}{4a} \displaystyle\int_0^{\pi/2} \sin 2\theta [1 - \exp(-at \sec \theta)]d\theta \\[2mm] = (n/4a) \ I \ \text{say.} \end{array} \right\} \qquad 6.12$$

This requires numerical integration if the term $\exp(-at \sec \theta)$ is appreciable; that is, if at is less than about 3. The result of this numerical integration is shown in Fig. 39. For thicker layers we may simply neglect the exponential term. Then the expression for n_t simplifies to

$$n_t = \frac{n}{4a} \int_0^{\pi/2} \sin 2\theta d\theta$$

$$= \frac{n}{8a}(-\cos 2\theta)_0^{\pi/2}$$

$$n_t = 0 \cdot 25n/a \ \text{cm}^{-2} \ \text{sec}^{-1} \qquad 6.13$$

$$= 0 \cdot 360nr_{\frac{1}{2}}$$

which is numerically equal to the number of γ-quanta emitted into the upper hemisphere by all of the decaying atoms within a distance of $1/2a$ or $0 \cdot 72r_{\frac{1}{2}}$ of the surface.

If we measure the total γ-flux outside the surface, and know the absorption-coefficient or the half-range, we can tell quantitatively from 6.12 what is the number of γ-emitting disintegrations per second per unit volume of the material. For convenience we may then rewrite 6.12 as either

$$n = 4an_t/I \text{ or } n = 2\cdot 77n_t/rI \qquad\qquad 6.14$$

where n_t is the total γ-flux in quanta/sec/unit area outside the surface, a is the absorption coefficient and $r_{\frac{1}{2}}$ the half-range of the γ-rays, n is the number of γ-decays per unit volume per second and $\frac{1}{2}I$ is the value of the integral shown in Fig. 39. For thick samples $I = 1$.

It has been assumed that the detector will record only the primary γ-quanta, as will be the case if a scintillation counter with a suitable discriminator is used. It must be noted too that, if equation 6.13 is to be used, the detector must be sensitive to γ-quanta from the sides

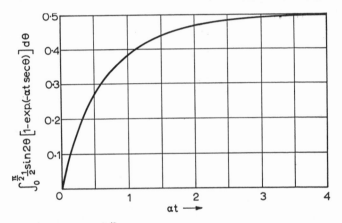

Fig. 39   Graph of $\displaystyle\int_0^{\pi/2} \frac{1}{2}\sin 2\theta[1 - \exp(-\alpha t \sec\theta)]d\theta, \; = \frac{1}{2}I$.

This function is useful in the determination of γ-emission from a moderately thick source. (See equation 6.12.)

as well as the front. The emergent flux will be far from isotropic, the flux per steradian being proportional to $\cos\theta$, as was the case for a thick source of α-particles. The flux will not vary at all with distance from the surface of the active material so long as this distance is small compared to the distance to the nearest edge of this material.

If we use a general detector of γ-rays, rather than one detecting a particular energy only, we shall observe a variety of secondary radiations as well as the primary ones. These are usually absorbed much faster so that the count recorded is unlikely to be increased by more than 10 to 20%, so long as β-particles are excluded.

An interesting case arises if we examine the emitting surface of a thick γ-source with a Geiger counter. This will have an efficiency of 1% or less for γ-rays but of nearly 100% for electrons. Most γ-emitters are the immediate product of β-emitters so we should naturally expect any observed counting rate to be due mainly to β-emitters. This interpretation must not be accepted without check. In the interior of a thick piece of γ-active material, far (i.e. many half-ranges) from any surface, we must clearly have on the average equal numbers of γ-rays emitted and absorbed per unit volume. Every process of γ-absorption involves the production of one or more fast secondary electrons. Hence, even in the absence of any β-emitters at all, the rate of fast-electron production at any point in the interior will be greater than the rate of γ-production.

Near the surface, with γ-rays coming from one side only, the rate of production of fast electrons per unit volume will drop to half that in the interior, but is still comparable with that of the γ-rays.

Thus a Geiger counter outside a thick slab of a pure γ-emitter will give much the same counting rate as it would outside a slab containing the same activity per unit volume of a pure β-emitter, in spite of its much lower efficiency for the primary γ-rays themselves. Hence a separate experiment is required to show what it is that we are observing.

If we have only a thin layer of material, the equilibrium of emission and absorption of γ-rays will not be established and we shall get very few secondary electrons. We can easily produce a slice of material which is very thin for most γ-rays but which is several half-ranges for most β-particles. The count from this will thus give—with only a small correction—the true β-emission of the material, while the thick slab will give something close to the $\beta + \gamma$ emission of the material.

Another plan is to put, over the active slab, a layer of inactive material of the same atomic-number distribution thick enough to stop all the true β-particles. If it is not thick enough to attenuate the γ-rays seriously, the rate of production of secondaries will be nearly the same as in the top layer of the active slab and our Geiger counter will record only the electrons due to the γ-flux.

If most of the γ-rays have energies in the region of 2 to 5 MeV, where there is little difference in the mass-absorption of different elements, it does not make much difference whether or not a similar material is used for the inactive layer so long as the mass-absorption coefficients for the secondary electrons do not differ too much. For softer γ-rays the material matters a great deal, and a layer of Perspex, for example, in which relatively few γ-rays are absorbed, will give a much smaller yield of secondary electrons than will most active substances which are likely to be investigated. On the other hand the insertion of a thin sheet of lead between the active slab and the counter might considerably *increase* the count because the small extra loss of γ-quanta would be more than compensated by the increase of efficiently detected electrons. This phenomenon provides a trap for the unwary but a useful method of increasing sensitivity for the well informed. It is known as the *transition effect* and may occur across the boundary between different materials whenever we have a mixture of primary and secondary radiations with different absorption-characteristics.

CHAPTER SEVEN

MEASUREMENT OF HALF-LIFE AND ESTIMATION OF RAPIDLY DECAYING SUBSTANCES

Of all of the properties of the thousand-odd known radioactive nuclides, the half-life is the most characteristic. If this can be measured to 10%, there will often be only a single possibility for the material responsible and rarely more than two or three possibilities in the entire list of active isotopes.

There are several methods of measurement, depending on the order of magnitude of the half-life concerned.

7.1 Measurement of Half-life by Direct Counting

The simplest is to take a sample of convenient activity and to measure the counting rate at intervals for a suitable period. The characteristics of the equipment used for counting must remain constant. If the source is moved between readings, it must be accurately replaced each time.

This method may be used for measuring half-lives from a few seconds to a few years; a factor of about ten million and, as this covers most of the experimentally important nuclides, we will deal with this first and in greatest detail. We shall assume the use of an end-window Geiger counter to measure the half-life of a β-emitter, but most of what is said would apply equally well to any kind of detector or radioactive nuclide.

At first sight, it might be thought that little need be said. The counting rate, after correction for background and the dead-time of the detector, obeys the law (see Chapter 2):

$$n = n_0 \exp(-\lambda t)$$

whence
$$\ln n = \ln n_0 - \lambda t$$

where n_0 is the initial counting rate, n is the counting rate at time t and λ is the decay constant. Hence if n is measured for several values of t, a graph of $\ln n$ against t should give a straight line with slope λ. It may be more convenient to plot $\log_{10} n$ against t, when the slope will be $0 \cdot 434 \lambda$. Much time is saved by using semilogarithmic graph-paper on which the counting rate can be plotted directly.

An example of a decay curve for a single radioactive nuclide is shown in Fig. 40.

Having found λ, the half-life is given by $\tau_{\frac{1}{2}} = 0.693/\lambda$ or can be read directly from a semilogarithmic plot.

The first point of practical detail to be discussed is the length of time that should be spent on each individual count. If this time is long compared to the half-life, the counting rate will change during the observation. On the other hand, if the source is decaying rapidly,

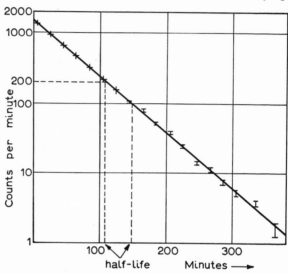

Fig. 40 Decay of a small source of Cl³⁸. The vertical line shown for each count shows the standard deviation for that count. The half-life may be read directly from the graph as indicated.

it may be impracticable to count only for very small fractions of a half-life.

In Appendix VI we give a quantitative investigation of the effect of counting for a time comparable with the half-life in obtaining each point.

It is proved there that to a first approximation it is correct simply to plot the average counting rate over a fairly short time interval against the mid-time of the interval. If the counting interval is not short, the mid-time should not be used, but rather a time t' changed from the mid-time by an amount Δt.

This is always an earlier time and can be obtained from the equation

$$\Delta t = -\tau_{\frac{1}{2}} f(\overline{t_2 - t_1}/\tau_{\frac{1}{2}}) \qquad 7.1$$

where $\overline{t_2 - t_1}$ is the counting interval. The function $f(\overline{t_2 - t_1}/\tau_{\frac{1}{2}})$ is plotted in Fig. 41. Its use is shown in the following example. Suppose we have counted 6000 counts in 50 minutes from zero time from a source which is decaying with a half-life of 40 minutes. Then $\overline{(t_2-t_1)}/\tau_{\frac{1}{2}}$ is 1·25. From Fig. 41 we find that the quantity $f(1·25)$ is 0·044. Then $\Delta t = 0.044\tau_{\frac{1}{2}} = 1·76$ minutes. Then on our graph we plot the average rate, 120 per minute, at time $t_{\frac{1}{2}} = 25 - 1·76 = 23·24$

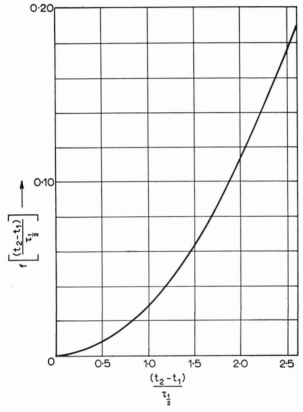

Fig. 41 Graph of the function $f[(t_2-t_1)/\tau_{\frac{1}{2}}]$ calculated in Appendix 6 and used as described in section 7.1 to find the effect of counting a short-lived activity for a long period.

minutes from zero time. In spite of so long a count, this is not a very large correction. When the counting interval is $\frac{1}{3}$ of the half-life the correction is about $\frac{1}{3}\%$ of the counting interval, which is very rarely important. Consequently, for a count over any period less than $\frac{1}{3}$ of the half-life, no correction at all is needed, so long as we remember to plot the average count against the mid-time. It is worth while to stress

F

this because it has not always been appreciated by previous authors.

This fortunate property of exponentials means that we can use practically every recordable count from the source, without excessive labour in plotting huge numbers of points. As we saw in an earlier chapter, it is important for good statistical accuracy to use the largest possible number of counts. With a rapidly decaying source, we cannot do better than to record every single count up to the time when the source becomes indistinguishable from the background. We thus switch the counter off for only the shortest possible times required to record the scaler reading, at intervals of $\frac{1}{4}$ to $\frac{1}{3}$ of a half-life. If the latter is not even approximately known (which rarely happens in applied work), we begin by taking two successive counts as short and close together as practicable. If the second one is less than the first by 30% or more, we continue with the same short counts; if not, we can lengthen the counts until there is a 20 to 30% drop between successive values, 26% corresponding to $\frac{1}{3}$ of a half-life.

It may turn out, of course, that we have a mixture of substances with different half-lives. The procedure suggested will not be ideal if there is only a little of the short-lived material, but when the proportions are roughly known it is easy to work out an appropriate modification. When the half-life of the source is a minute or less, or when the counting rate is small, it may be best not to switch the scaler off at all but to take "running readings" while it continues to operate. It needs little practice to note the reading at a time defined with an error of less than a second. It does not matter if the lights indicating "units" are changing too fast to be recorded. If the counting rate is high enough for this to be so, the standard deviation of the number of counts will automatically be large enough to make unimportant an error even of several tens in each recorded count.

Several automatic devices have been built which record every hundred or thousand counts on a uniformly moving tape, together with time signals at defined regular intervals. From this the counting rates at a suitable series of times can be derived by measurement. This may be of particular advantage when counting must be continued all night or when a large number of counters have to be operated by a single experimenter. It should be noted, however, that the actual working time of the experimenter is increased, not decreased, by this system as it takes longer to find each counting rate from the marks on the tape than directly from a scaler display. If practicable, therefore, it is more economical to deal with a large number of counters by fetching in just as many observers as will be kept continuously busy in recording the readings of the scalers at appropriate intervals. This has also the advantage that any unexpected results or failure of equipment will be noticed as they occur while something can still be done about them.

If much work is to be done on sources with half-lives of less than a day, automatic recording-equipment is nevertheless worth its cost, for use in emergencies and at inconvenient times of day or night.

The accuracy with which one can determine a half-life by this direct-counting method depends on the statistical accuracy of each point on the semilogarithmic graph (cf. Fig. 40) and on the number of half-lives over which the decay can be followed. If any desired strength of source can be used, the limitation at the beginning of the experiment is the counting rate at which dead-time or other counter corrections become inaccurate. At the other end the size of the background determines the smallest rate that can usefully be recorded. With most systems the useful range between these limits covers a factor of 10^3 to 10^4, representing 10 to 14 half-lives. It is good practice to indicate the standard deviation of each point on the graph as has been done in Fig. 40. The probable error of the best straight line through the points can best be determined by inspection. A least-square method of determining the slope is sometimes used, "weighting" the points according to their statistical standard deviations, but in the author's opinion this is never worth the arithmetical labour involved. The result never differs from that obtainable from a careful drawing on a suitable scale by more than the probable error. Working to fractions of the probable error gives only a marginal chance of approaching more closely the correct result, and the time involved would be better spent in repeating the experiment to improve the statistical accuracy of the original data.

If we have a source which is known to be *radiochemically pure* (i.e. in which all detectable activity is due to the single nuclide with which we are concerned), there is little advantage in plotting the decay curve for more than six or seven half-lives. The statistical error of any later points will be so much larger that they will contribute nothing of value. In practice, however, we are rarely absolutely sure of radiochemical purity and it is usually advisable to continue counting until the background is approached in order to detect any residual long-lived contaminating activity which could affect the result.

It has been assumed throughout the discussion so far that the only important errors are the inescapable statistical deviations discussed in section 6.2. This will usually be true for sources which are counted continuously, as suggested, with half-lives from a few tens of seconds to a few hours. In this case we can calculate the standard deviation of the half-life determined as we have recommended, using the whole of the counts recorded by the detector. The calculation will not be given here but the result will be quoted. If n_0 is the initial counting *rate*, and the measured half-life is $\tau_{\frac{1}{2}}$, then the standard deviation of the half-life is

$$0{\cdot}89\sqrt{\frac{\tau_{\frac{1}{2}}}{n_0}}.$$

Hence the *percentage* standard deviation, $100/\tau_{\frac{1}{2}}$ times the absolute standard deviation, is $89/\sqrt{n_0\tau_{\frac{1}{2}}}$. Since $n_0\tau_{\frac{1}{2}}$ is simply the number of counts we should observe if we counted for one half-life at the initial rate n_0, this is easy to work out. The result is almost independent of the counting interval so long as all the counts are used; i.e. it makes no difference whether we plot on our graph a moderate number of more accurate points or a large number of less accurate ones.

The percentage standard deviation may also be written $107/\sqrt{N}$, where N is the total number of counts recorded during the entire decay. This then gives us the best result that we can possibly hope for, if no other errors exist at all. To get a standard deviation of 1% we should need to record in all about 12,000 counts.

For longer-lived sources, the total number of counts which can be recorded may run into millions, giving a statistical standard deviation less than one part in a thousand. In this case the variations of the efficiency of the counter, inevitable when counting continues over an extended period, ultimately limit the accuracy. Where half-lives of weeks or months are involved, it is not usually either possible or desirable to leave the source under the counter for the whole period. Regular checks of counter efficiency and background must be made and particular care taken to see that the source is replaced each time, undamaged and free from dust, in the same identical geometrical position with respect to the counter. Where soft β-particles are concerned it may even be necessary to check also the atmospheric pressure and temperature, which determine the density of the air between source and counter. Taking into account all of the various sources of error, a final accuracy of 1% is quite a creditable achievement and a great deal of work must be expected if anything better than $\frac{1}{2}\%$ is required.

Even for short-lived sources, in which the error appears to be dominated by the simple statistical standard deviation, this is not really so. Changes of the characteristics of the counting system may not be important but the initial counting rate n_0, on which the whole accuracy depends, is limited by the maximum counting rate at which the dead-time corrections are accurately known. This gives a large advantage in principle to the scintillation counter over the Geiger counter. Unfortunately, only a part of this advantage is realisable in practice as the scintillation counter is often liable to rather shorter-term fluctuation in efficiency than is the Geiger counter when both are counting β-particles.

A dilution method by which both counter dead-time and background are eliminated, so long as both remain constant throughout the experiment, has been used in the Birmingham University Physics Laboratory for the accurate determination of the half-life of F^{18}.

The method needs considerably more activity than does the direct method but this does not usually present any difficulty.

The original sample of F^{18} is obtained radiochemically pure by distillation and is in the form of fluoride in solution in distilled water containing very little else, either of ordinary inactive fluoride or of dissolved salts. First, a trace (about 0.1%) of magnesium hydroxide is added so that part of the solution can be evaporated to dryness to give a thin source without loss of fluorine, which is firmly retained as non-volatile magnesium fluoride. A second solution, about 100 times weaker in activity than the first but containing the same concentration of magnesium hydroxide, is next made up. The dilution is carried out by adding a weighed quantity of the original solution to a weighed quantity of further 0.1% magnesium hydroxide. The dilution factor can be found to an accuracy of 0.03%. Thorough mixing is ensured by a mechanical shaker. Several drops of about 200 mg of the diluted solution are weighed out carefully on to standard counting trays and evaporated there to give thin sources covering small circular areas. These are defined by a ring of a hydrophobic (silicone) grease on the tray which limits the spread of the solution. The original solution and the dilution factor are chosen so that these sources give a counting rate of about 6000 to 8000 per minute at about 2 cm from the window of a Geiger counter. Mechanical arrangements must be made to ensure that a succession of samples can be put into the same geometrical position with considerable accuracy. Each source is then placed under a separate counter and the decay followed down to about 1000 counts/minute. Each set of counts is plotted as a graph from which the best possible determination is made of the exact time at which each source reached a rate of 4000 counts/minute.

Meanwhile, a similar set of weighed drops of the original strong solution are evaporated down on an identical set of counting trays. These are, of course, initially far too strong to count in the standard position and are set aside for 14 to 15 hours. By this time the weak sources will all be removed from the counters and the strong ones will have decayed to about 6000 to 7000 counts a minute. These are then counted in the standard positions in exactly the same way as the weak ones had been. In the same way again, the exact time at which each reaches 4000 c.p.m. is recorded.

Then suppose that on a particular counter the interval between the time when the weak source reached 4000 c.p.m., and the time when the strong one did, was T minutes. Then if the dilution factor was F and the weights of the weak and strong samples were W_1 and W_2 respectively, the total amount of activity initially present in the strong samples was FW_2/W_1 times that of the weak sample at the same moment. Hence the strong sample must have decayed by FW_2/W_1 in T minutes. But if the half-life is $\tau_{\frac{1}{2}}$, in this time the strong sample will have decayed by a factor of $2^{T/\tau_{\frac{1}{2}}}$.

Hence $$(T/\tau_{\frac{1}{2}}) \log 2 = \log (FW_2/W_1)$$

whence we find $\tau_{\frac{1}{2}}$.

The accuracy of this method does not depend at all on counter characteristics provided only that these remain constant during the experiment, as can be checked with a uranium standard source. The statistical error in the final value of half-life can be calculated and, in the same terms as before, gives a *percentage* standard deviation of $\dfrac{204}{m\sqrt{N}}$, where m is the number of half-lives between the two sets of measurements and N is the total number of counts recorded for each sample. m can easily be made as high as six or seven giving us a standard deviation of about $30/\sqrt{N}$ %. Remembering that N can be made larger, since uncertainty of dead-time is unimportant, this means that we can hope for a gain of five or six times in final accuracy over the direct method. This makes 0·1 % obtainable without specialised equipment or large numbers of control experiments. Interference by contaminating activities is easily detected and allowed for if necessary. Shorter-lived contaminating activities can be looked for in a small sample of the original solution, examined before any dilution, when their contribution to the counting rate will be larger than during the actual experiment. Longer-lived activities can be looked for in a large sample of the original strong solution after most of the desired activity has died away. The method by which allowance is made for undesired activities follows simply from the method of analysis of complex decay curves which is explained in section 7.5 below.

When the half-lives of substances of interest are very short or very long, or the specific activity is very low, the direct counting method or the dilution method are not applicable. We shall consider first very short half-lives.

7.2 Measurement of Very Short Half-lives

Here we necessarily have a source in which the active substance is being continuously or almost continuously produced, either as the daughter of a longer-lived nuclide or by means of nuclear reactions produced by a pile or an accelerator. Much, therefore, depends on the method of production. Where production is by an artificial source, the target material may be passed through the neutron flux or ion beam in a continuous stream of gas or liquid passing along a thin-walled tube. The radioactivity which is produced will then decay as it travels along and can be recorded as it passes two points a measured distance apart. The distance that the stream must travel for the activity to drop to half is then found and this, divided by the velocity of the stream, gives us the half-life. Half-lives down to 10^{-3} to 10^{-4} sec may be investigated by this means.

When the beam of ions from an accelerator is used to produce the active nuclide, the target may be distributed round the periphery of a spinning disc or cylinder which has counters or even nuclear emulsions to record the activities at various points round the periphery.

If the ion beam can be produced in a series of short pulses, a scintillation counter or other high-speed detector may also be switched on by a delayed pulse for a known brief period at each of a succession of times after the original pulse of ions. A large number of pulses may be used per second. All the counts in a particular range of time following each pulse are added together and recorded in a single scaler or in one channel of a "kicksorter". Then, even if only a few counts are recorded after any one pulse, the final statistical accuracy can be satisfactory. This method can be used for half-lives less than a microsecond, the limitation usually being fixed by the shortness of pulse which it is practicable to apply to the accelerator.

When the short-lived material is produced as the daughter of a longer-lived one, a similar method can be used if the decay of the parent substance can also be observed. We then investigate coincidences between the counter (A, say) recording the decay of the parent and that (B, say) recording the decay of the daughter whose half-life we want to know. We adjust the resolving time of the coincidence-equipment so that it records any counts which occur in B within say $\frac{1}{3}$ of the expected half-life of that in A. We record the coincidence rate observed. A known delay of the order of the half-life which is expected is then introduced into the line from the counter A to the coincidence unit. We then record again the coincidence rate, the resolving power of the equipment being kept unchanged. This procedure is repeated for a succession of further delay times. A semilogarithmic plot of coincidence rate against delay time then gives the half-life. This method has been used successfully to determine the half-lives of γ-transitions down to 10^{-10} seconds.

7.3 Measurement of Very Long Half-lives

Long-lived substances need a quite different approach. If the rate of decay is too small to show any change in a reasonable time, it is necessary to measure directly the decay constant λ rather than the half-life, by finding the absolute rate of disintegration of a known number of atoms. This may be a matter of some difficulty. An instructive example is the measurement of the half-lives of uranium 238 and uranium 235, which form 99.3% and 0.7% respectively by weight of ordinary uranium.

The first step is to spread a known weight of a uranium compound in a thin layer on the inside of an ionisation chamber or on a nuclear emulsion. The layer must be uniformly thin compared to the range in the uranium compound of the α-particles emitted. Then the weight of the compound gives N_{238}, the total number of atoms of U^{238} and the total number of α-particles observed in unit time can be used to calculate λN. It must be remembered that half of the α-particles are lost in the base on which the source is spread. Most of them come from the U^{238}, but allowance must be made for α-particles emitted by the other isotopes of uranium. There are two of these, U^{234} and U^{235}. Uranium ^{234} has a much shorter half-life and is produced from the U^{238} according to the decay scheme

$$U^{238} \xrightarrow{\alpha} Th^{234} \xrightarrow{\beta} Pa^{234} \xrightarrow{\beta} U^{234} \xrightarrow{\alpha} Th^{230} \text{ etc.}$$

The two β-emitters have very much shorter half-lives still. In old uranium ores secular equilibrium will have been reached in which, on the average, one U^{234} atom decays for every U^{238} decay. Hence, although the equilibrium weight of U^{234} is very small, it produces just the same number of α-particles per unit time as the U^{238}. The Th^{230} and later members of the series are removed during chemical purification of the uranium and, since the Th^{230} has a half-life of 80,000 years, will take a very long time to grow back to equilibrium. Hence it and its descendants contribute no significant number of α-particles to those observed. (The Th^{234} and Pa^{234} are also removed, during the extraction and purification of the uranium, but are short-lived and grow back again too quickly for there to be an important hiatus in the supply of U^{234}.) The U^{235}, on the other hand, is an independent isotope with a half-life comparable to that of the U^{238}. Its half-life must be found before allowance for it can be made. This can be done indirectly, without having to solve the difficult technical problem of separating the uranium isotopes in the laboratory or the possibly worse problem of extracting some U^{235} of known isotopic purity from the military authorities. The trick is to examine the activity of one of its descendants. These are produced according to the decay scheme

$$U^{235} \xrightarrow{\alpha} Th^{231} \xrightarrow{\beta} Pa^{231} \xrightarrow{\alpha} Ac^{227} \xrightarrow{\beta} Th^{227} \text{ etc.}$$

In any uranium ore these will all be in secular equilibrium with the parent U^{235}. The mean number of decays per unit time is therefore for all of them the same as for the U^{235}. Hence if all of any one nuclide is extracted from a known quantity of uranium ore, its initial rate of emission of particles will be identical with the rate of emission by the U^{235} therein. The most convenient nuclide to extract is Pa^{231}. No other α-emitting protoactinium isotope is produced by the other uranium isotopes, it has a conveniently long half-life of $3 \cdot 3 \times 10^4$ years so that no allowance needs to be made for its own decay and its daughter Ac^{227} is long enough lived not to build up inconveniently quickly. Then the Pa^{234} gives us $\lambda_{235}N_{235}$ and the weight of uranium in the ore from which it was extracted gives us N_{235} and between them we have the U^{235} decay constant. From this we can calculate the proportion of α-particles contributed by U^{235} to the observed total emitted by all the uranium isotopes. This turns out to be about $2 \cdot 3 \%$. Thus $97 \cdot 7 \%$ of the α-particles observed in the first experiment are due in equal shares to the U^{234} and U^{238} and finally those which are due to the U^{238} alone are $48 \cdot 8 \%$ of the whole. This then gives us $\lambda_{238}N_{238}$, and the relation $\tau_{\frac{1}{2}} = \ln 2/\lambda$ gives us the half-life desired.

This example illustrates all of the main principles of measurement of the half-lives of long-lived nuclides, which not infrequently occur mixed with other active isotopes of the same element.

The method of determining the rate of disintegration λN by measuring the activity of a daughter-nuclide, even if this is not known to be in equilibrium with its parent, is also a powerful method of dealing with an activity of moderate half-life but very low *specific activity*.* For example, Ca^{47} is a β-emitting isotope of calcium which

---

* The specific activity is the ratio of the rate of disintegration of the active material to the total quantity of material; it may be measured in a variety of units such as counts per minute per ml or millicuries per gram. It usually, but not always, refers only to the quantity of the same element present. Thus if we had 1 millicurie of Na^{24} in 10 ml of solution containing $2 \cdot 54$ gm of sodium chloride, its specific activity might be given as $0 \cdot 1$ mCi per ml but would more probably be said to be 1 millicurie per gram. since the solution contains one gram of sodium element. This lack of system does not often lead to confusion but it is always well to state exactly what is meant when using the term.

is difficult to produce without at the same time producing larger quantities of the longer-lived β-emitter Ca^{45}. To find the half-life of the Ca^{47}, a solution is made up and at regular intervals the daughter-element scandium is extracted chemically. This procedure is described as *milking* scandium from the solution. Sc^{45} is stable but Sc^{47} is radioactive (3·4 days, β^-). The activity of the scandium at a standard time after each extraction is proportional to the amount of Ca^{47} which remained in solution at the time of the extraction. Hence a semilogarithmic plot of the scandium activities against the time of extraction will give the half-life of the Ca^{47}. The method remains practical even if a large quantity of stable calcium is present in the solution though this might well prevent the production of a satisfactory source for direct counting even in the absence of Ca^{45}.

In this particular case, the half-life could also be found by counting γ-rays which are produced after the decay of Ca^{47} but not after that of Ca^{45} but the γ-ray method is less sensitive even where it is possible.

This discriminatory counting of a particular type of radiation or particle is a general method often of use for measuring the half-life of one out of a mixture of isotopes.

7.4 Quantitative Measurement of Decaying Nuclides

In many applications, the half-life of the active material is well known and what we want to know is the activity of a sample at some specified time. We may for example want to know how much of some active nuclide has been made in a pile or what proportion is excreted of a dose of radioactive material which has been fed to an animal. When the active substance is decaying during the experiment the activity measured obviously depends on when it is measured. Allowance must then be made for the decay since the production of the radioactive material. When we are concerned with the distribution of active atoms between different parts of an animal or between different chemical compounds, the time of manufacture of the active nuclide is of no interest. In this case we can refer measurements to an arbitrary time so long as this time is the same for all samples in a given experiment. If the measurement is made at a time t_1 and the chosen reference time is t_2, then we simply multiply the observed counting rate by $2^{\overline{(t_1-t_2)}/\tau_{\frac{1}{2}}}$ or, which is the same thing, exp $\lambda(t_1-t_2)$. (It is worth noting that the value of 2^x can be read off very simply from any slide-rule with a log. log. scale.) For comparison of different samples this procedure is applied to each measurement. For absolute determination of the strength of the source at time t_2 we must divide also by the counter efficiency.

A convenient method of allowing for decay, which avoids calculation on a slide-rule but which uses up a good deal of expensive graph-paper, is to plot the measured points on semilogarithmic paper with time as abscissa as in Fig. 40 and then to draw through each point a straight line with the correct slope to represent the decay of the activity concerned. Then the activity of each source at time t_2 can be read off directly from the graph. It is always safer when convenient to make more than one observation of each source. In this case, if there is appreciable decay between readings, the advantage of the graphical method is increased. A check that all the observations for a single source lie on the same straight line confirms the accuracy of the measurements.

When large numbers of measurements are going to be made on one radionuclide, it may be well worth while to construct a correction table once and for all, covering as many half-lives as may be required. This can be based on a sample which has an activity of 1·00 at midday and gives the activity at regular intervals over a day, a week or whatever convenient period gives a fall of say 10^4.

Using this, any measurement made at any time can be calculated back to the activity which it would have had at the previous noon simply by dividing by the figure given in the table for the time of measurement. Thus all measurements are referred to the same time by a simple division. If measurements run beyond the end of the chart they can be brought back 24 hours on to it again by dividing additionally by the 24-hour figure. When making such a chart—for which a calculating-machine is almost essential—it is well worth while to undertake the extra labour needed to prepare it with sufficiently close intervals to make interpolation very easy or even unnecessary.

7.5 Analysis of Mixed Activities

Frequently in applied nuclear physics we have to deal with sources containing more than one kind of activity. Unless these arise from different isotopes of the same element, they can in principle be separated chemically, but this may often be difficult or inconvenient. Even without this, and even when the different nuclides emit similar radiations, it is possible to measure the amount of each in the presence of the other by following the decay of the mixed source for an adequate time.

In Fig. 42 is shown the decay curve of a mixture of two positron-emitters, F^{18} (110 mins) and C^{11} (20 mins). Such a mixed curve is always concave upwards. If it is followed sufficiently far, until all observable activity of the shorter-lived activity has gone, we obviously have left the simple decay curve of the longer lived F^{18}. This may be extrapolated back, as shown, to show the number of counts for which F^{18} was responsible during the time that both activities were being recorded. We can find the part due to the C^{11} alone by subtraction of values taken from the extrapolated line from the curved part of the graph. The remainder after subtraction is plotted as the dashed line and shows the simple decay of pure C^{11}.

Both lines can be extrapolated back still further to the time of production of the source, even if this is some minutes before the beginning of counting. The initial counting rate of the C^{11} is seen to be 6·4 times that of the F^{18}. If the efficiency of the counter is the same for each, this means that the ratio of number of C^{11} atoms to

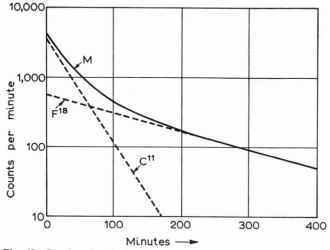

Fig. 42 Graph and analysis of the decay of a mixture (M) of F^{18} and C^{11}.

number of F^{18} atoms is $6·4 \times 20/110$ (since each counting rate is proportional to λN and λ is inversely proportional to the half-life).

In most practical cases the half-lives are known and it is sufficient to follow the decay of the longer lived material only far enough to check its identity, the slope of the line being drawn to suit the known half-life.

For moderate accuracy it is not even necessary to continue the decay curve until the shorter lived activity is entirely gone, so long as the identities of both are known and no other activities are present. A trial line of the correct slope is drawn by guesswork to represent a likely quantity of the longer lived material and subtracted from the mixed curve. If it has been guessed correctly, the difference can be replotted to give a second straight line with a slope corresponding to the half-life of the shorter lived component of the mixture. If too little of the long-lived component is subtracted, the difference curve will be concave upwards, showing that it still represents a mixture. If too much has been subtracted, the difference curve will be convex upwards. A second guess can then be made of the amount of long-lived material, and if necessary a third. After very little practice, two guesses will be found sufficient.

The method is of little value if the half-lives differ by less than a factor of $1\frac{1}{2}$ and is much easier to apply if they differ by at least a factor of 3. With large differences of half-life, mixtures of three or even more components can similarly be resolved. It may be necessary to make more than one source of the active material, otherwise the counting rate may be too great for the counter in the early stages and too close to background in the later stages.

It may be noted that exactly the same method of analysis can be used for separating the absorption curves obtained with mixtures of β- or γ-emitters. Here a factor of three or more in the half-range is desirable if useful results are to be obtained.

7.6 Very Weak Activities

We may conclude this chapter by considering briefly the minimum quantity of a known quickly decaying nuclide that we can hope to observe. The limitation here is due entirely to statistical fluctuation in the background. If we have a known average background rate of 10 per minute, we can detect an extra one count a minute with certainty if we can count for a few hours. If we are looking for a nuclide with a half-life of thirty seconds, on the other hand, an initial rate of 30 per minute will be difficult to detect with certainty. The total of counts to be expected in the complete decay of this is only 21.

The first thing to decide is how long each count shall be. If each is too short, the fluctuations due to the source itself are too great. In the example above, a three-second count would probably have only two from the source and half a chance of one from the background. The observation of two or three counts in 3 seconds will not carry conviction. On the other hand, if we count for too long, the cumulative fluctuation of the background is too great. In half an hour the total background count will be about 300 with a standard deviation of 17—which is almost the total that we could hope for from the source.

It turns out that the best period of observation is about two half-lives. In the example, we might hope for fifteen or sixteen counts in the first minute which, while still a *possible* fluctuation of background, has less than 0·01 % likelihood of happening by chance.

CHAPTER EIGHT

PRODUCTION OF RADIOACTIVE ISOTOPES

Few applied nuclear physicists are expected to make their own radio-isotopes. Many, however, will work in institutions with facilities for making some such isotopes and all can profit from some knowledge of the limitations of the various sources of supply.

Very few radioactive nuclides occur in nature in useful quantities. Nearly all must be made from existing stable isotopes. To do this we must be able to induce nuclear reactions in which nucleons can be subtracted from or added to a nucleus in much the same way as atoms are subtracted from or added to a molecule in chemical reactions.

Like chemical reactions, nuclear reactions may require an *activation energy* to bring them about. A neutron can enter any nucleus freely, but a proton or a second nucleus encounters a fierce electrostatic repulsion. This must be overcome before the particles can approach sufficiently closely for the still stronger, but short-range, nuclear forces of attraction to come into play. Charged particles must therefore be given sufficient kinetic energy to carry them through the electrical barrier. "Sufficient" kinetic energy means a million electron-volts (an MeV) or so, even for a singly charged proton or deuteron entering a light nucleus. To enter uranium a proton will require nearly 15 MeV and a doubly charged helium nucleus will need 30. In Fig. 43 is shown the variation of this *potential barrier* with the atomic number of the target nucleus for singly and doubly charged particles. To produce nuclear reactions, then, we must have a source of neutrons or a source of charged particles with high kinetic energies. There is an extensive literature available in this field; here we will give only a bare outline of the principles and capabilities of the main sources of particles.

By far the most important source of supply of radioactive isotopes is the nuclear reactor based on the fission of uranium.

8.1 Induced Fission

In Chapter 2 there is a brief mention of spontaneous fission as a rare form of radioactivity. In this process, a heavy nucleus breaks into two parts of comparable mass. While the process is slow in an undisturbed nucleus, it can be induced practically instantaneously if additional energy is added to the nucleus. Six or seven MeV may be enough just to push the nucleus over the edge, triggering the release of nearly 200 MeV. 6 MeV is near the binding energy of a single nucleon in this region of the table of elements. Consequently, capture of a single neutron, releasing this energy, may be enough to induce fission. The energy released when a neutron is captured by a nucleus with an odd number of neutrons is a little greater than in the opposite case. For example, the energies released when each of the more important isotopes of uranium capture a slow neutron are

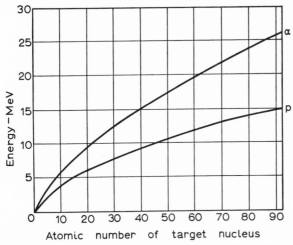

Fig. 43 Variation with atomic number of the potential barrier against protons and α-particles. The potential barrier for deuterons will be slightly less than that for protons and for He³ very nearly the same as that for α-particles.

as follows: U^{238}, 4·7 MeV; U^{236}, 5·4 MeV; U^{235}, 6·4 MeV; U^{234}, 5·2 MeV; U^{233}, 6·8 MeV.

The two odd isotopes, U^{235} and U^{233} will normally undergo immediate fission on capture of a slow neutron; the even ones will not. It is a sobering thought that our civilisation still exists because the threshold energy for fission in uranium is around 6 MeV rather than 4. If the binding energies of neutrons were 2 MeV greater or fission thresholds 2 MeV less, ordinary U^{238} (99·27% of ordinary uranium) would have the properties of U^{235}. Atomic bombs would probably have appeared during the 1914–18 war with a large-scale production cost of a few hundred pounds each and could easily be made in quantity, not only by the smallest independent countries but also by the undisciplined variety of private enterprise run by such characters as the late Mr. Al Capone.

We now come to the second important characteristic of fission, whether spontaneous or induced. This is that the two product nuclei have so great an excess of neutrons and so high an energy of excitation that they will usually evaporate one or more neutrons before settling down to their ground states. Hence the fission of one nucleus may lead to the induction of fission in further nuclei.

Consider a sphere of pure U^{235}. This has a partial half-life for spontaneous fission of 2×10^{17} years, giving about one fission every fourteen seconds per gram atom. The number of neutrons emitted following each fission event is not constant, but for a large number of fissions of a particular kind of nucleus we can measure the average number released, this number being characteristic of the nuclide concerned. For U^{235} it is 2·6 for induced fission and is probably little less for spontaneous fission. The fate of the neutrons produced will depend on the size of the U^{235} sphere. Since nuclei are so small compared with atoms, a neutron from fission of one nucleus is likely to pass through a large number of atoms before it happens to hit another nucleus. Furthermore, if it does hit one there is a large chance that it will bounce off before it produces any permanent effect at all. Hence, if our sphere is small, a neutron emitted from one fission will probably escape entirely without producing another.

If, on the other hand, neutrons start near the centre of a large sphere, each one will almost certainly produce a new fission. Knowing the target area for scattering and for fission of a U^{235} nucleus, we can work out the probability ϵ, averaged over all points in the sphere, that a neutron from any one nucleus will strike and cause fission in another before it can escape. Then, since the average number of neutrons produced per fission is 2·6, the average probability of one fission anywhere in the sphere inducing another is $2 \cdot 6\epsilon$.

Everything depends on whether $2 \cdot 6\epsilon$ is greater or less than unity. If it is less, a spontaneous fission, though it may be followed by a few induced fissions, will not lead to an indefinitely large number. If, on the other hand, $2 \cdot 6\epsilon$ is greater than one, each "generation" of fissions will give rise to a larger number of fissions in the following generation and the numbers will increase continuously. This process is known as a *chain reaction*, though the rapid expansion in all directions that can occur evokes a picture which is more like a kind of three-dimensional chain mail than it is like the linear structure usually referred to as a chain.

Naturally, this expanding process cannot continue for ever. Eventually the conditions must change and $2 \cdot 6\epsilon$ falls below one.

This happens as the result of either the using-up of the U^{235} or the evaporation and dispersion of the material of the sphere as the fission energy is transformed into heat. The quantity 2.6ϵ is called the *reproduction constant* and is usually given the symbol k. In the jargon of the nuclear engineer, "the k of the system must be equal to one for the fission rate to remain constant". It clearly depends on the shape as well as the size of a piece of U^{235}. Neutrons can escape more easily from sheets or wires than they can from a sphere which, having the smallest possible ratio of surface to volume, is the shape giving the largest reproduction constant possible for a given quantity of material. Even for a sphere of given radius, k will vary with the amount and kind of surrounding material. If this consists of an element such as carbon which will not absorb many neutrons but which can scatter or reflect them back, k will be considerably greater than it would be for the sphere in free space. For any particular shape, k will increase from zero at zero size to 2.6 (for U^{235}) at infinite size. There will be a critical size for which $k=1$. The radius of the sphere for which $k=1$ is of critical importance and the mass of material then contained is called the *critical mass*.

8.11 *The atomic bomb*

An atomic or nuclear bomb consists in principle simply of two or more pieces of fissionable material such as U^{235} which are separately below the critical mass but which immediately before the explosion can be combined into a single piece well above the critical mass. The Hiroshima bomb, for example, consisted of two hemispheres of U^{235} each a little less than the critical mass for its shape. One of these formed the "bullet" of a field-gun which could fire it at the other so as to combine the two very quickly into a sphere which then had more than twice the critical mass owing to its more efficient shape. The critical mass for a sphere of U^{235}, in the absence of a neutron reflector, is about $2\frac{1}{2}$ kg. Only a fraction of this is used. In the Hiroshima explosion about 450 gm of U^{235} actually underwent fission, yielding 5×10^{20} ergs or fifteen thousand megawatt-hours in less than a microsecond.

To ensure a successful explosion before the two hemispheres bounced apart or were broken up as a result of their mechanical impact, it is likely that an extra neutron source to start the fission chain was needed. The critical mass of U^{235} gives only about one spontaneous fission per second, and it would clearly be desirable to start the reaction much faster than this could be relied on to do. The only really enjoyable feature of the reports of the Hiroshima bombing was a picture in one of the London newspapers giving an artist's impression of the bomb, which was shown to include an operating cyclotron for the purpose of providing these extra neutrons. Nothing so complicated would be required to ensure this. Neutrons are obtained when α-particles strike beryllium, and little ingenuity would be required to ensure that a small quantity of beryllium would be exposed to the α-particles from, for example, a curie of Po^{210} at the moment of impact of the two hemispheres.

A chain reaction must also be established in a nuclear reactor for producing useful power or radioactive isotopes. We do not in this case want to establish a reaction which will increase as fast as possible but one which will continue steadily at a controllable rate. The use of a pure sample of U^{235} involves first a risk of explosion and second a certainty of great expense in production since it involves separation of the uranium isotopes.

8.12 *The nuclear reactor or pile*

The big reactors which are now producing power are fuelled with ordinary uranium or uranium only slightly enriched in U^{235}. This is made possible by the fact that slow neutrons are much more readily absorbed by U^{235} than by U^{238}. The neutrons resulting directly from fission have a mean energy around 1 MeV. These are moving too quickly to be affected by the attraction of a uranium nucleus until they get very close indeed. The sizes of all the uranium isotopes are much the same. Hence in ordinary uranium, these fast neutrons will be collected in proportion to the numbers of atoms present, i.e. 99·3 % will be collected by U^{238}, almost all without producing fission, and only 0·7% by the U^{235} with fission.

When the neutrons are moving slowly, the different interactions with the different isotopes have a large effect. The cross-section of the U^{235} is then some 200 times that of the U^{238}, and even in the natural mixture of isotopes rather more of such neutrons will be absorbed in the U^{235} than in the U^{238}. The essential requirement of a reactor running on natural uranium, therefore, is that neutrons emitted from the fission of one nucleus should lose the greater part of their kinetic energy before they have much chance of meeting another uranium nucleus. This is achieved by distributing the uranium in small pieces in a large mass of an inert element, i.e. one which does not easily absorb neutrons. Suitable elements are beryllium, carbon, oxygen, or the heavier isotope of hydrogen, deuterium. Fast neutrons entering a mass of any of these will make frequent collisions, in each of which some kinetic energy will be lost. The material used to reduce the energy of the neutrons in this way is known as a *moderator.* The lighter the atoms of the moderator, the more energy will be lost per collision, so that the best nuclide for this purpose is ordinary hydrogen, whose nuclei are of almost the same mass as the neutron. Unfortunately, single protons have a considerable absorption cross-section for neutrons and it is no use to slow these efficiently if they are then gobbled up before they can reach a new piece of uranium. The next lightest nuclide is deuterium, but this itself is expensive and cannot be made into a solid compound which will stand high temperatures. Carbon forms a good economic compromise. The power reactors now operating in Britain at Calder Hall, Chapel Cross and their heirs all use carbon, in the form of graphite, as moderator. The reactors consist of piles of graphite blocks through which are cut channels in which lie rods of natural uranium protected by thin aluminium cases and cooled by high-pressure carbon dioxide gas.

Like the atomic bomb, the power reactor has a critical size at which the chain reaction becomes self-sustaining. Since some neutrons are absorbed by the best of moderators and nearly half of those left are still taken by U^{238}, there is very little margin to spare for loss from the outside of the pile. The critical mass necessarily therefore runs into tons. It is necessary to make the pile rather larger than the critical size (i.e. with $k > 1$) to allow for small changes during running and to make it possible to raise the power output to the running value fairly quickly after a shut-down. The size may also be increased still further by intentionally changing the distribution of uranium and carbon away from the optimum, so as to spread the heat produced over a larger volume. The final size may use over 100 tons of uranium metal in two thousand tons of graphite. The excess reactivity is removed during steady running (i.e. k is reduced to 1) by control rods containing cadmium. This has a large cross-section for absorption of neutrons. The rods can be run into and out of the pile either automatically or manually by remote control.

We are not concerned here with the extensive complications of design needed for a pile which is to produce heat steadily at a rate up to a million kilowatts. A whole technology of nuclear engineering has grown up since the war for this purpose. We shall accept the fact that many piles are now running with heat-outputs of this order and shall concern ourselves only with their use for the production of radio-nuclides.

8.2 Fission Products

They can do this in two ways. First, the products of the fission of uranium are themselves a major source of radioactive materials. As has been stated, the initial nuclear products have always a very large neutron excess and are therefore violently β-active. (The neutron excess arises simply from the fact that the stable proportion of neutrons in uranium is far greater than the stable proportion among the elements which have close to half of its atomic number.) The fission usually is asymmetric, giving a most probable ratio of the mass of the heavy fragment to that of the light fragment of almost 1·5 : 1, but this ratio may vary in individual breakdowns from 1 : 1 to 2 : 1. Nuclides with mass numbers from 72 to 161 have been observed among the fission products. In Fig. 44 is shown the distribution of the fission products as a function of mass, showing the characteristic

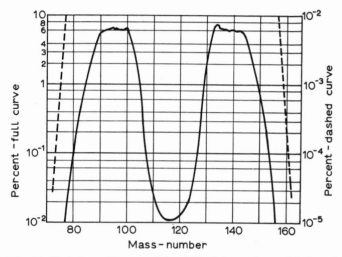

Fig. 44 Variation with mass of the yield of the products of fission of U^{235} by slow neutrons. Note that the sum of the yields of nuclides with different charges but the same mass number is plotted as a single point.

double-humped curve. Over most of this region, useful amounts of the various radioactive species are produced. Those which are made less efficiently, however, though probably produced in adequate absolute amounts for any existing experimental purpose, may be troublesome to free from the larger quantities of other radioactive nuclides. Even where this trouble is tolerable, purification may be difficult to do quickly. The spent uranium rods in which the activity is contained are so radioactive when they come out of the pile that they can be handled only by remote control behind heavy shielding screens. With the additional chemical difficulty that they initially contain active isotopes of every element from zinc (30) to terbium (65) it is impracticable to use them as sources of most of the shorter lived isotopes. Only isotopes of such elements as the inert gases or iodine, of which useful amounts can be removed by heating, can economically be extracted if their half-lives are less than a day or two. As sources of large quantities of the longer lived isotopes within the range shown by Fig. 44, reactors are, however, unequalled. A big pile will produce such isotopes as Sr^{90} (28 years) or Cs^{137} (30 years) at a rate of tens of grams (or thousands of curies) per day which is orders-of-magnitude faster than any other practicable method.

8.3 The (n, γ) Reaction

The second way in which a pile can be used to produce radioactive elements is by making use of the very large flux of slow neutrons in its interior. As has been said, a pile is designed with some excess reactivity and the extra neutrons can be used to practical advantage instead of being mopped up by control rods.

The simplest possible nuclear reaction is involved; that in which a free neutron is added to a nucleus. A slow neutron is attracted into all known stable nuclei except He^4 and absorbed with a considerable release of energy. If the initial kinetic energy of the neutron is small, this energy will not usually be enough to lead to the ejection of a proton or α-particle. (The only important exception is N^{14} which, on absorbing a neutron, will usually eject a proton, leaving the radioactive nucleus C^{14}). In the great majority of cases the energy will be released in the form of γ-radiation, leaving an isotope of the same element with an extra neutron. For example, ordinary phosphorus, P^{31}, will absorb neutrons, releasing γ-rays and giving rise to the radioactive isotope P^{32}. This reaction is usually written P^{31} (n, γ) P^{32}.

A large number of useful radioactive isotopes may be made by similar reactions; for example Na^{24} from Na^{23}; Co^{60} from Co^{59}. The method will often be preferred for the production of nuclides also available as fission products—e.g. Y^{90}—because no chemical separation is required.

The yield of radioactive isotopes which can be obtained in this way from a large pile is very large. The average flux of slow neutrons may be from $10^{12}/cm^2/sec$ up to 10^{14} or even more. A sample thick enough to absorb most of the neutrons falling on it can be introduced. Hence active atoms may be made by the (n, γ) process at a rate of 10^{12} or more per second per cm^2 of sample. The maximum permissible weight of sample depends on the excess reactivity of the pile, i.e. the proportion of pile neutrons which can be withdrawn without stopping the chain reaction. Since a big pile may easily be producing nearly 10^{19} neutrons per second, the number withdrawable can be 10^{14}/sec with very little disturbance. If the target element is left in long enough, therefore, we can build up the product of the (n, γ) reaction till we reach an equilibrium limit when the rate of loss of the new nuclide is also 10^{14} atoms/sec. For a short-lived product, the loss will mainly be due to its own radioactive breakdown, 10^{14} disintegrations/sec being about 3000 curies. For a long-lived product the main loss may be due to the removal of the desired isotope by further neutron absorption. For example, after we have made a large quantity of Co^{60} by the n, γ reaction with Co^{59}, we shall begin to lose it again at a serious rate by the reaction Co^{60} (n, γ) Co^{61}. It is uneconomic to increase the yield of any isotope to the point where we are losing a large part of any further increase, but it is clear that something of the order of 1000 curies can readily be obtained.

A serious drawback of the method is that, though the total activity produced may be very high, the specific activity is often low. Since the base material from which the active nuclide is made is the same element, chemical separation is not possible. One exception to this has already been mentioned; C^{14} is produced from nitrogen by slow neutrons by the reaction N^{14} (n, p) C^{14} and can therefore be separated without difficulty free of stable carbon isotopes. Several other nuclides may be produced, though less efficiently, by the same reaction if the smaller flux of faster fission-neutrons are used. For example, P^{32} may be made from sulphur according to the scheme S^{32} (n, p) P^{32} by neutrons with energy over about 2 MeV. Another important special case is tritium. This can be made by the action of slow neutrons on Li^6, the reaction being Li^6 (n, α) H^3. The tritium can readily be freed from lithium by heating and, if necessary, from the other product, helium by adsorption into zirconium or passage through heated palladium.

Another difficulty is that most elements consist of a number of isotopes and only one of these will yield the particular radionuclide desired. Even where the appropriate stable isotope forms a good proportion of the whole, this may be serious. We have seen that in a nuclear reactor the U^{235} can absorb about as many slow neutrons as can U^{238} even though there is about 140 times as much of the latter present. Similar variations occur among the isotopes of other elements. The relative effectiveness of different nuclides in absorbing slow neutrons (which form by far the most important part of the flux inside a reactor) is given in the nuclear chart in terms of a *cross-section* σ.

This represents the effective target area presented to a neutron with a velocity of 2·2 km/sec. This velocity is chosen as the appropriate mean for a Maxwellian distribution of neutrons in thermal equilibrium near to room temperature. The value of σ is given in *barns*, one barn being 10^{-24} cm^2. It is convenient to remember that the mean-free-path for capture of slow neutrons in solids and liquids is of the order of $10/\sigma$ centimetres. The free path for scattering may easily be much less than this, so the formula cannot be used for estimating the attenuation of a flux of slow neutrons unless σ is fairly large, when the probability of absorption will be bigger than that of scattering.

Using the values of σ for the different stable isotopes of an element, we can see quickly whether the slow neutrons of a pile will give us a useful source of the active isotope desired. If we want Cu^{64} (12·8 hrs, β^-, β^+, e-cap), for example, we see from the chart that ordinary copper consists of 69 % Cu^{63} for which σ is 4·1 barns and 31 % Cu^{65} for which σ is 2·0 barns. The relative neutron absorptions will then be proportional to $69 \times 4\cdot1$ or 283 and 31×2 or 62 respectively and we see that 82 % of the neutrons absorbed will go into producing the nuclide we want. Furthermore, Cu^{66}, the nuclide produced from the Cu^{65}, has a half-life of only 5 minutes, while Cu^{64} has a half-life of 12·8 hours. Hence, six hours after exposure of our copper sample in the reactor, the Cu^{64} will have lost by decay only about 40 % of its activity, while the whole of the Cu^{66} will have decayed away, and we shall be left with a radiochemically pure sample. Finally, since the cross-section (4·1 barns) is fairly large, the activity required can be produced in a fairly small piece of copper, giving a good specific activity.

On the other hand if we wanted the nuclide Cl^{38} (37·5 mins, β^-) the situation is much less favourable. Here Cl^{37} forms only 24·5 % of ordinary chlorine and has a cross-section of 0·56 barns while Cl^{35} forms 75·5 % and has a cross-section of 42 barns. Hence little over 0·4 % of the neutrons absorbed will be used to give the nuclide we want. Less favourable examples still can easily be found.

8.31 *The Szilard-Chalmers process*

When an adequate total activity can be made, the problem of low specific activity can sometimes be solved by an ingenious method independently proposed by Szilard and Chalmers. This requires the neutrons to be absorbed, not in the pure element but in a compound of the element. Furthermore, the element must be held into the compound in a non-ionic form. For example, if a high specific activity is required of I^{128} (25 mins, β^-), the stable nuclide I^{127} is bombarded as an organic compound, such as ethyl iodide. (It is of course desirable that the other elements in the compound should not absorb too many neutrons; in the case under consideration the iodine will take 80% of the total.) The slow neutrons absorbed have too little momentum to disturb the compound, but the energy released by their capture is almost instantaneously emitted as a γ-ray. This will cause the I^{128} nucleus to recoil and the recoil energy will be large enough to tear it out of the compound. The free iodine atom may recombine with a neighbouring molecule when it is brought to rest, but a large fraction of atoms will not do so and will remain as the free element.

If the ethyl iodide is now shaken with water, with which it is immiscible, the free iodine can be dissolved out into the water phase, practically free of unchanged stable iodine. By this means, even if tens of grams of iodine have been used to produce only a few millicuries of activity, the latter can be obtained with a most satisfactory specific activity.

The Szilard-Chalmers process can be used for most non-metals but it is difficult to use for metals* only a few of which, such as manganese, can be made to form suitable compounds.

8.4 Particle Accelerators

There are many important radioactive nuclides left which cannot be made in a reactor, either as fission products or by slow-neutron absorption. They can be made with the help of particle accelerators. These cannot be used to produce anything like the yield of which a reactor is capable in favourable conditions, but are more versatile.

As has already been said, if nuclear reactions are to be produced by charged particles, the latter must first be accelerated to considerable energies. Fig. 52 gives a good idea of the minimum energy likely to be needed in different parts of the periodic table. It must be remembered, however, that many reactions are highly endothermic —i.e. they absorb a lot of energy—so that a great deal more than this minimum may be needed in a particular case.

By far the most important accelerator for the production of radio isotopes is the cyclotron and this alone will be discussed here.

The first cyclotron was built over thirty years ago, by Professor E. O. Lawrence and Dr. M. S. Livingston, who thus laid the foundation of one of the major fields of modern physics. We shall describe briefly how it works.

A particle of mass M and charge e moving in a field of magnetic

---

* It has recently been shown by W. Parker that active silver ions in solution can be obtained by recoil from colloidal particles of metallic silver dispersed in water. The yield is lower than in the cases described above.

induction B with velocity v will follow a circular orbit of radius r given by

$$r = \frac{Mv}{Be} \qquad 8.1$$

The periodic time will be given by

$$\tau = \frac{2\pi r}{v} = \frac{2\pi M}{Be} \cdots$$

and the angular frequency of rotation in the orbit, ω_p, will be

$$\omega_p = \frac{Be}{M}. \qquad 8.2$$

The frequency is then independent of the particle velocity at speeds sufficiently low for M to be unaffected by relativistic effects.   To show the order of magnitude of energy obtainable with this restric-

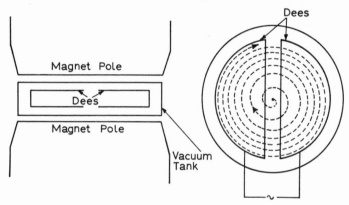

Fig. 45 The essential components of a cyclotron;
A—elevation;   B—plan.
The dotted line shows the path of one accelerated particle.

tion, a 1 % increase of mass represents about 10 MeV for a proton and 20 MeV for a deuteron, but only 5 keV for an electron.   The fact that the rotation frequency is independent of velocity represents the basic requirement of the cyclotron.   From the figures just given, this kind of accelerator is therefore suitable only for heavy particles.   The principle of operation is as follows.

In a uniform and constant magnetic field an alternating electric field with constant angular frequency ω_E is used to accelerate particles, the position of which in their orbit bears a constant relation to the phase of the electric field.   The essential components of a cyclotron are shown in Fig. 45.   Two hollow dee-shaped electrodes, connected to a high-frequency voltage supply, are mounted in a vacuum box in a strong magnetic field perpendicular to the planes of symmetry of the dees. Then a charged particle initially at rest at the centre of symmetry of the system will follow a path of the form shown by the dotted line if $\omega_E = \omega_p$ (see equation 8.2).

If the gap between the dee-edges is small, the particle will pick up an energy $eV \cos \phi$ at each transit of the gap, and $2eV \cos \phi$ in each complete revolution, where $V \cos \omega_E t$ is the instantaneous voltage-difference between the dees and $\omega_E t = \phi$ gives the phase of the electric field at which the particle starts.

The frequency relation given in equation 8.2 shows that the electric field will be exactly reversed while the particle travels between one gap-transit and the next, so that it is similarly accelerated on each occasion. Each dee acts in effect as a Faraday cage with no internal high-frequency field, so that except when crossing a gap the particle is acted on only by the magnetic force and pursues a semicircular path. After N complete revolutions the particle will have total energy $eV \cos \phi (2N - k)$, where k represents the fact that the full gap-energy may not be obtained in the first revolution or so, when the orbit is comparable in size to the dee gap. Then, since we are at this stage limiting ourselves to the non-relativistic region

$$\tfrac{1}{2} M v^2 = eV(2N - k) \cos \phi,$$

so that

$$v = \left(\frac{2e}{M} \right)^{1/2} \{ (2N - k)V \cos \phi) \}^{1/2}$$

From (8.1)

$$r_N = \frac{1}{B} \left(\frac{2M}{e} \right)^{1/2} \{ (2N - k)V \cos \phi \}^{1/2} \qquad 8.3$$

Hence

$$\tfrac{1}{2} M v^2 = \frac{e^2 B^2 r N^2}{2M} \qquad 8.4$$

and

$$N_{\max} \simeq N_{\max} - \tfrac{1}{2} k = \frac{E_{\max}}{2V \cos \phi},$$

where $E_{\max}$ electron volts is the maximum energy obtained.

Equations 8.3 and 8.4 show first that as the energy and radius increase the orbits are more closely crowded together, and second that the final energy depends only on B and $r_{\max}$ for a given particle, as is of course clear from 8.1.

The first requirement of cyclotron design for the production of particles of any desired energy is a magnet giving a field of sufficient extent and magnitude to provide the required $(Br)_{\max}$. A particle whose orbit reaches the design radius $r_{\max}$ will then necessarily have the correct energy, but a number of subsidiary conditions must be satisfied before a particle can in fact reach such a radius. These conditions lie outside the scope of this volume and will not be further discussed except to say that their satisfaction costs a lot of time and money.

The second requirement is a vacuum tank of 1 to 2 metres diameter, according to the particle energy required. This must be kept at a pressure of 10^{-5} mm of mercury or less in spite of a steady input of gas for the ion source, air leaking in from outside and gas given off from target or dees. It is not easy to construct a leak-free tank of the size required with all the necessary inlets and outlets. The original small tanks were fabricated but for the bigger ones of 150 cm (60 in.) or so built after the war this was at first thought difficult.

The first solution to be tried was to make the main part of the tank as a single bronze casting with steel lids top and bottom, and it is interesting to see how this same idea was applied in three different countries. In U.S.A. and in Britain (Birmingham) a complete bronze casting was made of the form of the finished tank and was then machined to the exact shape required. Unfortunately, in both cases the bronze proved not to be as solid as it seemed, but rather to be a kind of metal soufflé with a reasonably good outer skin. Where this skin had been broken by machining, air found its way into the soufflé and leaked out again through another machined surface into the interior of the tank. The British worked away piecemeal at this with vacuum grease and plasticine for about nine months and eventually got the leak-rate down to a workable value. The Americans, after sufficient work to appreciate the scale of the problem, drilled thousands of small holes through the outer skin and then dipped the whole tank into a huge bath of molten solder in the hope of filling up all the cavities. The casting absorbed some hundred kilograms of solder but continued to leak. Its owners then threw it away and fabricated a complete new one of stainless steel sheet.

The Russians, starting a little later, had heard of the difficulties resulting from the use of a casting. They therefore had a single forging made (forgings, unlike castings, being non-porous), nearly 2 metres square and 30 cm thick, and cut the interior out with a kind of glorified dentist's drill. All three systems work; the American machine was finished first, and the Russian has the lowest leak-rate.

Even with a low leak-rate, pumping speeds of the order of a thousand litres per second are needed.

The third requirement is a high-power oscillator to supply radio-frequency power to the dees. A large machine may need a good fraction of a hundred kilowatts at ten megacycles or more per second.

This was a serious difficulty in the early days. Now the only trouble is the cost, though there is still much scope for ingenuity in reducing this.

The final requirement is a suitable source of ions. A great variety of sources have been developed. Practically all of them consist of a heated tungsten or tantalum filament supplying an ampere or so of electrons to ionise a column of the appropriate gas in a narrow tube along the axis of the magnetic field at the centre of the machine. A slot or hole in the side of the tube enables ions to escape into the tank where they can be picked up and accelerated by the electric field of the dees.

Having built a successful cyclotron, we have still to arrange for the particles to hit the required elements. A large cyclotron will give beams of hundreds of microamperes at tens of MeV, so that many kilowatts may have to be dissipated by the target. This power will be delivered to an area of perhaps a square centimetre, providing a serious engineering problem in heat removal, especially as the layer in which the heat is liberated is a millimetre or less thick and as, for efficient use of the beam, the target must be placed inside the vacuum tank. Water-cooled cooper blocks can be designed to stand some 10 kilowatts in these conditions and various high melting-point metals can be plated or soldered on to these to take comparable powers. For metals which melt or sublime more easily, rapidly rotating targets have been successful.

Non-metals and compounds which have a low thermal conductivity are much more difficult to make into satisfactory targets for use inside the vacuum.

To make radioisotopes from such materials, the particles which have been accelerated by the cyclotron must be brought out of the machine into the air through a thin metal foil "*window*". The extraction of the beam in this way is not easy to do efficiently; 20% is generally regarded as a reasonable figure. Hence 100 μA of protons or deuterons and 20 to 50 μA of He^3 or α-particles is regarded as satisfactory. This is more than enough for most liquid or powder targets, although these can now be cooled by an air blast. For many such targets, 200 watts (representing for example 10 μA at 20 MeV) is as much as they can stand. Often this question of what input power the target will survive without destruction is what determines the choice of reaction to produce a particular radioactive nuclide.

Cyclotrons have been designed to accelerate several kinds of particles, but for the purpose of producing radioisotopes the important ones are deuterons, followed by α-particles, protons and, less frequently, He^3 and tritium.

8.41 *Deuteron reactions*

The deuteron is a relatively loosely bound combination of a neutron and a proton and either of these may alone be captured when the deuteron strikes another nucleus. The resulting interactions are known as (d, p) or (d, n) reactions according to whether the proton or the neutron escapes as a free product of the reaction.

For example, if carbon^{12} is bombarded by deuterons, we can get either C^{13} by the reaction C^{12} (d, p) C^{13} or N^{13} by the reaction C^{12} (d, n) N^{13}. C^{13} is stable and forms 1 % of ordinary carbon so this is not of practical value. The N^{13}, however, is radioactive, emitting positrons with a half-life of ten minutes and could not be made in a pile. It will be noticed that the net effect of the (d, n) reaction is to add a proton to the nucleus concerned. By means of this reaction therefore we can produce all those radionuclides which have a single proton above the series of stable nuclei. The (d, p) reaction, on the other hand, has the net effect of adding one neutron, which can usually better be done in a pile unless the target nuclide has an extremely small cross-section for absorption of neutrons.

The cross-section of a nucleus for reaction with charged particles, whose energy is well above the potential barrier, depends mainly on the geometrical size of the nucleus. There is very little variation between different isotopes of the same element.

More complex deuteron reactions can also be used. For example, the standard method of producing the longest lived positron emitter, Na^{22} (2·6 years), is by bombardment of magnesium by deuterons. The reaction concerned is then

$$Mg^{24} \text{ (d, } \alpha) \text{ } Na^{22}.$$

When the deuterons have high energies (15 MeV or more), reactions in which more than one particle is ejected begin to give fair yields; for example (d, 2p) or (d, 2n) reactions. For example our old friend P^{32} can be made according to the scheme S^{32} (d, 2p) P^{32}. This will give a much smaller yield than will the method of making it by absorption of pile neutrons in P^{31}. It can, however, give a higher specific activity since the phosphorus can be chemically separated from the sulphur target.

Protons may also be accelerated by the cyclotron and used to give radionuclides by similar reactions. For example, C^{11} (20 min, β^+) may be obtained by bombarding boron with protons according to the reaction B^{11} (p, n) C^{11}. Exchange of a proton for a neutron does not usually need much energy, the potential barriers for protons are low and protons are easy to accelerate, all of which favour the use of this kind of reaction. Another useful proton reaction is the p, d reaction which in effect *subtracts* a neutron from the target.

For example, we can obtain Cl^{34} from Cl^{35} by the reaction Cl^{35} (p, d) Cl^{34}. Like the (n, γ) reaction, this does not give a high specific activity.

8.42 *Alpha-particle reactions*

The simple (α, n) or (α, p) reactions rarely produce a nuclide which *cannot* be made by a deuteron or proton reaction, but may make it possible to use a more convenient element as target. Thus, P^{30} ($2\frac{1}{2}$ min, β^+) can be made by bombarding silicon with protons according to either of the schemes Si^{30} (p, n) P^{30} or Si^{29} (d, n) P^{30}, but silicon is chemically rather intractable and Si^{29}, Si^{30} form only 4.7% and 3.1% respectively of natural silicon. On the other hand, aluminium makes a satisfactory target and consists of a single isotope. It can give the same product, by the reaction Al^{27} (α, n) P^{30}, with much greater efficiency.

There are a few nuclides, such as F^{18} (110 min, β^+), which can best be produced by α-particles as in the reaction O^{16} (α, d) F^{18}. The alternative method using protons, O^{18} (p, n) F^{18} is unsatisfactory since O^{18} forms only 0.2% of ordinary oxygen and to use the reaction F^{19} (p, d) F^{18} by bombarding a fluoride target gives a low specific activity.

8.43 *He^3 and tritium reactions*

Helium^3 has not yet been used extensively for isotope production but is likely to be used more as time goes on, since it is perhaps the best particle available for producing highly neutron-deficient isotopes such as Fe^{52} (8.2 hrs, β^+) or Zn^{62} (9.3 hrs e.cap., β^+).

Tritium, being itself radioactive and thus adding yet one more hazard to those attached to an accelerator, has as yet been used only at low energies. Its great value in producing such biologically important isotopes as Mg^{28} (21.4 hrs, β^-), by the reaction Al^{27} (t, 2p) Mg^{28} may increase its use in the future.

8.44 *Accelerator yields*

The yields which can be obtained from an accelerator are much smaller than those from a pile. A cyclotron target can be made to take perhaps 500 μA of charged particles at 10 MeV or more. This represents 3×10^{15} singly charged particles per second, which is fully comparable with the number of neutrons per second that can be taken from a large pile. Even at 20 MeV, however, only about one deuteron in a thousand will produce a nuclear reaction of any sort.

The rest fritter away their energy in ionisation and similar processes without ever coming close enough to a nucleus for interaction. Other particles and deuterons with lower energies, will have a lower efficiency still. *Every* slow neutron which is captured by a target will produce a reaction. Again, the charged particle of high energy can give many different reactions so that a proportion only of the reaction products will be of the particular kind desired. With the exceptions mentioned, *every* slow neutron will produce the same (n, γ) reaction.

The result is that the limiting activity of any particular nuclide that can be produced in a cyclotron is at the best a few curies and often only a few tens or hundreds of millicuries.

8.5 Alpha-Beryllium Sources

The high efficiency of utilisation of slow neutrons makes possible the use in certain cases of a quite different and much simpler "two-stage" source. In this we start with a neutron source consisting of beryllium mixed with an α-particle emitter. This produces neutrons according to the reaction Be^9 (α, n) C^{12} or Be^9 (α, n) 3α. The efficiency is not very high, a curie of Ra^{226} mixed with beryllium, producing about 10^7 neutrons/sec. Where the large γ-radiation from the decay products of radium would be troublesome, Po^{210} may be used instead of Ra^{226}. A one-curie Po-Be source gives $3-4 \times 10^6$ neutrons/sec., but decays with the half-life of Po^{210}, 138·4 days.

Although such a source gives few γ-rays directly (about one quantum for every ten neutrons, from a rare mode of disintegration of the Po^{210}), it must be remembered that every nucleus which captures a neutron will at once emit one or more γ-quanta. Since in any practical arrangement most of the neutrons are captured by the hydrogen in the moderator, there will still be some millions of 2·1 MeV γ-quanta from the H (n, γ) D reaction even in a "γ-free" source.

A source of smaller but comparable efficiency can be made by mixing a γ-ray emitter such as Sb^{125} with the beryllium, utilising the reaction Be^9 (γ, n) $Be^8 \rightarrow 2\alpha$. In either case the neutrons are reduced to thermal velocities by immersing the source in a moderating tank of water or block of paraffin wax of 20 to 30 cm radius. Some loss of neutrons occurs owing to capture by hydrogen to form deuterium, but enough are left to produce some microcuries of the product of an (n, γ) reaction from a 1 curie Ra-Be source.

In cases where the Szilard-Chalmers (see section 8.31) or similar processes can be used to concentrate the product into a convenient source this is enough for most experiments in teaching laboratories and for some research purposes. The source involves no maintenance, no operating staff and, if left in its moderating tank, no health hazard. When it is needed for only a short time, Pu-Be sources which, like Po-Be sources are "γ-free", can be hired.

8.6 Small Accelerator Sources of Neutrons

Quite recently, a new kind of laboratory neutron source has become available, intermediate between the α-beryllium source and the nuclear reactor in convenience, cost and performance.

This consists of a continuously pumped deuteron accelerator with a tritium target. A small electrostatic generator can give 150 to 400 kV with an output of some hundreds of microamperes d.c. This can be used to accelerate deuterons in a demountable vacuum tube to strike a target of zirconium or titanium into which tritium gas has been allowed to diffuse. Such targets can now be obtained commercially at a cost from £10 upwards, and will last from ten to a few hundred hours, according to the power input, in practical running conditions. Neutrons will be produced, with an initial energy of some 14 MeV, by the reaction T (d, n) α. The total yield of fast neutrons may be of the order of 10^{10}/sec, some thousands of times the yield from a 1-curie Po-Be source. These can be moderated with water or paraffin wax to give 10^7 to 10^8 slow neutrons/cm²/sec. The upper figure can be used only for short bursts if rapid deterioration of the target is to be avoided. The total cost of such a system, ready made, might be in the region of £10,000.

Having discussed the installation and some of the reactions by which radionuclides are made, we shall conclude this chapter by considering the production of active sources to meet specific practical requirements.

8.7 Choice of Production Processes for Radio Nuclides

The reasons for making a radioactive nuclide may be divided into two classes, the first being for the study of the nuclide itself and the second for its use in other investigations.

In the first class we have no problem in deciding on which isotope of an element to make; we have to make the one we want to study. The aim of the experiment affects a good deal, however, the choice of method of making it.

For example, if we want to study the γ-rays emitted, a fairly high total activity will be needed; perhaps some millicuries; but the specific activity is not very important unless extremely soft γ-rays are concerned. Little self-absorption will usually be produced by a gram or so of the stable element. Consequently, production by the (n, γ) reaction in a pile is, whenever possible, the best method of production. Alternative methods should be used if (a) the half-life is inconveniently short, when the best producer may simply be that which is nearest (in transit time) to the research worker's laboratory or (b) if several stable isotopes of the original element exist which will produce inconvenient additional γ-activities. With luck, even if these are produced, they may have a much lower yield or may have a shorter half-life so that they can be simply allowed to decay away before beginning the main experiment. It will be equally good if they have an extremely long half-life; for example K^{40} will be produced from K^{39} some 22 times faster than is K^{42} from K^{41} but, neglecting the fact that 0.01 $\mu Ci/gm$ of K^{40} is present already, this would not matter. The half-life of K^{40} is 5×10^{11} times greater than that of K^{42} so that the rate of decay of the additional K^{40} at the end of a bombardment, though 22 times as many atoms have been made, will be 2×10^{10} times *less* than that of the K^{42}.

There are plenty of less simple cases. For example, suppose we wish to study Ge^{75} (82 mins, β^-). This can be made in adequate quantity from ordinary germanium in a pile according to the reaction Ge^{74} (n, γ) Ge^{75}. At the same time we shall make about 10% as much of Ge^{77} (12 hours, β^-) according to the reaction Ge^{76} (n, γ) Ge^{77}. Ge^{77} gives some 25 different γ-rays and, since its half-life is longer than that of the Ge^{75}, the proportion of them will get steadily larger with time. X-rays from the electron-capturing Ge^{71} (12 days), which is produced in larger quantities than either of the others, may also be troublesome. One solution is, of course, to put a pure sample of the single isotope Ge^{74} into the pile, but separated isotopes are very expensive and fair quantities would be needed since the neutron-absorption cross-section is only 0.55 barns.

A much better method will be to bombard As^{75}, the only isotope of arsenic, with deuterons from a cyclotron. Then we shall get our Ge^{75} from the reaction As^{75} (d, 2p) Ge^{75} and it will be impossible to produce Ge^{77}. If we use deuterons of moderate energy we shall not get Ge^{71} because the *threshold energy* of the reaction As^{75} (d, α 2n), or the minimum energy required to make the reaction possible, is not forthcoming.

The use of the threshold energy, illustrated by this example, is valuable and it is worth while to discuss its application in detail.

In several reference books and charts; for example, William H. Sullivan's *Trilinear Chart of Nuclides*, the masses of most known isotopes are given to $1/10,000$ of a mass-unit. (A mass-unit is now defined as $1/12$ of the mass of the nucleus C^{12}. The figures below are, however, taken from the Sullivan Chart which is based on $1/16$

of the mass of O^{16} as unit. The difference is only about 3 parts in 10,000.) By adding up the masses of the initial and final products of a reaction, we can find the gain or loss of mass involved. In the case we have been considering, the initial masses are those of As^{75}, which is 74·9455, and of the deuteron, 2·0147. Hence the total initial mass is 76·9602. If the reaction goes to the desired products $Ge^{75} + 2p$, the total final mass is $74·9466 + 2 \times 1·0081(5)$, which comes to 76·9629. On the other hand, if it goes to $Ge^{71} + \alpha + 2n$, the total final mass is $70·9476 + 4·0039 + 2 \times 1·0090$ or 76·9695. The final mass is therefore greater than the original in the desired case by 0·0027 mass units and in the undesired case by 0·0093 mass units. The mass increase, according to Einstein's law $E = mc^2$, requires that before either reaction can take place, extra kinetic energy equivalent to this mass must be supplied. The energy equivalent of one mass unit is 931 MeV. Hence the extra energy needed is, for the desired reaction (d, 2p) $931 \times 0·0027 = 2·5$ MeV, and for the undesired reaction (d, $\alpha2n$), $931 \times 0·0093 = 8·6$ MeV.

The energy change can be shown in the equation for the reaction. If nuclide X is bombarded with particle x to produce nuclide Y with emission of particle y, and if in the reaction an energy Q is liberated, we can write the equation of the reaction as $X(x, y)Y + Q$. The two reactions that we have been considering can then be written

$$As^{75} \text{ (d, 2p) } Ge^{75} - 2·5 \text{ MeV}$$

and $$As^{75} \text{ (d, } \alpha2n) \text{ } Ge^{71} - 8·6 \text{ MeV.}$$

Hence, if we bombard our arsenic with deuterons whose energy is a little less than 8·6 MeV, the unwanted reaction is impossible while the wanted one is quite easy. In practice, an energy which is greater than 8·6 MeV by several MeV would still be safe as the α-particle and the two neutrons will always take away some kinetic energy with them so that more than the absolute minimum would have to be supplied.

One thing still needs to be remembered. Although we shall have no undesired isotopes of germanium itself, there will be a large yield of radioactive arsenic and selenium isotopes. The Ge^{75} must therefore be separated chemically before use; this is laborious rather than difficult and is usually worth while to avoid having to use a separated isotope and to obtain a high specific activity.

If we wish to study the α- or β-activity of a nuclide, we do not usually need such large sources; a few microcuries will be enough for most experiments. On the other hand, we need sources which are very thin compared with the range of the particles and this necessitates a high specific activity. For this, therefore, the (n, γ) reaction from a pile is unsuitable unless by a trick such as the Szilard-Chalmers process we can separate the active atoms from those of the stable elements. When this can be done, a weak neutron source such as a one-curie source of radium-beryllium or its equivalent may be adequate for a limited rate of production of sources.

Most β-sources must, however, be made by a particle accelerator from some neighbouring element so that a chemical separation of the active nuclide can be made.

When sources are needed as tools in other investigations, a somewhat different set of requirements must be satisfied. The first difference is that we may choose which isotope to use of the element needed and often we may even have a choice between two or three elements in the same molecule.

8.71 *Choice of isotope*

Important factors are half-life, type of radiation and ease of production. The choice is often quite wide. Iodine, for example, has sixteen radioactive isotopes with half-lives of more than 20 minutes. Twenty minutes is short for most tracer experiments, but the half-lives of eight isotopes are more than twelve hours, which is a convenient time. Generally speaking, it is best always to choose the isotope with the shortest half-life usable in the experiment proposed. This means firstly that activities left over from one experiment are less likely to interfere with the next one and, secondly, that a conveniently observable activity of a short-lived isotope requires fewer atoms than does a long-lived one and is, therefore, likely to be cheaper to produce. In medical work too it may be desirable to use the shortest possible half-life to reduce to a minimum the radiation dose received by the patient.

When an active nuclide of long half-life *must* be used, great care has to be taken to avoid contamination of counting apparatus, as this is often difficult to clean. Contamination by an active material whose half-life is only a few hours, on the other hand, is less serious. Even if it is quite impossible to clean away, a few days will see its activity drop to an unobservable level.

The minimum half-life that can be accepted depends a great deal on the experiment to be performed. A chemical experiment on solubilities or partition coefficients may be completed in an hour while one on the retention of heavy elements in bones may take several years. Since we can allow for the loss of activity due to radioactive decay, it is unnecessary that the half-life should be much greater than, or even equal to, the length of the experiment. It is rarely objectionable to employ an initial activity which is ten times that with which the experiments will conclude. Hence a half-life of a third or even less of the length of the experiment is quite acceptable. It must be remembered, however, that in this case the advantage in cost of a shorter-lived material may disappear. The minimum activity required is naturally defined by that which can be counted to sufficient accuracy at the end of the experiment. If more of the shorter-lived material must be used, it may cost *more* than would a longer lived one made by a similar process. Other things being equal, a half-life around half the total time of the experiment is about the best one to choose.

After the half-life, the next consideration is the kind of radiation produced. If we are studying the amount of iodine taken up in the thyroid of a human patient, a pure β-emitter is of little use, as it cannot be observed without cutting out a piece of thyroid or surgically inserting a thin-walled counter. In such cases a γ-emitter must be chosen. For this purpose all positron emitters can be counted as γ-emitters, since they will all produce the 0·511 MeV annihilation radiation.

In the case of iodine this still leaves us with a wide choice. This leaves availability as the final determining factor. The most practical approach to this is to begin by looking up iodine isotopes in the lists produced by the Radiochemical Centre at Amersham, Bucks (Great Britain), or Oak Ridge (U.S.A.) and then comparing the prices and delivery times. If none of the prices is satisfactory, it is rarely useful to consider unlisted isotopes; if they are unlisted they may be perfectly possible to make, but they will not be cheaper. Only if none of the listed nuclides is technically suitable should the possibilities of others be examined. The probable feasibility of production of such nuclides can then be considered in the light of the principles given in the earlier part of this chapter. Advice can be obtained in Great Britain on isotopes likely to be made in a nuclear reactor from the Radiochemical Centre at Amersham, and on isotopes likely to be made in the cyclotron from the Physics Department of the University of Birmingham.

A final feature of the production of radioisotopes used as tools is that the specific activity is usually of great quantitative importance. In experiments on the rarer elements used in living systems and in tracer analysis, the highest possible specific activity is needed. In other experiments the specific activity may not need to be particularly high but must be accurately known. When a choice exists of the target element to be used to produce a given radio nuclide, it may therefore be important to favour the one likely to have the lowest percentage of the stable element concerned as a normal chemical impurity.   To take one extreme and rather unlikely case, small quantities of pure K^{43} (22·4 hrs, β^-) could be made from argon by the reaction A^{40} (a, p) $K^{43} - 3·3$ MeV better than from calcium by the reaction

$$Ca^{43} \text{ (d, 2p) } K^{43} - 3·3 \text{ MeV.}$$

Potassium is a normal impurity in calcium but is not found in argon. Furthermore, if the argon is bombarded by a-particles of less than 10·8 MeV, the threshold for the (a, d) reaction, the K^{43} will be uncontaminated by the other radioisotope of potassium K^{42} (12·5 hrs, β^-).

In a deuteron bombardment of calcium the latter could be made according to the reaction Ca^{42} (d, 2p) $K^{42} - 5·3$ MeV which requires very little more energy and comes from an isotope of four times higher abundance.   (Ca^{43} is 0·145% and Ca^{42} is 0·64% of ordinary calcium; Ca^{43} is thus also a rare isotope which will give a poor absolute yield.)

8.72 *Use of radioisotopes*

Part of the content of this paragraph is a repetition of observations which have already been made.   It is useful, however, to give it as a whole without too many cross-references.

In all radioactive experiments it is particularly important not to use more active material than is really useful.   Reasons of economy apply no more than in other work, but to these are added the need for safety and the need to reduce the risk of contamination of counting equipment.

Careful planning is therefore necessary and it may be well worth while to do a pilot experiment with very small quantities of active material before deciding the amount to be used in the definitive experiments.

In planning, it is usually best to start at the end and work back. For example, if it is proposed in a particular experiment to use a particular nuclide and to finish with samples on counting trays to be recorded by end-window Geiger counters, the first figure to consider is the counting rate required from each of such samples.   If only a few samples of a long-lived nuclide will be needed, a counting rate of a few tens of counts a minute above background may be sufficient for each.   (Ten counts per minute above background corresponds to 14,400 per 24 hours, giving a standard deviation of a little under 1%; the standard deviation of the background must also be allowed for, but if the counter is well shielded the combined s.d. should not much exceed 1%.)

On the other hand, if many samples or short-lived nuclides are involved, so that only a few minutes' counting of each will be possible, some thousands of counts per minute may be needed from each sample.

Secondly, having settled the final counting rate needed, allowance must be made for the efficiency of the counter and the probable self-absorption of the sample.

Thirdly, we must estimate the proportion of the initial active material which will be likely to appear in each sample. This may be fairly easy, if it depends on the roughly known chemical yield of a known aliquot of a uniformly mixed material, or very difficult if the biological yield of some complex organic compound is involved. In such a case the first estimate may be little better than a guess and a pilot experiment is practically essential.

Fourthly, the length of the experiment from the time of production or dispatch of the radioactive material must be considered in relation to the half-life of the nuclide used, so as to allow if necessary for radioactive decay. If a pilot experiment is performed, this will automatically be taken into account.

Finally, the number of times the experiment is likely to be repeated must be decided.

At this point a preliminary estimate of the quantity needed of active material can be made. This must be looked at first from the point of view of safety and second from that of cost and availability. The tables given in Chapter 14 will give an indication of the hazard involved. This may make necessary anything from reasonable care and cleanliness to the need for a fully equipped radio-chemical laboratory with remotely controlled handling equipment. If anything more than the former is needed, it is well worth while to spend a lot of time and thought on possible alternative procedures for the experiment, to reduce the amount of active material required. Reconsideration should include the possibility of using a different isotope than that first chosen, or even of ways of getting the same information without using radioactive materials at all. Given an adequate factor of safety, the ordinary rules of planning apply to the consideration of cost; if the material is readily available and the cost is only a small part of the total cost of the experiment, it may well be worth multiplying the provisional answer by a small factor to ensure efficient use of the more expensive equipment. On the other hand, if the radioactive nuclide represents the main cost it may again be worth while to reconsider procedure to see whether the proposed amount is really necessary before sending out the orders.

The planning of a radioactive experiment does not stop when it has been decided how best to get the result. Safe disposal of the radioactive material which survives the experiment must also be arranged before the material is ordered. Short-lived nuclides can conveniently be stored until their activity has dropped sufficiently and then washed down the sink in the ordinary way, but this cannot be done with long-lived nuclides. When these are involved the procedures recommended in Chapter 14 or in the references given therein *must* be followed. If for any reason no satisfactory procedure for disposal of waste is possible, this must be accepted as tantamount to saying that the whole experiment is impossible.

Even when the amounts of activity are well below those which could represent a health hazard, disposal of long-lived waste products must be carefully thought out. It is terribly easy for disregarded waste solutions to get spilt or dried up and distributed as dust. There may be little indication of this having happened except for a gradual rise of counter backgrounds but accidental pick-up by important samples with small activities can lead to disastrously wrong results. In handling even the smallest quantity of long-lived activities it is good practice to cover the working bench with disposable sheets of paper. Newspapers are cheap to renew and are perfectly adequate if inelegant, and, if chosen to fit the political views of the head of the department, can even help to underline the maturity and sound judgment of the experimenter.

CHAPTER NINE

RADIOCHEMICAL METHODS

This chapter is not intended to remove the need for a good textbook of radiochemistry. That subject has now a literature and a "know-how" nearly as extensive as has nuclear physics itself. Apart from full-scale textbooks, a series of sizeable monographs on the radiochemistry of individual elements is being produced by the National Research Council of the U.S. Academy of Sciences in their Nuclear Science series. Authors are perhaps rather too ready to describe as indispensable works that they have themselves found useful, but it is safe to say that anyone embarking on the radiochemistry of an unfamiliar element would at least be unwise to dispense with the appropriate member of this series.

The object of this introduction is the humble one of reminding the student of some basic principles and of helping him to know what questions he should ask of a real radiochemist.

The fundamental assumption of radiochemistry is that all isotopes of a given element will show the same chemical behaviour. Except for a small difference in the *rates* of reactions, which may affect the balance-point of a dynamic equilibrium, this is reliably true.

Even as regards the rate of a reaction, a radioactive isotope is expected to differ from other isotopes of the element only by an amount proportional to the square-root of the mass. This is usually only 1 to 2% and may well be less than the differences between the stable isotopes of the same element. In reactions which go to completion quickly, differences of rate will have no influence at all on the result. In a reaction which does not go to completion but reaches a dynamic equilibrium there may be a very slight sorting by weight of the atoms of an element which is taking part but, except in the case of hydrogen, the resulting differences between the behaviour of different isotopes are too small to detect. Even tritium, with a mass three times that of ordinary hydrogen, shows only a barely observable difference of distribution between different phases or chemical groups.

Only when a whole series of reactions occurs, each involving a dynamic equilibrium between the initial and final components, is there a measurable change in the isotopic distribution. Such a case is by no means uncommon in biological processes and then the small change in isotopic distribution may actually be valuable in indicating something of the complexity of the processes concerned.

Having shown that radioisotopes of an element behave just as do the stable isotopes, it might be thought that the rest of radiochemistry could be taken unchanged from any good textbook of ordinary chemistry. This unfortunately is far from true. The reason is simple; in the main chemical processes we are interested in constituents which form a considerable part of the system. In favourable conditions we can observe the disintegration of a few thousands of radioactive atoms present in several grams of material. Hence we may be in a position to detect a nuclide which is present in a proportion of one part in 10^{20}. This may change completely its chemical behaviour.

9.1 The Precipitation Process

Let us consider, for example, the very basic chemical process of precipitation of salts from solution. It is probable that no salt is absolutely insoluble in water. Salts in dilute solution are almost fully ionised. There is, however, a dynamic equilibrium between the ions and the unionised molecules; for example in a solution of sodium chloride we have an equilibrium represented by

$$NaCl* \rightleftharpoons Na^+ + Cl^-$$

The proportion of each component depends on the concentration for the following reason. The rate at which the forward reaction goes depends only on the probability per unit time of dissociation of the NaCl molecules. Like the decay constant of a radioactive process, this probability, in a dilute solution, is affected little if at all by other molecules in the environment. The number of NaCl molecules which ionise per second is therefore directly proportional to the number of molecules present, as given by the concentration of NaCl. We shall write concentrations in gram-molecules or gram-ions per litre as [NaCl], [Na$^+$], etc.

The rate at which the reverse action goes, on the other hand, depends on the frequency with which the Na$^+$ and Cl$^-$ ions meet each other. This is proportional to the concentration of each of the ions and hence to the *product* of their concentrations [Na$^+$] × [Cl$^-$]. When equilibrium is reached, the rates of the two reactions must be equal, so that

$$[NaCl] \propto [Na^+] \times [Cl^-].$$

---

* The "molecule" of a salt such as sodium chloride is to be regarded as only a temporary association of a pair of oppositely charged ions, not involving any serious electronic rearrangement.

G*

In dilute solutions, when almost all of the material is ionised, this means that [NaCl] is proportional to the square of the total concentration in gram-molecules per litre. In a saturated solution, the molecules of NaCl are also in equilibrium with the crystalline solid sodium chloride. Again, the equilibrium is a dynamic one in which molecules are leaving the solution at the same rate as they are dissolving.

The actual solubility of the unionised NaCl molecules may be exceedingly small, since even in a saturated solution there are far more ions than complete molecules.*

Whether it is small or large, there is a critical value above which the rate of crystallisation will exceed the rate of solution so that a rapid fall in concentration will continue until the critical saturation value is reached. This critical value of [NaCl] must evidently correspond to a critical value of $[Na^+] \times [Cl^-]$. This value is known as the *solubility product, K_s.*

We can extend these considerations to salts containing more than two atoms or groups, such as Na_2S or LaF_3. To form a molecule of Na_2S we must collect together two ions of sodium as well as one of sulphur so that the sodium-ion concentration appears twice. The solubility product will then be the critical value of $[S^{--}] \times [Na^+]^2$ at which precipitation begins. Similarly for LaF_3 it will be the critical value of $[La^{+++}] \times [F^-]^3$.

Now let us go back to sodium chloride. Suppose that to a saturated solution are added further supplies of either kind of ion. For example, suppose we add HCl. Then the value of $[Cl^-]$ will be increased and the product of the ionic concentrations $[Na^+] \times [Cl^-]$ will be momentarily increased above the value of the solubility product. [NaCl] will then also be increased momentarily to a value above that at which it is in equilibrium with the solid and salt will be driven out of solution. This will continue until the value of $[Na^+] \times [Cl^-]$ has fallen again to the solubility product.

A similar situation arises if we mix solutions of two different salts. Suppose, for example, we have solutions of sodium chloride and silver nitrate. The solubility product values $[Na^+] \times [Cl^-]$ and $[Ag^+] \times [NO_3^-]$ are both quite large; measuring all concentrations in gram-molecules or gram-ions per litre, the first is about 40 at room temperature and the second about 100.

The solubility product of silver chloride, however, is very small, about 4×10^{-11}. Hence if we mix solutions of sodium chloride and

---

* It is not necessary to assume that NaCl molecules exist at all. The results of this and the following paragraphs can be obtained by considering directly the equilibrium between Na^+ and Cl^- and solid sodium chloride.

silver nitrate, each containing a few gram-molecules per litre, we shall momentarily have concentrations of silver and of chlorine ions the product of which is some 10^{12} times the solubility product. The result is the very rapid loss of solid silver chloride from the solution until $[Ag^+] \times [Cl^-]$ reaches the solubility product of 4×10^{-11}. Whichever ion is present in the smaller amount is practically eliminated from the solution. If enough chloride is added to give an excess of one gram-ion per litre over the concentration of silver ions, the latter will be reduced to a concentration of 4×10^{-11} gram-ions/litre or about 4 parts by weight in 10^{12}.

The solubility product for sodium nitrate is quite large, approaching 100, so that all of the sodium and nitrate ions will be left in solution together with the excess chlorine ions.

Since 4 parts in 10^{12} is unobservable for any stable element we can, for most practical purposes, say that *all* the silver in solution can be removed by precipitation with an excess of chloride or *all* the chloride can be removed by precipitation with an excess of silver.

While this is true enough for ordinary chemical purposes, in radio-chemical work we must remember that it is not exactly true. A concentration of only 4×10^{-11} gram-ions/litre still means $2 \cdot 4 \times 10^{12}$ ions per litre since there are 6×10^{23} atoms in one gram-atom. Far fewer ions than this can be detected if they are radioactive.

9.2 Precipitation of Trace Quantities

Let us now start from the other end, with a practical case in which radioactive silver is involved. Suppose that we want a small source of a few microcuries of Ag^{112} (3·2 hr, β^-). To make this, we can bombard silver-free cadmium with deuterons of moderate energy in a cyclotron. Cadmium has a number of isotopes and from each of these a silver isotope will be made by the (d, α) reaction. Some are very short-lived, and after half an hour or so we shall be left with the Ag^{112} that we want, a comparable amount of stable Ag^{109} and a little of the long-lived (270 days, β^-) isomer of Ag^{110}. If we have 10 μCi of Ag^{112}, corresponding to $3 \cdot 7 \times 10^5$ disintegrations per second, we must have $3 \cdot 7 \times 10^5/\lambda$, or $3 \cdot 7 \times 10^5 \times (3 \cdot 2 \times 3600)/\ln 2$ total atoms of Ag^{112}. This is 6×10^9 so that, with the other silver isotopes made at the same time, we have a total of perhaps 10^{10} atoms of silver, or $1 \cdot 6 \times 10^{-14}$ gram-atoms.

The active silver must be separated from the bulk of the cadmium, both to make a thin source for β-particle work, and to get rid of interfering cadmium and indium activities; for example, Cd^{117} (two.

isomers, 52 min and 3 hrs, β^-) which would be made in large quantities by the (d, p) reaction, or In^{111} (2·84 days, e-capture) made by the (d, n) reaction. The normal chemical procedure would be to dissolve the cadmium (with the silver contained in it) in nitric acid and to precipitate out the silver as silver chloride. Suppose that after we have dissolved the material in acid, and added chloride we have 100 ml of solution, with the high chloride concentration of 10 gram-ions/litre. Our silver concentration, $1·6 \times 10^{-14}$ gram-atoms in 100 ml, is then $1·6 \times 10^{-13}$ gram-ions/litre so that the product $[Ag^+] \times [Cl^-]$ is $1·6 \times 10^{-12}$. This is still a long way below the solubility product 4×10^{-11} so that no silver chloride at all will be precipitated.

Our large-scale chemical knowledge of how to separate silver from cadmium has therefore proved useless. Worse is yet to come. If we can't remove the silver from the cadmium, perhaps we can remove the cadmium from the silver. The usual procedure in elementary qualitative analysis is to precipitate it as sulphide with H_2S in acid solution.

There is no difficulty in this, as we have plenty of cadmium present and, since the solubility product is around 10^{-10}, it can be removed pretty completely, stable and radioactive isotopes alike. Unfortunately the silver may not be left in solution, although the solubility product for silver sulphide will not be approached. The few silver ions will be readily adsorbed on the mass of growing crystals of cadmium sulphide and locked in as the crystals grow larger still. By the time the CdS precipitate is fully formed, much of the silver may have been swept out of the solution. To all intents and purposes, therefore, our Ag^{112} has behaved chemically as though it were an isotope of cadmium and not of silver at all. Even with considerably larger quantities of silver, this behaviour is still likely to occur. Suppose we had made 100 millicuries, which would have given us a concentration of silver 10^4 times as great, leading to a value for the product $[Ag^+] \times [Cl^-]$ of some 400 times the solubility product. This would still represent only $1·8 \times 10^{-8}$ gm in 100 ml and, though all but 1/400 would eventually be precipitated, it would take a long time for the molecules to find each other and form solid crystals. Even when equilibrium was reached, the material is likely to be dispersed through the 100 ml in large numbers of such tiny crystals that some might pass through a filter and take an abnormally long time to separate in a centrifuge. They would, however, readily adhere to and be removed with the cadmium sulphide crystals if attempts were made to remove the cadmium. Hence, although a partial separation *might* be achieved, it would be likely to be inefficient.

9.3 Use of Carriers

This misbehaviour can be overcome very simply if we are prepared to dilute our Ag^{112} with a small amount of stable silver. If we add ten milligrams of ordinary silver nitrate to the original active cadmium-silver solution, the concentration of silver in 100 ml will be about 10^{-3} gram-ion/litre and the product $[Ag^+] \times [Cl^-]$ will exceed the solubility product by several million times. The precipitate will still be a little thin and will need care in collection but should carry with it practically all the silver including practically all the Ag^{112} since in the formation of the silver chloride crystals there will be no discrimination between the stable and unstable isotopes. A stable salt used in this way to take with it a radioactive isotope is known as a *carrier*.

Carriers are so generally used to ensure the proper behaviour of any radioactive nuclide that it is quite usual to remark the fact if one has *not* been used, when the nuclide is said to be *carrier-free*.

A carrier can also be used in the opposite way.

9.31 *Hold-back carriers*

Suppose that we wish to study Cd^{107} (6·7 hours, e-capture). This can conveniently be made by bombardment of a silver target with protons, by the reaction Ag^{107} (p, n) Cd^{107}. We have then to remove it from the bulk of the silver and from any Ag^{106} (8·3 days and 24 min, e-capture and β^+) formed by the reaction Ag^{107} (p, d) Ag^{106}. Now, if we dissolve the target in strong nitric acid and then attempt to remove the silver by precipitation as silver chloride, we shall get rid of the silver all right—this time there will be hundreds of milligrams of silver present—but we are liable to lose much of our Cd^{107} by adsorption on the silver chloride precipitate.

We can avoid this by adding a few milligrams of cadmium nitrate before the chloride is added. The absolute amount of cadmium adsorbed in the precipitate of silver chloride is doubtless increased, but it will now form only a minute fraction of the total cadmium in solution and a correspondingly small proportion of the Cd^{107} activity will be lost.

A carrier used in this negative way to prevent precipitation of a nuclide is known as a *hold-back carrier*. To obtain a nuclide radiochemically pure (i.e. without any other observable activities) it is usually essential to add hold-back carriers for every element of which active isotopes could be present. In the Ag^{112} and Cd^{107} separations described above, for example, hold-back carriers of indium and palladium respectively would also have been used during the precipitation of the desired nuclide. When a weak activity must be freed from a strong unwanted one, it may be necessary to repeat a separation several times, adding fresh hold-back carriers each time. For example, when the long-lived isomer of the aluminium isotope Al^{26} (740,000 y, β^+) is obtained by bombarding magnesium with protons —Mg^{26} (p, n) Al^{26}—a much larger activity of Na^{22} (2·6 y, β^+) will also be made—Mg^{25} (p, α) Na^{22}. Even when some hundreds of milligrams of sodium chloride are added as hold-back carrier, an appreciable amount of Na^{22} will follow the precipitate of the few milligrams of aluminium hydroxide carrier for the Al^{26}.

This precipitate is a gelatinous one best removed by centrifuging rather than by filtration. After washing, it is redissolved in dilute hydrochloric acid containing a further few hundred milligrams of sodium chloride and reprecipitated with sodium hydroxide. The same small fraction of the total sodium is again carried down with the precipitate, but with only a small fraction of the previous amount of Na^{22}. This procedure is repeated until the solution of sodium chloride decanted from the precipitate shows no further trace of activity.

With elements other than the alkali metals and alkaline earths, the carrier may be useless if it is not in the correct chemical form. This fact is made use of in the Szilard-Chalmers process described in section 8·31 above. The same fact, however, is often a nuisance. For example, if we have a small quantity of an active isotope of iron in solution in the ferric state, it will be no use to add a ferrous salt as carrier. In this case the treatment is simple, even if we do not know in which state our active atoms will be. We first add a ferrous salt as carrier and then oxidise it to the ferric state. If the active atoms were in the ferrous state they will be oxidised too and in whichever form they were at the beginning, both radioactive and carrier atoms will be in the same state at the end and the carrier will operate according to expectation. If our further plans for the activity require the active atoms in the ferrous state, we reverse the procedure, first adding a ferric salt as carrier and then reducing all ferric ions to the ferrous state.

Elements such as manganese or phosphorus need more complex procedures to get all of the active and carrier atoms into the same chemical state. It may often in fact be preferable simply to add carrier samples of *each* of the possible chemical forms and then to extract a particular one of them again chemically. Most of the activity will usually be found in a single chemical form. It may often be sensible to discard the other forms rather than to take the trouble to treat them separately, unless the distribution of the active atoms between the different possible states is an object of research in its own right.

9.32 Non-polar compounds

When we come to non-polar compounds it is useless to talk about carrier elements and we must instead consider carrier compounds. The radioactive carbon atoms in active carbon dioxide, $C^{14}O_2$ for example, will not exchange at all, in solution or otherwise, with the carbon atoms in another compound such as alcohol. Indeed, we can make a compound in each molecule of which the radioactive carbon is in a particular position; for example, ethyl alcohol could be made with C^{14} either as $C^{14}H_3 . CH_2OH$ or as $CH_3 . C^{14}H_2OH$, and the carbon atoms inside the same molecule will not exchange positions but can be relied on to stay in the group into which they were first built. In organic chemistry it is the structure of a compound rather than the list of elements comprising it that is of interest so that this is a major advantage rather than a limitation. When we do want to extract the total carbon activity from a mixture of compounds, regardless of its distribution between them, it is quite easy to do so by oxidising the whole mixture at high temperature in a copious stream of oxygen. All the carbon will then appear as carbon dioxide which can readily be removed from the gas stream, as $BaCO_3$, by bubbling the gas through a solution of barium hydroxide.

9.33 Non-isotopic carriers

While a carrier of a radioactive nuclide is normally the stable form of the same element, in the same chemical form, this need not necessarily be so. When the original activity is to be extracted from a very large bulk of material, a considerable weight of carrier may be needed to ensure effective precipitation and this may interfere with the making of a satisfactorily thin source.

For example, strontium 90, derived from the fall-out from bomb-tests, is now found in milk in all parts of the world and it is very desirable to measure its amount. It occurs—fortunately—in exceedingly small quantities, of the order of a millimicrocurie or less per litre. A kind of two-stage carrier procedure is then used. The first stage is to use a calcium carrier to separate the strontium from the bulk of the milk. Calcium, like strontium, is an alkaline-earth metal. When the milk is dried and then burnt to ash to remove all organic material, the radioactive strontium remains with the calcium salts rather than being adsorbed on smoke-particles and lost. The total ash is of the order of 1 %, of which about a tenth is calcium; this can readily be separated as phosphate or carbonate, carrying the strontium with it, from the more soluble alkali-metal salts. In this particular case, extra calcium need not be added as the natural calcium can be used as the preliminary carrier of the radioactive strontium.

The next stage, of separating the strontium from the bulk of the calcium, employs a normal added carrier of stable strontium. This is separated from the calcium dissolved in only a few millilitres of solution, by precipitation as the nitrate with fuming nitric acid. Then we have essentially all of the Sr^{90} from some tens of litres of milk in a few milligrams of material, a concentration of nearly 10^7 times, which could not easily be achieved in a single process.

Although for clarity in explanation we have supposed that each carrier is added only when it is to be used, it may be simpler and better in practice to add both calcium (if enough is not naturally present) and strontium carriers in the appropriate quantities to the original milk. The successive use of the two carriers can then be regarded as the use of two successive processes with an increasing chemical resolving power, so as to have at each stage a convenient quantity of material to handle.

This represents the use of a particular non-isotopic carrier chosen for its chemical similarity to the nuclide of interest. Even this is not always necessary. Several precipitates, particularly ferric hydroxide, will carry with them a large variety of active nuclides with very remote degrees of chemical similarity. This can be particularly useful either as a first stage when very high specific activity is needed or when no stable isotopes exist of the nuclide concerned. Such materials may also be used to remove unwanted radio-active isotopes of several different elements from the solution at the same time. They may then be referred to as *scavenger precipitates* or simply as *scavengers*.

We shall not consider in detail any chemical processes other than precipitation, but carriers or hold-back carriers are often needed also in procedures such as oxidation, reduction or electrolysis.

If no carrier is added, not only is the behaviour of an active nuclide liable to be quite different from that of the bulk element but it may not even be the same in successive, apparently identical, experiments. The general reason for this is not difficult to see. The best of analytical reagents are often known to contain a few parts per million of several impurities, although particularly objectionable contaminants may have been reduced to much lower levels. At the level of 1 part in 10^{10} or so the unknown impurities will include nearly all the lighter elements up to calcium, in a great variety of compounds, together with a good selection of the heavier elements.

Hence our newly made radioactive nuclides, which may form only a few parts in 10^{12}, might be able to find an adequate weight of almost any known compound with which to react.

9.34 *Radio colloids*

The effect of impurities on active nuclides is not confined to their chemical interaction. The active element may well behave as a colloid rather than as a true solution, even though its concentration is far below that required to reach the solubility product of any of its known compounds. It is presumed that the active atoms are adsorbed on the surfaces of minute particles of silicaceous dust or similar materials, the presence of a few of which can hardly be avoided. The resulting highly active solid particles are often known as *radio colloids*.

The effects of impurities are of minor importance for the alkali metals and alkaline earths.* Both of these groups will remain as simple positive ions in practically all watery environments, although, as indicated above, serious amounts of them may be adsorbed and removed from solution by the bulkier precipitates such as $Al(OH)_3$ or $Fe(OH)_3$. For elements with any capacity for complex formation, however, there is a serious chance that the radioactive atoms will be present as some unsuspected compound or compounds, which may respond to chemical treatment in unexpected ways. The results are liable to vary from one experiment to another simply because the kinds and quantities of the invisible impurities are liable to vary from one experiment to another. Fortunately, in many applications the situation is better than this discussion may suggest. Most work is done with nuclides which have half-lives longer than a few hours and often with activities which are to be measured in millicuries rather than microcuries. Ten millicuries of carrier-free Fe^{59} (45 days, β^-), for example, represents nearly $\frac{1}{6}$ of a microgram which, while too little for observation by ordinary macrochemical methods, is too much to be affected by unsuspected impurities in reagents of analytical quality.

Enough has been said to show that many of the processes of ordinary chemistry, quick and convenient though they are, may be unsatisfactory when carrier-free samples of short-lived activities are required. In biological and chemical work, the addition of even small amounts of carrier is to be avoided. Several effective though sometimes laborious techniques are available for separating them from other elements or compounds.

* Although care must be taken with the latter if there is any reason to suppose that chelating agents are, or may have been, present.

9.4 Special Methods for Carrier-free Separations

9.41 *Ion-exchange systems*

The most effective general method employs the differential adsorption of ions on a specially made ion-exchange resin. The process is similar to that used in a water-softener.

Resins have been made for both cation and anion adsorption. The cation-exchange resins, such as Dowex-50 or Amberlite IR-1, mostly consist of water-insoluble polymers containing sulphonic acid groups. According to the acidity and other characteristics of the solution in contact with these, particular cations may be bound firmly, loosely or not at all. Anion-exchange resins, such as Dowex-1, mostly contain quaternary amine groups which can similarly hold anions. For any particular kind of ion in solution a dynamic equilibrium will exist between the bound and free states, the proportion in each state depending on the composition of the solution. The proportion will depend on the kind of ion and this fact can be used for the separation of one from another. The great value of the process for use with carrier-free activities results from the fact that each individual ion has a definite average *time of dwell* on the surface of the resin and a definite expectation of free-floating time in a given thin layer of solution. This time is nearly independent of the presence of other ions, if enough resin is used for plenty of surface to be available for all. Consequently, the behaviour of a particular radioactive ion is practically unaffected by the presence or absence of carrier ions of the same element. The reliability of this statement has been most strikingly shown in work with the transuranic elements, where a few atoms of a particular element have been successfully separated from 10^{18} or so atoms of closely similar elements in several ml of solution.

The usual method of operation is as follows:

A vertical glass or plastic tube, some 10 to 50 cm long and 2 to 10 mm diameter, is filled with finely divided particles of resin, and the solution is run through it from the top. If the flow-rate (controlled by air pressure if necessary), quantity and composition of the solution are correct, the ions to be adsorbed will be held in a narrow band at the top of the column. In the simplest case, some kinds of ions will be so held and others will not be held at all, but will run straight through. For example, divalent nickel and cobalt ions may be separated from a mixed solution in 12-molar HCl run through an *anion*-exchange Dowex-1 column. In these conditions the cobalt forms negatively charged complexes with the chloride present and is adsorbed, while the nickel remains as a cation and passes through unchecked.

More frequently, all of the ions present will at first be adsorbed together in static equilibrium in a short length at the top of the column. They are then separated by passing through the column a second *eluting solution*, or *eluent*, in which a dynamic equilibrium is set up between adsorption and desorption for each ion. Such exchange between solution and column will take place many times per second. While the ion is in solution, it will be moving down the column, at the same rate as the eluting solution itself, and its mean rate of progress therefore depends on the ratio of its mean free-floating time to its mean time-of-dwell on the resin. If the eluent is properly chosen (from one of the large number of chemical cookery-books now available; see bibliography), this ratio will differ for the different ions present and they will therefore move down the column at different rates. The preliminary firm adsorption of all ions at the top of the column will have prevented any from getting a start on the rest during the passage of the original solution, so that as the various ions move down the column each type remains in an adsorption band which becomes further and further separated from the other bands until they can be separately collected in successive fractions of the effluent from the bottom of the column. The width of a band will increase if large quantities of one ion are present, owing simply to the filling of all available adsorption sites over a significant length of column, so that the quantity of material that a given column can handle in one run is limited to something of the order of 10 mg of ions per gram of resin. The minimum width of band is fixed by a process analogous to the straggling of α-particles and depends on the number of times an ion is adsorbed or resorbed in the course of its passage down the column. In good conditions the band may be remarkably narrow, the whole of a particular carrier-free nuclide appearing in one or two successive drops of the effluent.

The method is capable of giving clean separations in an hour or less even of ions so similar as those of successive rare earths or actinide elements, and as already indicated, works as well with carrier-free solutions as it does in the presence of carriers—so long, of course, as the chemical state of the active nuclide is adequately controlled.

9.42 *Solvent extraction*

It is standard procedure to purify some elements by extraction from a watery solution with an organic solvent. For example the extraction of uranyl nitrate with ether from acid solution will separate uranium effectively from nearly all other metals; this is of particular importance in removing it from a mixture of fission products. Where the compounds or complexes are in the same state of association in the two phases, the *partition coefficient* (the ratio of concentration in the two phases at equilibrium) will often be almost independent of the absolute concentration down to 10^{-15} gram-mol/litre so that carrier-free nuclides can be extracted. On the other hand, if complexes are associated in pairs or higher polymers in the organic phase, the concentration in the latter will be proportional to the square or higher power of the concentration in the water phase. Carrier-free extraction from water to organic phase will then be impossible, though extraction in the reverse direction will still be satisfactory.

The process is not so generally applicable as the ion-exchange method, and two or three repetitions may be needed to give clean results. It is quicker however, and by using appropriate compounds —see the cookery-book again—quite a large number of elements may be separated from particular contaminants.

9.43 *Paper chromatography*

The principle of an ion-exchange column can be combined with that of solvent extraction to give a process known as *partition chromatography*. In this, two liquid phases are used, one of which is mobile and the other held by sorption on a suitable fixed support. A small quantity of material soluble in both liquids is carried by a thin layer of the mobile phase over the fixed phase. The solute is then exchanged frequently between the two phases and while in the fixed phase has a definite average time-of-dwell just as do the ions adsorbed on a resin. The mean rate of movement depends on the partition coefficient between the two phases; clearly if this favours the mobile phase the solute will move fast, and vice versa. The discriminating power of the method was first shown by Martin and Synge, using silica gel to hold water as the fixed liquid phase.

Filter-paper, which has a strong affinity for water, is now almost always used. The paper takes no direct part itself, forming merely a convenient base with a large effective surface to which the water clings and through which capillary attraction will draw a thin layer of the mobile phase at a suitable slow speed. The full name of the process is then *paper-partition-chromatography* but the "partition" is usually omitted.

In practice it is not necessary to add water and the other solvent separately. A water-saturated solvent is allowed to travel over the filter-paper, which takes up automatically the appropriate (very minute) quantity of water. The substances to be separated are usually placed near one end of a strip of filter-paper and the solution applied to the same end. Capillary attraction then takes solution and solutes along the paper. As this process is mainly of importance for analysis, the details of technique will be left until the next chapter.

9.44 *Electrochemical and electrophoretic methods*

Carrier-free elements often behave in the same way as do the bulk materials in either of these processes. Isotopes of the easily plated metals, such as copper and silver, are particularly easily extracted by electrolysis from solutions which contain large quantities of their neighbour elements zinc and cadmium, at the other end of the electrochemical series. The method is particularly valuable for the production in one step of thin uniform sources of β-particle emitters.

Some elements may be removed from solution by direct ion exchange, without the use of any supply of electric current, the mechanism being the same as that which produces a thin plating of copper on a steel knife-blade when it is dipped into a solution of copper sulphate. For example, pure Po^{210} may be deposited on a piece of silver or copper which is agitated for a few minutes in a solution containing the polonium together with its precursors Pb^{210} and Bi^{210}, both of which fall below silver in the electrochemical series and which therefore remain in the solution. The silver can be thoroughly washed without risk of removing the Po^{210}.

Electrophoresis uses an electric field to separate different ions without causing their actual deposition on the electrodes. The separation occurs in any mixed solution but it is difficult to prevent remixing in a free liquid. The electric field is therefore usually applied between the ends of a piece of filter-paper wetted with a dilute electrolyte. Fields in the range from 3 to 50 volts/cm are usually employed. The solution containing the mixture of ions to be separated is then placed near one end of the strip. If cations (positive) are to be separated, the solution would be placed near the anode. Then positive ions of different mobility will move at different speeds along the strip, being easily followed by observation of their radioactivity. Anions, on the other hand, would be placed near the cathode. If it is desired simply to separate anions from cations the solution may be placed near the middle, when the two kinds of ions will move in opposite directions.

It must not be imagined that by such simple means we could achieve a separation of electric charges. In the absence of some mechanism for neutralising excess charges, the smallest radiochemically detectable separation of positive from negative ions would require millions of volts.

A self-adjusting supply of H^+ and OH^- ions is everywhere available to keep down the net charge per unit area of the strip whatever is the local concentration of added ions.

The resolving power of the method for simple ions* is not very high. It can be much increased by introducing a complexing agent into the solution. Ion complexes can then be formed which will spend much of their time absorbed on the filter-paper. We have then a situation closely analogous to that which is used in paper chromato-

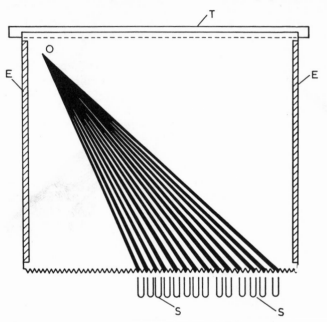

Fig. 46 Continuous separation of rare-earth elements by electrophoresis.

T—trough containing 0·05% lactic acid; *E-E*—electrodes; *S-S*—collecting tubes; *O*—point to which the active mixed solution is fed.

graphy, with the additional factor that the speeds of different ions in the mobile phase are different and are controlled by the electric field rather than by the rate of flow of the mobile phase itself.

The progress of the radioactive ions may be stopped and fixed at any point simply by drying the paper and the region containing useful activity can be cut out and extracted.

A very effective method of combining electrophoresis with paper

---

* Even the simplest ions, such as Na^+ or Cl^-, will normally be hydrated to an extent which much affects their mobility.

chromatography is illustrated by the method used by Dr. Pučar in Zagreb to effect a continuous separation of the rare earth metals. It has the great advantage over chromatography that any "tail" left by one nuclide does not get mixed up with the following one.

In this, a large sheet ($\sim 50 \times 60$ cm) of chromatographic paper is suspended vertically in a glass tank with its top edge in a trough of inactive 0·05% lactic acid solution. This solution runs steadily down the paper in the usual way, finally dripping off the bottom. A solution containing the active rare earths to be separated is fed continuously at a very slow rate (a few ml per hour) on to the paper near one top corner.

A vertical conducting metal strip is connected to each side of the sheet and a potential difference of some 1500 volts is maintained between them, thus giving a uniform field of some 30 volts/cm across the sheet.

Under the combined influence of the flow of solution and the electric field, the metallic ions travel along oblique paths, the slope being different for each element. The arrangement is shown schematically in Fig. 46. If the voltage and the rate of flow are kept constant, each element will flow steadily towards, and finally drop from, the same region of the sheet. The bottom of this is cut as shown into a series of "teeth" defining the points from which the drops of liquid fall. A series of test-tubes can then be arranged each to catch one, and one only, of the rare earths, the proper points to put the tubes being found by following the progress of activity with a small counter. By this means a complete set of fission-product rare-earth elements can be separated in a few hours.

The process as described here is unsuited to the separation of very large quantities of material but can be very useful in analytical or other experiments with small to medium activities.

9.45 *Other physical methods*

Such methods as sublimation and distillation may often be used for carrier-free separations, though recrystallisation is—for the same reasons as chemical precipitation—rarely effective. Thus mercury may be extracted from gold by melting the gold in a vacuum under a collection plate cooled by liquid oxygen. The mercury can be evaporated from thin foils without melting the gold or causing an appreciable amount of gold to evaporate, as it diffuses readily through the hot metal, but this process is slow for thick pieces. The extraction can then be done in two stages. In the first, the gold is melted and the mercury removed from the bulk of it and deposited on a cooled tungsten or molybdenum foil. This in its turn is heated, to perhaps 700°C, and the mercury alone transferred to a second cooled plate.

It is one of the harmless ironies of history that, after the alchemists had spent hundreds of years trying to transmute base metals into gold, the first useful transmutation of a stable element was that of gold into mercury.

The mercury discharge-tube is an efficient source of light and much of its energy is concentrated into a few spectral lines. These are intrinsically very sharp, but when ordinary mercury is used, each line is a close multiplet due to the fact that different isotopes give slightly different wavelengths. For experimental work on interference, the sharpest possible line is desirable and the light from a single isotope of mercury would be extremely suitable. For the small quantities required, it turns out that it is cheaper to make it from gold in a pile than it is to separate the isotopes of existing mercury. The pile actually produces Au^{198} by absorption of slow neutrons—Au^{197} (n, γ) Au^{198}—but the Au^{198} is β-active with a half-life of 2·7 days, giving Hg^{198} which is stable. One gram of gold foil in a large pile giving a neutron-flux of $10^{12}/cm^2/sec$ would produce Hg^{198} at a rate of about a milligram a week; a milligram is plenty for a small discharge tube.

Distillations and sublimations are efficient for carrier-free materials though in many such processes (for example, steam distillation) a good deal of other non-isotopic material may come over with the nuclide desired. Having once obtained radiochemical purity of a carrier-free nuclide, however, it is not usually very difficult to separate this from the bulk of extraneous materials.

9.46 *Isotope separators*

We conclude this section with a reference to the heroic method of carrier-free extraction, by use of an isotope separator. This is an expensive and inefficient method, but is the only effective one when one of several active isotopes of the same element is to be extracted. Where volatile compounds can be used, enrichment of a particular isotope can be obtained by various diffusion processes, making use of the fact that lighter molecules at a given temperature travel slightly faster than heavier ones. For example, uranium is enriched in this way in U^{235} from 0·7 to 15% or more, to make possible the construction of nuclear reactors with a small critical mass.

Diffusion processes are slow and usually employ considerable quantities of material, so that they are not very suitable for the separation of radioactive isotopes with a high specific activity. For this a magnetic, electrostatic or electromagnetic mass-separator can be used, though they too are far from perfect. In these, positive ions are produced in a near-vacuum from a low-pressure arc or from a heated solid. The ions are then accelerated by a negative electric field to a potential of some tens of kilovolts and at the same time focussed into a well-defined beam. All have the same energy, derived from the accelerating potential, and hence particles of different mass have different velocities and momenta. The beam is then deflected by either an electrostatic, magnetic or high-frequency electromagnetic field. Ions of different mass and velocity will suffer different deflections which allow them to be separately collected on a cooled metal plate or in cooled cavities behind suitably placed slits.

A small magnetic mass-spectrometer is not very difficult to build with sufficient resolving power to give 1% mass-resolution, i.e. to separate successive isotopes up to a mass of about 100. Large instruments, such as those designed at the Nobel Institute in Stockholm, give a good separation between successive isotopes of the heaviest elements.

It is not easy to make an ion source with a high efficiency for a small apparatus. Most small instruments do well to deposit on the collector one part in 10^5 of the material supplied to the source. They are therefore of little use for providing useful quantities of carrier-free radioactive isotopes. It is neither safe nor economic to put 100 curies of activity into the ion source to obtain one millicurie of a useful nuclide, free of contamination. One of the big Stockholm mass-separators can give an efficiency of up to 15% for most solid materials and even more for the inert gases, and will be worth while to use in special cases if it is conveniently available. In cost and complexity, however, it is quite comparable to a small cyclotron and it is not worth installing for routine radiochemical separations. It has proved valuable in the determination of the mass numbers of new radioactive nuclides, but the number of people wishing to identify new active nuclides is small.

It is worth while to make great efforts, and to accept low yields, to use a method of production of a desired nuclide which produces it free of unwanted activities in the first place. In difficult cases it is better to use a separated stable isotope as target material rather than to try to separate several active isotopes after production.

9.5 Maximum possible Specific Activity

To end this chapter, we will consider quantitatively, for a carrier-free nuclide, what will be the specific activity, or number of disintegrations per second per gram, of the element concerned. If a nuclide is carrier free, then the number of disintegrations per second is simply the number of atoms N multiplied by the decay constant λ. The specific activity is therefore $\lambda N/NM$, where M is the mass of one atom in grams, and hence NM is the total weight. The maximum possible specific activity is thus simply λ/M disintegrations per second per gram. If A is the atomic weight, this may more conveniently be written as

$$1 \cdot 13 \times 10^{13}/A\tau_{\frac{1}{2}} \text{ curies per gram}$$

where $\tau_{\frac{1}{2}}$ is the half-life in seconds. The specific activity of a short-lived element, carrier-free, can thus be extremely high. For example, pure O^{15} (2 min, β^+) would have a specific activity of 6×10^9 curies per gram and P^{32} (14·3 days, β^-) about 9 million curies per gram. On the other hand the long-lived nuclide thorium 232, even carrier-free, has a maximum specific activity of only a tenth of a microcurie per gram.

In all cases, the specific activity can be reduced to any extent below the maximum by the addition, intentionally or otherwise, of the stable element.

CHAPTER TEN

USE OF RADIOACTIVE TRACERS IN CHEMICAL ANALYSIS

Radioactive methods may be of assistance as a tool in any field of chemistry, inorganic, organic or physical.

Their use makes it possible, and perhaps necessary, to re-examine the whole of the first two of these fields at such low concentrations of the active elements that two atoms of such an element can rarely or never react with each other. An entirely new subject, "hot-atom" chemistry, has grown up for the study of interactions of atoms which enter the system with energies well above the normal thermal range.

Since it is impossible to cover the whole subject, we will take as an example only the limited field of analysis.

The ease with which radioactive molecules can be detected makes them of very great value in some kinds of analysis. Great care must be used if accuracies better than a few per cent are required, but the limits of sensitivity to small traces of certain elements are far lower than those obtainable by any other means. Broadly speaking, there are two modes of approach. In one, a radioactive isotope of the interesting element is introduced in known quantity into the raw materials and followed through to the final product whose composition is to be determined. In the other, radioactive nuclides are added to or produced from the substance to be analysed.

The first group is of less general applicability, but may be useful in determining particular impurities.

10.1 Radioactive Control of Purification

For instance, suppose that a manufacturer of an analytical grade of chemicals wants to know just how good is his method of making a metallic chloride as free as possible from fluoride. All of the basic raw-material sources of chloride contain a little fluorine and removal of the last traces is difficult. A quick and satisfactory procedure would be to add radioactive F^{18} (110 min, β^+), to the crude salt before purification and then to look for it again in the final product. [Suppose that at first there are w grams of fluorine per gram of chlorine and that a measured quantity of F^{18}, with a negligible weight of carrier, has been added per gram of chlorine. When the purification procedure is completed the F^{18} activity per gram of chlorine is re-measured. The natural radioactive decay of the F^{18} during the purification must be taken into account. If then it is found that only 1/1000 part remains of the activity to be expected if the chloride had not been purified, the final fluorine impurity has been reduced to $w/1000$ grams per gram of chlorine. If the process consists of several operations taking 12 hours or more, it may be necessary to do this investigation in several stages, measuring the surviving activity and adding known amounts of fresh F^{18} each time.

The main difficulty in such investigations—and in the purification itself—is to avoid adding more ordinary fluoride with the reagents used in the operation.

A radioactive isotope used in this way to follow the chemical behaviour of the stable isotopes of the same element is known as a *radioactive tracer*. An element or compound, to which a radioactive version of the same element or compound has been added as a tracer, is said to have been *labelled* with the tracer. Most of the rest of this book will be concerned with various uses of tracers.

In most analytical problems we are interested in the composition of an existing mixture of non-radioactive materials and our second mode of approach is necessary. Several techniques are available.

10.2 Tracers as Monitors

In the simplest case a radioactive tracer is used qualitatively to confirm that a normal chemical analysis is going according to plan. For example, suppose we went to determine the amount of silver in a piece of medieval lead. We dissolve a sample of the lead in strong nitric acid as usual and, before carrying out any other operation, add to the solution 10 to 100 μCi of any radioactive isotope of silver which emits γ-rays; for example Ag^{105} (40 d, e-capture, γ) or Ag^{106} (8·3 d, e-capture, γ). This should be carrier-free if possible; if not, the weight of silver added must be known. Then at each stage of the normal separation we can use an external γ-counter to check that no significant fraction of the silver has remained in the wrong solution or precipitate. This technique is particularly valuable where an unfamiliar separation has to be done, when things often do not go according to the book.

10.3 Activation Analysis

The most direct method of quantitative analysis, using a radioactive tracer, is to change a known fraction of the interesting element into a radioactive form, when it becomes easy to observe in extremely small quantities. The change is accomplished by exposing a sample of the material to be analysed to a large flux of bombarding particles which are capable of producing nuclear reactions in the element to be detected. In the great majority of cases neutrons are used, as a flux of some 10^{12} per cm^2 per sec is conveniently available in many large nuclear reactors.

More convenient, and adequate for many purposes, is the small laboratory accelerator described in Chapter 8, which gives 10^7 to 10^8 slow neutrons/cm^2/sec. This makes it possible to use radio-isotopes with much shorter half-lives than can be employed if samples must be sent to and returned from a large reactor.

The principle of the method is that neutrons are absorbed by the element of interest to give a radioactive isotope whose decay can be detected.

10.31 *Gamma-spectrum analysis*

The quickest method of doing this is to observe the characteristic γ-rays which are produced by most radioactive nuclides. If a multi-channel pulse-height analyser (kick-sorter) is available, two or three different elements can be investigated simultaneously. Since neither carbon nor hydrogen emits delayed γ-rays after absorption of slow neutrons, samples may be both irradiated and counted in the same sealed polyethylene vessels. Oxygen gives a small yield of O^{19} from the 0·2% isotope O^{18} but the half-life of O^{19} is only 29·4 sec. This will be of no importance a few minutes after irradiation, and so the samples need not be dried and may even be in solution in water without disadvantage.

A good illustration of the method is the estimation of arsenic in sulphuric acid. Sulphuric acid is used in many processes in the pro-duction of food and forms an important part of the raw material of most detergents, so that it is important that the proportion of arsenic be kept low. When a sample of sulphuric acid is placed in a flux of slow neutrons, some of them will be absorbed by arsenic atoms to give As^{76} (26·6 hr, β^-, γ). This gives a strong γ-ray at 0·56 MeV. Sulphur, unlike oxygen and hydrogen, gives a conveniently observ-able radioactive isotope, S^{35} (87·1 d, β^-) but this gives no γ-rays. Hence any γ-rays emitted by irradiated sulphuric acid will be derived from the impurities. A typical spectrum displayed on a 100-channel kick-sorter is shown in Fig. 47. Besides the main 0·56 MeV γ-peak of As^{76} which masks the weaker 0·64 MeV peak, there are peaks due to Mn^{56} (2·58 hr, β^-, γ) and weaker ones due to Na^{24} (15·0 hr, β^-, γ) produced by absorption of neutrons by manganese and sodium respectively.

The relative heights of the peaks do not give directly the relative amounts of arsenic, manganese and sodium present. Firstly, the number of active atoms formed will depend on the thermal-neutron cross-sections. These are as follows: As^{75}, 4·2 barns; Mn^{55}, 13·2 barns and Na^{23}, 0·53 barns.

Secondly, the three isotopes decay at different rates and if equal numbers of atoms are formed, the initial rates will be proportional to the decay constants. Thirdly, the proportion of disintegrations which give γ-quanta varies from one nuclide to another and finally the efficiency of the scintillation counter depends on energy. Fortunately none of these factors need be measured separately if a known quantity of each element is irradiated under the same conditions and

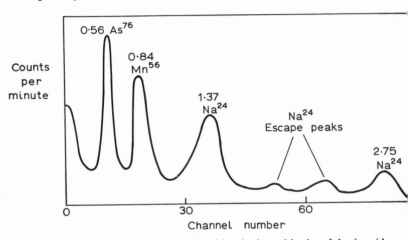

Fig. 47 γ-ray spectrum produced by the impurities in sulphuric acid after irradiation with slow neutrons.

counted with the same equipment and the same interval after irradiation. Then $W_1/W_2 = N_1/N_2$ where W_1 is the weight of impurity to be measured and N_1 the number of counts in the γ-peak from the test sample and W_2, N_2 are the corresponding figures for the control.

As in all radiochemical work, short irradiations followed immediately by short counts will emphasise the peaks which result from short-lived activities while long irradiations followed by a delay before counting may make possible their complete elimination. Periodical observation of the spectrum makes it possible to measure the half-life of the nuclide responsible for each peak—an important method of confirming their identity.

As was shown in an earlier chapter, a flux of γ-quanta which have all the same energy E will not give a spectrum consisting of a single peak, but may give also a quite complex spectrum of lower-energy peaks, due to the escape from the phosphor of some of the secondary radiations. High-energy γ-rays from one nuclide may thus conceal lower-energy quanta from a second nuclide. When the nuclide giving

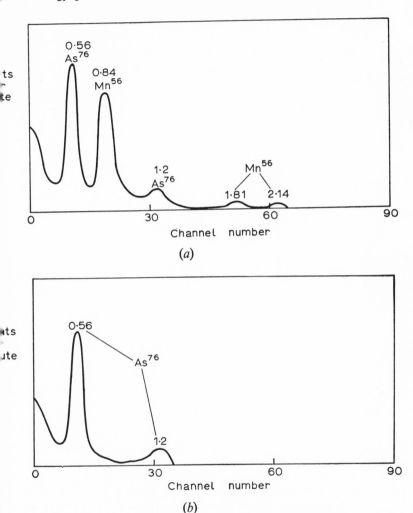

Fig. 48 (a) Spectrum of Fig. 56 after subtraction of the spectrum of Na^{24}. The 1·81 and 2·14 MeV peaks of Mn^{56} are now observable. (b) Spectrum of Fig. 57a after subtraction of the spectrum of Mn^{56}. A simple spectrum of As^{76} remains, the 1·2 MeV peak now being clear.

the higher energy has the longer half-life, so that its peak cannot be eliminated simply by waiting, the process known as *spectrum stripping* may be useful. For this, a reference spectrum is produced with the same equipment from a radiochemically pure sample of the higher-energy nuclide. This is scaled down to give the same number of counts in the highest good peak as occurred in the corresponding peak in the mixed spectrum. The whole reference spectrum is then subtracted from the mixed spectrum. In many commercial kick-sorters the subtraction can be done automatically. The general background must be carefully accounted for. If the total numbers of counts are large enough in each channel for statistical errors to be small, the subtraction will remove not only the peak taken as reference but the whole contribution of counts on all channels due to the nuclide concerned, including both escape-peaks and the main peaks due to other γ-rays emitted by the nuclide.

In Fig. 48a is shown the effect of "stripping" the spectrum due to Na^{24} from that of Fig. 47. The smaller Mn^{56} peaks are now clear, confirming that the 0·84 MeV peak is due to Mn^{56}.

In Fig. 48b is shown the further effect of stripping the Mn^{56} spectrum from Fig. 48a. The subsidiary peak of As^{76} at 1·2 MeV is now discernible.

The sensitivity of this method of analysis, even with a small neutron source, is high. In Table 10·1 are shown the detection limits for several elements, assuming an exposure to 10^8 neutrons/ cm^2/sec for 30 minutes or three half-lives, whichever is the shorter. The quantities quoted can be detected in samples of a few grams except in the case of cobalt which depends on a very soft γ-ray which will show considerable self-absorption in a few millimetres of material.

If irradiation can be carried out in a pile, much higher sensitivities can in principle be obtained, as a neutron flux of 10^{12}/cm^2/sec or more may be available and exposures of the order of a month are practicable. Such elements as aluminium or vanadium will usually be ruled out because of the short half-lives of the radioisotopes produced from them. The sensitivity for the longer lived elements might be expected to be greater by 10^5 or even 10^6 times when a pile is used but such an improvement can rarely be achieved when we use the quick and convenient method so far described. At the level of 10^{-9} gm or less, large numbers of impurities occur and impossibly complex spectra may be produced. The sample itself may not be responsible for this. The mere handling of polyethylene bottles with metal forceps may wipe off enough metal to be registered. Zinc (from brass) is particularly troublesome.

10.32 *Activation analysis using chemical separation*

Because of these factors, use of a pile gives an improvement in sensitivity of only a hundred times or so even for a million times increase in neutron-flux. To do better than this, a more complex procedure than γ-analysis is necessary. This begins with a chemical separation of the desired active element from other radioactive substances. A large reduction of volume can usually be achieved in the course of

TABLE 10.1

Sensitivity for Activation Analysis for a selection of Elements using γ-ray Spectrometry, assuming 10^8 neutrons per cm^2 per sec and exposure for up to 30 minutes

Element	Observable quantity (μ gm)	Half-life of isotope observed
Aluminium	30	2·3 minutes
Antimony	30	2·8 days
Arsenic	700	26·6 hours
Bromine	100	18·5 minutes
Chlorine	400	38 minutes
Cobalt	15	10·5 minutes
Copper	400	5·2 minutes
Indium	1	54 minutes
Iodine	40	24·99 minutes
Manganese	10	2·6 hours
Rhenium	4	17 hours and 91 hours
Sodium	300	15·0 hours
Tungsten	300	24 hours
Vanadium	4	3·8 minutes

extracting the desired active nuclide, so that the β-activity, rather than the γ-activity, of the resulting small source can be measured with an ordinary Geiger counter. In this case, the chemical procedure rather than the characteristics of the radiation is used to identify the nuclide concerned. We shall explain the method with the help of a specific example; again the estimation of arsenic, but this time in a biological material. (We are interested here in the normal amount present, which is often a tenth of a part per million or less. The analysis of stomach contents, so dear to readers of whodunits, is of no forensic interest until the concentration reaches

H

a few hundred parts per million, which is easily measured either by the γ-analysis method already described or by conventional chemical methods.)

The procedure is as follows. A few grams of the material to be analysed is sealed into a small polyethylene or silica ampoule and irradiated in a pile, along with a second sample of the same size to which arsenic has been added to a level of say 1 μg/gm.

In both unknown and control samples, the same proportion of arsenic will be converted into As^{76} (26·6 hours, β^-, γ) by the (n, γ) reaction. This radioactive arsenic will usually be accompanied by several high activities due to such nuclides as Na^{24} (15·0 hours, β^-, γ). Some of these (but not the Na^{24}) may be eliminated simply by waiting 24 hours, which will cause the loss of less than half the As^{76}. After this, the active arsenic must be separated from the other activities with the help of a stable arsenic carrier. The quantity of carrier added must be known. Fifty milligrams per gram of sample would be suitable. Since the active arsenic may be in the trivalent or pentavalent state, each mixture must be put through an oxidising or reducing process to ensure the efficiency of the carrier. The arsenic is extracted chemically from each sample and then redissolved and re-extracted two or three times to obtain a reasonably pure product which can be dried and weighed to determine the chemical yield. The activity of the arsenic extracted from each sample is then measured in a standard β-counting assembly, both activities being followed for a few half-lives to detect any traces of contaminating activities. Then if the chemical yields of the unknown and control samples were the same, the ratio of the counts from the two samples is equal to the ratio of the quantities of arsenic that they contained at the beginning of the irradiation. (It is better to make both counts with the same counter, correcting for decay, than to make both at the same time and to attempt to correct for the differences between counters.)

Quantities of arsenic of about one part in 10^8 in samples of the order of grams can easily be observed by this method. Besides the great sensitivity, there are two major advantages. The first is that untreated original material is irradiated so that no harm is done by any later addition of traces of arsenic with reagents. After the irradiation, arsenical impurities in the reagents would have to be comparable with the 50 mg of carrier before they could affect the results.

The second advantage is that the efficiency of chemical separation does not matter so long as the final yield is known. Differences of yield between unknown and control samples are easily allowed for.

10.33 *Conditions for and limitations of activation analysis*

For activation analysis to be useful, it is desirable for the following conditions to be satisfied.

1. The cross-section for slow neutrons of a suitable isotope of the element to be detected must be large enough to produce a measurable activity. By "a suitable isotope" is meant one that will give a radioactive isotope on absorbing a neutron and not another stable one. It must also be present as a reasonable percentage of the parent element. Thus, the method is not very suitable for iron owing to the low abundance ($0 \cdot 3\%$) and poor cross-section ($0 \cdot 98$ barns) of Fe^{58} which gives the only observable active isotope.

2. The isotope produced must be reasonably easy to detect; those decaying mainly by electron capture are much less easy to observe than β-emitters.

3. The half-life of the active isotopes employed should fall within a range determined by the availability of the pile. If the chemical work is done in the laboratory attached to the pile, a few minutes may be enough. Otherwise the minimum usable half-life will depend on the time taken for transport between pile and laboratory. A half-life of an hour or so should suffice for most British laboratories, if air transport can be used.

Long half-lives must be avoided because of the length of time required to build up a satisfactory activity. For a given disintegration rate the number of atoms which have to be made is proportional to the half-life. Except for elements with particularly large cross-sections, a few months is probably the maximum half-life which can be tolerated. It should be remembered that the cost of irradiation of a sample increases with the time for which it is irradiated.

Such elements as fluorine (whose only stable isotope, F^{19}, gives rise to the $11 \cdot 6$-second F^{20}) or carbon (giving the 5700-year C^{14}) cannot usefully be measured by radioactivation.

4. A large flux of neutrons must be available. The sensitivity is directly proportional to this.

5. It must be impossible for significant amounts of the observed isotope to be produced from other elements. Thus we detect lithium by means of tritium ($12 \cdot 3$ y, β^-) which is produced by the reaction Li^6 (n, α) H^3. In a water solution the sensitivity is limited by the reaction H^2 (n, γ) H^3. The low cross-section and concentration of deuterium make this important only if there are some 10^9 times as many atoms of hydrogen as of lithium, so in this case the effect is not very serious.

More important is the difficulty that arises from the fact that a pile contains fast neutrons as well as slow ones, and that fast neutrons can give rise to (n, p) or even (n, α) reactions as well as to the usual (n, γ) reactions. Then if, for example, we are searching for traces of cobalt in nickel, we shall make cobalt isotopes by the reactions Ni^{58} (n, p) Co^{58} and Ni^{60} (n, p) Co^{60}, etc., as well as by the intended reaction Co^{59} (n, γ) Co^{60}. To correct for this, we may make a blank measurement on nickel, as free as possible from cobalt, in the same region of the pile, or we may irradiate the sample in a region well away from the pile centre, shielded from fast neutrons by a column of moderator. Either way, ultimate sensitivity is seriously reduced.

Even where a pure flux of slow neutrons is available confusion can sometimes arise. For example, suppose that we are looking for gold in the presence of platinum. The reaction we are expecting is Au^{197} (n, γ) Au^{198} (2·7 d, β⁻), but we also find the reaction Pt^{198} (n, γ) Pt^{199}, and Pt^{199}, with a half-life of half an hour, decays into Au^{199} (3·15 d, β⁻). The yield of Au^{199} from platinum is 1/300 of that of Au^{198} from gold and, although the two isotopes have different β-and γ-energies, it would be very difficult to approach the theoretical sensitivity for gold in the presence of any significant amount of platinum. Cases as awkward as this are rare, but the possibility of such interference should be remembered.

The best length of time for irradiation depends on a balance between two conflicting processes in operation; the production of the desired active nuclide from the stable element and the loss of it by radioactive decay. After a sufficiently long period these processes will balance and the rate of decay will just equal the rate of formation. There is no gain in irradiating for a longer period.

The useful activity produced after irradiation for a time t can be shown to be

$$n = g\sigma F \ [1\text{-exp}\ (-\lambda t)] \times 0\text{·}602 \text{ disintegrations per second}$$

where g is the number of gram-atoms present of the operative isotope of the stable element, i.e. the isotope which is the parent of the active nuclide observed; σ is its cross-section in barns; F the flux of neutrons per square centimetre per second averaged through the specimen and λ the decay constant. The value of g is of course $W\theta/A$ where W is the total weight in grams of the element being investigated, A is its atomic weight and θ is the abundance of the operative isotope expressed as a fraction of the total quantity of the element.

It will be seen that there is little advantage in continuing irradiation for a time longer than that which will reduce exp $(-\lambda t)$ (or $2^{-t/\tau_{\frac{1}{2}}}$) to about 0·1. This time is a little over three half-lives.

The efficiency of chemical extraction and of the counter which it is proposed to use must be considered. Where we wish to use such an isotope as S^{35} (85 d, β^-), which gives particles of only 0·17 MeV, the counter efficiency may be low owing to the quantity of carrier needed to remove the desired activity from the bulk of material. A thick source with considerable self-absorption is almost unavoidable.

Since F can be in the region of 10^{12}, isotopes for which $g\sigma$ is as small as 10^{-10} to 10^{-11} can be observed in good conditions, i.e. if the half-life is suitable and if there are no special difficulties in the radio-chemical purification of the active extract.

A list is given in Table 10.2 of elements which can be detected in good conditions in concentrations of less than one part in 10^8 with a flux of 10^{12} n/cm^2/sec. It is assumed that irradiation of the sample is continued either to saturation (i.e. until exp $(-\lambda t) \ll 1$) or for a month, whichever is the shorter. By "effective" cross-section in column 4 is meant $\sigma\theta$, where σ and θ are as defined above. The final column, showing the minimum actual weight of each element that can be determined is calculated on the assumption of a counter re-cording 40% of all disintegrations and a counting rate of 1000 per minute. The figure of 40% is typical for a small source close to the window of a good β-counter and could not easily be improved upon. Especially for the longer-lived nuclides, much less than 1000 c.p.m. could be observed easily enough, but the initial counting rate must be high enough for it to be possible to count through several half-lives to detect residual radioactive contamination. The table assumes 100% chemical extraction and negligible self-absorption in the sources to be counted. A factor of ten in sensitivity may easily be lost here if difficult chemical separations are involved. The time taken between irradiation and counting has also been neglected. This may be important for the shorter-lived nuclides.

In perfect conditions with an isotope the half-life of which is some days, ten or even a hundred times less than is shown in column 6 might be detectable, but it would be unwise to assume this to be the case without one or two experimental trial runs.

A few elements omitted from the table such as chlorine, sulphur, iron or the alkaline earth metals, may be determined by activation analysis, with limiting sensitivities between 0·1 and 10 μg, but for most of them alternative methods of comparable or better sensitivity are available.

With the help of an ingenious trick, oxygen may be determined by neutron activation analysis. The trick is to add an oxygen-free salt of lithium to the mixture to be analysed. Slow neutrons give 2·8 MeV tritons from the lithium—Li^6 (n, α) H^3—and these tritons can react with oxygen —O^{16} (t, n) F^{18}—giving an observable fluorine isotope

TABLE 10.2

Sensitivity for Activation Analysis using Chemical Separation and a β-particle Counter, assuming 10^{12} neutrons per sq.cm per sec and exposure for up to a month

Element	Isotope produced	Half-life	Effective cross-section barns	% saturation 1 month	Sensitivity m μg of element
Antimony	Sb^{122}	2·8 d	3·89	100	2
Arsenic	As^{76}	26·6 hr	4·3	100	1
Bromine	Br^{80m}	4·58 hr	4·09	100	2
Bromine	Br^{82}	35·8 hr	1·11	100	5
Cobalt	Co^{60}	5·27 y	34	1·08	10
Copper	Cu^{64}	12·8 hr	2·97	100	4
Dysprosium	Dy^{165}	139 m	738	100	0·02
Erbium	Er^{171}	~7 hr	1·0	100	10
Europium	Eu^{152}	9·2 hr	659	100	0·02
Gadolinium	Gd^{159}	18 hr	0·9	100	10
Gallium	Ga^{72}	14·3 hr	1·35	100	4
Gold	Au^{198}	2·69 d	96	100	2
Hafnium	Hf^{181}	45 d	3·5	34·2	10
Holmium	Ho^{166}	27·3 hr	60	100	0·2
Indium	In^{116m}	54 m	139	100	0·05
Iodine	I^{128}	25·0 m	7·0	100	1·0
Iridium	Ir^{192}	74·4 d	285	25	0·2
Iridium	Ir^{194}	19 hr	80	100	0·2
Lanthanum	La^{140}	40 hr	8	100	1·0
Lithium	H^3	12·3 y	71	0·5	10
Lutecium	Lu^{176m}	3·7 hr	19·5	100	0·5
Lutecium	Lu^{177}	6·8 d	91	95·2	0·2
Manganese	Mn^{56}	2·59 hr	13	100	0·3
Osmium	Os^{191}	16·0 d	1·4	75	10
Palladium	Pd^{109}	13·6 hr	3·1	100	2
Phosphorus	P^{32}	14·3 d	0·029	76·7	10
Praseodymium	Pr^{142}	19·2 hr	10	100	1
Rhenium	Re^{186}	3·8 d	37	99·5	0·4
Rhenium	Re^{188}	16·9 hr	47·2	100	0·3
Samarium	Sm^{153}	47 hr	36	100	0·3
Scandium	Sc^{46}	85 d	14·4	21·5	1·0
Sodium	Na^{24}	15 hr	0·4	100	4
Tantalum	Ta^{182}	112 d	21	16·2	4
Terbium	Tb^{160}	73·5 d	22	26	2
Thulium	Tm^{170}	129 d	100	15·0	1
Tungsten	W^{187}	24·0 hr	9·9	100	2
Ytterbium	Yb^{175}	102 hr	15	99·3	1
Yttrium	Y^{90}	2·64 d	1·24	100	5

(110 min, β^+). In good conditions the limit of sensitivity may be around 1 μg.

Activation analysis is sometimes useful even where there is a fair quantity of an element present and chemical analysis presents little difficulty, for example in non-destructive analysis of valuable objects, such as ancient coins, and in work on biological structures.

10.34 *Determination of the distribution of elements*

An irradiation with neutrons followed by autoradiography may give an excellent picture of the distribution of a particular element, which is often as important as its quantity. Because different elements produce isotopes with very different half-lives, a single irradiation may be able to show the distribution of two elements. An immediate short exposure to a photographic emulsion may be made to give a picture of the distribution of, for example, chlorine—using Cl^{38} (38 min, β^-)—while an exposure started a few hours after irradiation would show the distribution of sulphur, using S^{35} (85 d, β^-).

10.35 *Activation analysis by charged particles*

Particles other than neutrons have been little used for activation analysis, largely because of experimental difficulties. Samples must be exposed directly to the beam of an accelerator and must be able to dissipate a considerable amount of heat. This is particularly troublesome when soft biological tissues are involved. There is, however, another serious drawback; that there is usually an ambiguity in the origin of the isotope involved. Thus bombardment of a sample by deuterons can produce P^{32} from phosphorus, sulphur and chlorine by the reactions

$$P^{31} (d, p) P^{32}, \quad S^{32} (d, 2p) P^{32} \text{ and } Cl^{35} (d, p\alpha) P^{32}.$$

The last reaction might be avoided by using deuterons with low energies, but the other two could not be reliably separated in this way.

There are some advantages to be exploited. Some elements, such as vanadium, do not produce a useful isotope by neutron irradiation (V^{52} has a half-life of only 3·76 minutes which is too short for most analyses). Proton bombardment can give the reaction V^{51} (p, n) Cr^{51} and Cr^{51} has a convenient half-life of 28 days. It decays, unfortunately, by electron capture but emits an observable γ-ray. The reaction Cr^{52} (p, d) Cr^{51} may also occur if there is chromium in the sample, but requires a threshold energy of 9 MeV and can be prevented by using protons of lower energy than this.

Another example is the estimation of carbon in steel using the 20-minute positron-emitter C^{11}. C^{11} can be produced by any charged particle of sufficient energy, the best being He^3. This will give a good yield of C^{11} by the reaction C^{12} (He^3, a) C^{11} with a bombarding energy of 5 MeV and, since the potential barrier of iron against He^3 is about 10 MeV, reactions between them at this energy will be rare. Hence, unless other light elements are also present, no chemical separation will be needed.

This technique could be used for the detection of any light-element contamination among heavy elements though it may be difficult to avoid ambiguities among the light elements themselves. Thus in the example just cited, C^{11} could be produced from either boron or beryllium at the energy stated. This is usually, however, unimportant as beryllium and boron are rare impurities if they have not been introduced on purpose. If boron is suspected it could be confirmed and its quantity determined with the help of a proton bombardment —B^{11} (p, n) C^{11}. The threshold for this is only about 2 MeV, while that for the reaction C^{12} (p, d) C^{11} is over 10 MeV so that for protons of say 8 MeV there is no risk of confusion.

The method has one more characteristic which might be of use in investigating the distribution of elements. This is that at suitable energies the ranges of charged particles are very small. If the particular particles are allowed to strike a surface near glancing incidence, their penetration may be only a few microns. This may improve the resolution of autoradiographs a good deal, without the trouble of cutting and mounting sections, as the autoradiograph of the bombarded surface of a thick specimen will be identical with that of a section only a few microns thick.

Since charged-particle beams are generally limited to 10^{13} or 10^{14} particles per second and since less than 1 in 10^4 produces any nuclear reaction at all, the sensitivities of activation analyses by use of such particles are usually thousands of times lower than those of analyses carried out by neutron activation.

10.36 *Activation analysis during bombardment*

If a sample is arranged so that its radiations can be observed during its actual bombardment by a beam of charged particles, several very specific methods of analysis for light elements become possible.

For example, the reaction F^{19} (p, a) O^{16} can be used to detect fluorine in samples of a few milligrams down to concentrations of 10 parts per million (ppm.). Advantage is taken of the fact that the reaction is strongly exothermic and usually leaves the O^{16} nuclei in an excited state. They then emit γ-rays including some of high energy* which can be detected with a scintillation counter in the pre-

* 6.06, 6.14, 6.92 and 7.12 MeV.

sence of a considerable background of lower-energy γ-rays. A further advantage is that there are sharp reaction-resonances at 0·340, 0·874 and 0·935 MeV.* Protons of conveniently low energy have therefore a fair cross-section for reaction in spite of the fact that they cannot pass over the top of the potential barrier of the fluorine nucleus. They will, however, be able to produce very few reactions with other elements and the background will be less. Nitrogen may produce a little interference, owing to the possibility of the reaction N^{15} (p, γ) O^{16} from which some of the same γ rays may be produced, but the sensitivity to fluorine will be thousands of times as great as to nitrogen.

The method is a comparative one, the apparatus being calibrated by finding the rate of emission of γ-rays from a sample with a known fluorine content bombarded by a known current of protons, the same proton-energy being used in all experiments. If the proton-beam has a sharply defined energy (above 0·340 MeV) the fluorine content will be determined only at sharply defined depth in the target; where the protons have energies corresponding to the resonances this is rarely troublesome and may be useful if it is desired to find the distribution of fluorine over small ranges of depth.

The same equipment can be used to detect nitrogen if 2 MeV deuterons are used instead of protons, by the help of the reaction N^{15} (d, n) O^{16}, again counting high-energy prompt γ-rays from the O^{16}. This would not distinguish well between nitrogen and fluorine, which could undergo the reaction F^{19} (d, αn) O^{16}. A modification of the procedure, however, gives an absolutely unambiguous measurement of the nitrogen present. At the same time as the reaction above, the reaction N^{15} (d, p) N^{16} will also occur. N^{16} has a half-life of 7·4 seconds, emitting β-particles up to 10·30 MeV and γ-rays of 7·12 and 6·14 MeV. Consequently, if 6-7 MeV γ-rays are counted immediately after the accelerator is switched off no prompt γ-rays will be recorded and very low percentages of nitrogen can be observed in the presence of any other element. In practice the accelerator would be switched on and off for perhaps two seconds each way, the counting equipment being automatically switched on whenever the accelerator was off.

10.4 Isotope Dilution

In ordinary chemical analysis two requirements must be met; not only must 100% of each constituent be extracted but it must be extracted in the form of a pure, known compound. Either of these

* As measured in the laboratory. The centre-of-mass energies are 5% lower.

H*

requirements may easily be fulfilled but to achieve both together is often difficult. With the help of radioactive materials the requirement of 100% extraction can be avoided.

Suppose that we want to estimate the chloride in a complex mixture. We first add, with as little carrier as possible, a known activity, say n counts per minute at time zero, of Cl^{38} (38 mins, β^-) in the chemical form of chloride. We then extract the mixed chloride by any method desired and purify it, finishing up perhaps with a clean, dry precipitate of silver chloride. Loss of part of the chloride during the purification process does not matter at all so long as the precipitate contains nothing besides silver chloride. The precipitate is counted until the desired statistical accuracy has been obtained and is then weighed. Suppose that the activity, after correction for radioactive decay back to zero time, corresponds to $\frac{1}{4} n$ counts per minute. Then clearly during the extraction and purification we have managed to keep only $\frac{1}{4}$ of the radioactive chlorine atoms. It can be inferred that we have also kept only $\frac{1}{4}$ of the stable chlorine atoms. Hence, if our weighed precipitate contained w mg of chloride the total weight of chloride in the original mixture was $4w$ mg.

To summarise, the isotope-dilution method consists in doing a good chemical separation and purification of a convenient fraction of the element investigated and weighing this fraction, while the chemical yield is measured by a radioactive tracer.

The same process can be applied to groups such as sulphate or phosphate, or compounds such as oxalic acid, by using corresponding labelled groups such as $S^{35}O_4^{--}$, $P^{32}O_4^{---}$ or $C^{14}OOH \cdot COOH$.

High accuracy is difficult to obtain where self-absorption effects are high, as with S^{35} (85 d, β^-, 0·17 MeV). It is better in such cases not to determine the original activity simply by counting a sample of the carrier-free radioactive sulphate. It is best, instead, to make up two identical carrier-free solutions, adding one (A) to the unknown mixture and preserving the other (B) until after the weighing of the final sample. Suppose that this is in the form of barium sulphate, and contains w mg of sulphate. Then w mg of sulphate (as sodium sulphate or any other soluble sulphate) is added to the preserved active carrier-free solution B and the whole of the sulphate in B precipitated with an excess of barium. This precipitate is washed, distributed over just the same area as the precipitate derived from the unknown mixture, dried and counted by the same counter. If the thicknesses and areas of the precipitates are identical, this will enable us simultaneously to compensate for self-absorption and for counterefficiency.

It will be seen that this method, unlike the method of activation analysis, is not more sensitive than standard chemical methods, although it may be more accurate and is often a great deal easier.

10.41 *Inverse isotope-dilution analysis*

It is possible in some cases to use an *inverse* isotope-dilution method. This may in some circumstances have a higher sensitivity. It has the important advantage that an unlabelled compound is used as additive. The method is mainly of use in biochemical analysis. Here there are many compounds of interest which it is difficult to prepare in any quantity in labelled form. The method of operation is as follows. Suppose that we wish to determine the quantity of a particular amino acid per gram of total protein from a particular plant. First, the plant is grown for a period in an atmosphere containing labelled carbon dioxide —$C^{14}O_2$—until it can be expected to have incorporated a convenient amount of C^{14} in the amino acid of interest. A sample of protein is separated from the plant material, and divided into two parts, each of weight W gm, say. A pure specimen of the amino acid is extracted from one part of this protein. Suppose this weighs w mg. and has a measured activity of n counts per minute in standard counting conditions.

To the second part of the protein are added 100 mg of the *unlabelled* amino acid. After thorough mixing, a pure sample, weight w' say, of the amino acid is again extracted, and its activity, n', measured in the same conditions. Suppose that $n/w = an'/w'$, i.e. the specific activity has been reduced a times. No allowance need be made for decay, as the half-life of C^{14} is 5700 years, but allowance may have to be made for the thickness of the source as the β^- energy is only 0·156 MeV. If either w or w' is over a milligram, it may be best to make them equal by rejecting the excess of the larger sample, and also to ensure that they cover equal areas before measuring the activities.

If 100 mg of added unlabelled amino acid has reduced the specific activity a times, the total weight of the labelled compound in the original W gm of protein must have been $100/(a-1)$ mg.

In this case the sensitivity is not materially different from that of the direct method; the advantage lies in the fact that the plant itself has done the expensive part of the experiment, that of making the labelled compound.

A modification of this method is possible if the plant has for its whole life used labelled CO_2 of constant specific activity as its sole source of carbon. It can then reasonably be assumed that all carbon compounds in the plant have the same fraction of their carbon in the form C^{14}. In this case the first part of the experiment described above is unnecessary. A sample of any part of the plant tissue can be used to establish the value of n/w, so long as the carbon content is known. Allowance can be made for any difference of carbon content between it and the sample w' of active amino acid.

In this case the sensitivity is considerably increased since there is no need to have enough of the compound to give a weighable extract before unlabelled material is added, although there may be a risk of error owing to a slight degree of isotope separation which is known to occur in living organisms. This risk is unlikely to be significant and can be minimised by using compounds of the same nature as that being investigated—in this case protein rather than carbohydrates for example. Attempts to get very high sensitivity may, however, be risky as they entail very high specific activities throughout the plant tissues which may upset the plant's normal metabolism.

For this reason, and because a great deal of the active element is lost in irrelevant compounds, the inverse isotope-dilution method is of much less general value than the normal method.

10.42 *Special methods of isotope-dilution analysis*

Several ways exist, at least in principle, of increasing greatly the sensitivity of the normal method and hence of making it more sensitive as well as quicker than conventional chemical methods. These all stem from the fact that the limits of sensitivity are set, not by the radioactive measurements but by the gravimetric ones. Most chemical balances require at least a milligram of material for an accuracy of 10 to 20% Commercial torsion balances exist to weigh a microgram to this accuracy but are by no means so easy to use. Such methods as electro deposition, solvent extraction, ion exchange or selective adsorption might be considered instead of weighing for determining the quantity of material in an active sample. As an example we will describe a method which has been used to measure small quantities of fluorides. Fluorides are of importance in the prevention of dental decay and concentrations of a few parts per million close to the tooth surfaces may be significant.

Instead of extracting a weighable sample of the fluoride after addition of tracer F^{18} (110 min, β^+), a sample is obtained by adsorption on a small area of glass. The size of the sample may be of the order of 10^{-12} gm and must be determined by a control experiment. In practice, the method is therefore a comparative one, in which a solution of the material to be analysed is compared with a series of standard solutions, made up in a suitable range of strengths.

The procedure is as follows. The unknown sample is dissolved in normal nitric acid to give a pH close to 1. A series of standard fluoride solutions, bracketing in strength the expected value of the unknown, are also acidified to the same degree. To the same volume of each standard solution and of the unknown is then added the same amount of carrier-free F^{18} also in solution.*

The solutions are then applied to equal areas of a clean glass surface, and are left for equal times, ten minutes to an hour being suitable. At the end of this time the solutions are made alkaline. As soon as the pH rises above about 5, the adsorbed fluoride on the glass is fixed and can be added to or removed only with considerable

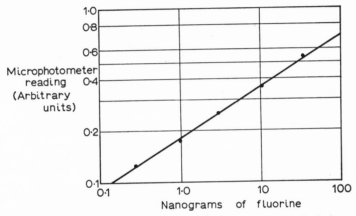

Fig. 49 Variation of F^{18} activity adsorbed by glass from $1\,\mu l$ of solution, with the stable fluorine concentration. The activity is measured by means of X-ray film and a microphotometer and is given in arbitrary units. 1 nanogram in $1\,\mu l$ is 1 part per million.

difficulty so that the glass can then be washed and dried without disturbing the adsorbed activity.

This can then be measured for each area, either by cutting up the glass slide and counting each area in a standard position under a counter or by making an autoradiograph of the whole slide, and measuring its density by means of a microphotometer.

In the conditions employed the amount of fluoride per unit area adsorbed by glass is approximately proportional to the square root of the fluoride concentration [F]. Each F^{18} atom has the same chance of adsorption as an F^{19} atom, and the *proportion* of F^{18}

* Carrier-free F^{18} can very conveniently be produced already in solution as fluoride, by bombarding distilled water with α-particles—O^{16} (α, d) F^{18}—but α-particles of 30 MeV or more are required to give a useful yield.

atoms is inversely proportional to the total fluoride concentration, since the absolute concentration of F^{18} atoms is exceedingly small and is the same in all samples. Hence the amount of F^{18} adsorbed per unit area is proportional to $\sqrt{[F]} \times 1/[F]$ or $1/\sqrt{[F]}$. Measurement of the activity of each area of glass therefore gives the value of [F] for the unknown sample. In Fig. 49 is shown the experimental variation of optical density of the autoradiographs from a series of 1 μl samples of solution as a function of fluoride content. Quantities down to less than 10^{-9} gm are measurable. The range of concentrations lies between 0·1 and 1000 parts per million. Over this range, an accuracy in the region of $\pm 15\%$ is quite possible to obtain.

At concentrations of fluoride below 0·1 p.p.m. the amount of F^{18} taken up by the glass begins to change less rapidly then as $1/\sqrt{[F]}$ and the sensitivity of the method declines.

One microlitre (1 mg approx) of solution is as small an amount as can conveniently be handled without evaporation difficulties.

If all work were done under a layer of a fluid such as benzene, immiscible with water, doubtless much smaller quantities could be used, with a corresponding increase of sensitivity, but the method is already sufficiently complicated to make such additional difficulties very discouraging.

10.5 Radiometric Analysis

This can be thought of as a development of the gravimetric method in which the quantity of precipitate is determined by radioactive measurement rather than by weighing.

For example, small quantities of silver may be determined by precipitating it as silver iodide using an iodide labelled to a known specific activity with I^{131} (8·06 d, β^-, γ). Then a precipitate which is far too small to weigh or even to see with certainty can be filtered off and measured by finding its β-particle activity. The method also has been used successfully for the measurement of small quantities (down to 10^{-8} gm) of calcium, which is not easy to determine by activation analysis.

Here the calcium is precipitated as fluoride using F^{18} (110 min, β^+) as indicator. The calcium sample in acid solution is mixed with an excess of labelled fluoride and placed on chromatographic paper previously washed in ethylene diamine tetracetic acid. The pH is raised to 9 or more by exposure to ammonia vapour, when calcium fluoride is precipitated. The paper is then irrigated with inactive fluoride solution (0·1 %) to remove excess of F^{18} and the active spot of precipitate is dried and counted. A control series of known samples are treated in the same way to find the efficiency of the process, which is usually under 100% but is practically constant for any particular grade of paper.

As in many measurements, the limit of sensitivity is not determined by the difficulty of detection but by the presence of "background" phenomena. Many elements could be precipitated in quantities of 10^{-10} gm or less from a millilitre of solution, combined with enough active material to be readily detectable. Except in very fortunate cases, however, larger quantities of activity will be taken up by other solid impurities in the solution or on the filter itself. Thus in the calcium analysis mentioned, the activity per square centimetre of clean chromatographic paper corresponds to some hundredths of a microgram of calcium and sets a corresponding limit to what can be observed. In the absence of this incidental activity there would be no difficulty in detecting 10^{-10} gm of calcium by this method.

Radiometric analysis may be useful even where great sensitivity is not required, because of the speed with which measurements can be made. For example, an effective method has been worked out for the determination of magnesium oxide in Portland cement, which can be completed in one hour as compared with 28 hours for the standard chemical method. Calcium and iron would interfere with the method and must be removed by an ion-exchange column, or otherwise, beforehand. A 1-gm sample of cement is dissolved and mixed in acid solution with a known quantity of ammonium hydrogen phosphate labelled with P^{32} (14·5 d, β^-). In acid solution the magnesium will not be precipitated. A standard volume of the solution is placed in a liquid-type Geiger counter (see Chapter 3) and its counting rate recorded. The rest of the solution is then made alkaline with ammonia, when the magnesium is precipitated as magnesium ammonium phosphate. The precipitate is removed and a standard volume of the residual liquid is put in the same counter and its activity recorded. The proportional loss of activity compared to the first solution gives directly the proportion of phosphate precipitated and hence the quantity of magnesium. The washing, cleaning, drying and weighing which are needed in the gravimetric method have been eliminated.

For this difference method to be successful the concentration of magnesium should be roughly known as for good accuracy the loss of activity should be at least 30 but less than 100%.

A combination of radiometric and activation analysis may be used. For example, in the determination of thallium the first stage is to precipitate the thallium, in the thallous state, as iodide. This is best done by precipitating from an aqueous alcoholic solution in the presence of sulphite, a reducing agent. The precipitate is collected on Perspex and dried. It can then be irradiated in a neutron flux for an hour, together with a standard consisting of a known quantity of thallous iodide. The I^{128} (25 min, β^-) activities of sample and standard can then be compared.

10.6 Neutron-absorption Methods

A few substances with very high cross-sections for absorption of slow neutrons can be estimated by observing the attenuation of a slow-neutron flux in a sample of standard thickness. Such elements as cadmium, boron or gadolinium can be determined in this way, so long as it is known that no other element with a high cross-section is present. Again, measurement is by comparison with a known sample.

Many variants exist of the methods described, but enough has been said to indicate the power and specificity of the new analytical methods made possible by radio isotopes. The equipment required is more complex and expensive than most analytical apparatus and it is important not to assume that radioactive methods are necessarily better because they are newer and more frequently discussed. Nevertheless, where any difficult analytical problem arises, radioactive methods are always worth serious consideration.

CHAPTER ELEVEN

RADIOISOTOPIC DATING

The presence in any material of radioactive nuclei, or of stable daughters known to be derived from these, may be used in several ways to indicate the age of the material. The mere fact that a particular radioactive nuclide is observable in a structure in which it could not have been created gives a limit to the age of the structure. For example, the element technetium has been observed spectroscopically in certain rare stars. No stable isotope of technetium is known and the longest lived active isotope, Tc^{97} (e-capture), has a half-life of 2·6 million years. New technetium cannot be made in steadily burning stars so it must have been incorporated before the beginning of the stars' lives in their present state. It might have been incorporated as long as ten million years ago, since when Tc^{97} will have decayed to about one-fifteenth of its original amount. It certainly hasn't lasted as long as 100 million years, although this is only a fiftieth of the age of our own planet. In that time Tc^{97} would have been reduced by a factor of 6×10^{11} so that even if the star had originally consisted entirely of this single isotope it would be impossible to detect it spectroscopically now. Thus the qualitative observation of a single radioactive element gives a firm numerical result which is exceedingly important for our understanding of the evolution of the stars.

We can even make quantitative deductions from the qualitative observations that we *haven't* made. Thus we have not found on the earth any primary radioactive nuclides with half-lives less than 10^8 years, though some which have been made in the laboratory, such as Pu^{244} (7×10^7 y, a) should be very easy to observe. All known isotopes with half-lives equal to or greater than that of U^{235} ($7 \cdot 1 \times 10^8$ y, a) have, on the other hand, been found in nature. The implication is strong that the material composing the earth has existed long enough for all of the primordial Pu^{244} to have decayed beyond the limits of observability; at least some 3000 million years.

We will now consider various means of gaining more definite quantative information, not merely top or bottom limits, for the ages of various objects.

11.1 Geological Dating

11.11 *Uranium and thorium*

The classical method for determining the age of the oldest rocks consists in comparing the amount of uranium present with the amount of its decay products, which have been accumulating at a calculable rate since the rocks were formed. By the time any known rocks were laid down, only two uranium isotopes survived in any quantity: U^{238} (4.5×10^9 y, α) and U^{235} (7.1×10^8 y, α). The relative abundances of these are now 99·3 % and 0·7 % respectively and each, after a series of α- and β-decays, gives rise to a stable isotope of lead.

11.111 *Lead.* U^{238} gives Pb^{206} and U^{235} gives Pb^{207}. If an old uranium-bearing rock had initially contained no lead at all, measurement of the ratio of either of these two lead isotopes to the uranium would give the age of the formation.

Suppose that the age of the rock is T years. Then if we now have N atoms of U^{238} in a sample of rock,

$$N = N_{-T} \times 2^{-T/\tau_{238}}$$

where N_{-T} was the number of U^{238} atoms in the same sample T years ago and τ_{238} is the half-life of U^{238}. This may be rearranged to give

$$N_{-T} = N \times 2^{T/\tau_{238}}.$$

The number of uranium atoms which have decayed in the sample in time T is then given by

$$N_{-T} - N \text{ or } N[2^{T/\tau_{238}} - 1].$$

When T is large compared to 10^6 y, nearly all of the U^{238} atoms which have decayed will have attained their final stable state as Pb^{206}.* Then the number of lead atoms will be equal to the number of decayed U^{238} atoms, or

$$N_{Pb} = N[2^{T/\tau_{238}} - 1].$$

Hence
$$2^{T/\tau_{238}} = (N_{Pb}/N) + 1$$

or
$$T = \tau_{238} \frac{\ln [(N_{Pb}/N) + 1]}{\ln 2} \qquad 11.1$$

$\therefore$
$$\left. \begin{array}{l} T = 6.5 \times 10^9 \ln [(N_{Pb}/N) + 1] \text{ years} \\ = 15 \times 10^9 \log_{10} [(N_{Pb}/N) + 1] \text{ years}. \end{array} \right\} \ 11.2$$

---

* This assumption will lead to an underestimate of the age of the rock by about half a million years. The error could easily be corrected by finding the amount of the intermediate products U^{234} and Th^{230} present, but it is usually quite negligible compared to other errors.

If N_{Pb} is less than $1/10$ of N, this may still more simply be written

$$T = 6 \cdot 5 \times 10^9 \ (N_{Pb}/N) \text{ years}$$

with an error of less than 5%.

Similarly, from the decay of U^{235} we obtain

$$\left. \begin{aligned} T &= 1 \cdot 03 \times 10^9 \ \ln \ [(N'_{Pb}/N') + 1] \text{ years} \\ &= 2 \cdot 26 \times 10^9 \ \log_{10} \ [(N'_{Pb}/N') + 1] \text{ years} \end{aligned} \right\} 11.3$$

where N'_{Pb} and N' are the number of atoms respectively of Pb^{207} and U^{235} present today.

In most cases the amount of Pb^{206} produced from U^{238} is much greater than that of Pb^{207} produced from U^{235} so that it will give the more accurate results. It is worth noting, however, that when the oldest known rocks were laid down, around 3000 million years ago, the percentage of U^{235} in uranium was some twenty times greater than now and the rate of production of Pb^{207} was comparable with that of Pb^{206}. A rock 5200 million years old would contain equal quantities of the two kinds of lead derived from uranium and still older rocks would contain more Pb^{207}.

The ratio of Pb^{207} to Pb^{206} gives, therefore, a method of determining the age of an old rock. For rocks over 1000 million years old it is probably the best of all the uranium-lead methods as it does not require quantitative determination of the absolute amounts (often very small) of lead and uranium present. It does, however, require the measurement of the proportion of each lead isotope in the sample. This can be done by means of a mass-spectrograph. A variety of designs of these instruments have been published and any of them could be built by a well-equipped laboratory. To do so is, however, a considerable undertaking and it will usually be best to arrange for the mass-analysis of samples by one of the laboratories specialising in such work.

Radioactive methods of dating can be applied also to rocks containing thorium. Here there is only one long-lived isotope, Th^{232} ($1 \cdot 39 \times 10^{10}$ y, α) which gives rise to Pb^{208}.

The relation between age and lead-thorium ratio is then

$$T = 2 \cdot 01 \times 10^{10} \ \ln \ [(N''_{Pb}/N'') + 1] \text{ years}$$
$$= 4 \cdot 62 \times 10^{10} \ \log_{10} \ [(N''_{Pb}/N'') + 1] \text{ years}$$

where N''_{Pb} and N'' respectively are the numbers of atoms of Pb^{208} and Th^{232} in a sample of the rock.

In practice the situation is much complicated by the fact that few uranium- or thorium-bearing rocks were free of lead to begin with. This primeval or aboriginal lead can be allowed for, if we know what was its isotopic composition, by observation of the isotopes which are *not* produced by radioactive decay.

The percentage composition of ordinary present-day lead is: Pb^{208}, 52·3%; Pb^{207}, 22·6%; Pb^{206}, 23·8% and Pb^{204}, 1·5%. Where the sample is free of thorium, the Pb^{208} present can be used to determine the primeval amount of Pb^{206} and Pb^{207}. Where the sample contains thorium but is free of uranium, Pb^{206} and Pb^{207} can be used to determine the primeval amount of Pb^{208}. When both uranium and thorium are present the Pb^{204} may be used, but its percentage is so small that this is liable to give less accurate results and it is better to solve the necessary set of simultaneous equations, using both the uranium and thorium results, to find both the age of the sample and the primary lead content.

A drawback to all of these procedures, apart from their technical complexity, is that we cannot be sure that the isotopic composition of the primeval lead in the rock was exactly the same as that of modern lead. An appreciable percentage of all the lead in the crust of our planet is derived from the decay of uranium or thorium. Any non-uniformity of distribution of these two elements, therefore, must result in an inhomogeneity of isotopic composition of lead from different regions. The extent to which such inhomogeneities have been ironed out, at various periods of our geological history, is a matter for cautious hope rather than confident dogmatism.

11.112 *Helium.* A method which is free of the difficulty arising from the possible presence of aboriginal lead is to examine instead the ratio of helium to uranium or thorium. During the series of decays from U^{238} to Pb^{206}, eight helium atoms are evolved; as U^{235} passes to Pb^{207} seven and as Th^{232} passes to Pb^{208} six. Uranium with its present isotopic constitution is producing helium at the rate of 120 cubic millimetres at S.T.P. per gram per million years while thorium is producing it at the rate of 29 cubic millimetres per gram per million years. If a rock is impervious to gas, this helium will be retained. Very sensitive methods of measurement are available which can determine the helium content even of relatively recent rocks, and we can be quite sure that no primeval helium was incorporated in the rock when it was laid down. The main drawback to the method is that we can *not* be sure, even if the rock is now impervious to gas, that it has been so throughout its history. Helium diffuses through almost all substances when hot and few old rocks have escaped some degree of heating during their history. The method must therefore be regarded as giving a reliable *lower limit* to the age of any particular rock rather than a reliable actual value. The age of some meteorites has thus been shown to be over 2000 million years. In this case there is a risk of additional helium being produced by cosmic rays and though the proportion so produced may be indicated by the presence of helium 3, which is not produced in radioactive transformations, the need for a mass-spectrographic analysis makes the method a great deal less attractive.

11.113 *Particle tracks.* When charged particles pass through certain crystalline minerals, such as mica, they cause permanent damage to the crystal structure. The disturbed atoms can be removed slightly more easily by a weak etching solution than can the atoms in the undisturbed lattice. If thin flakes of irradiated mica are etched with hydrofluoric acid the tracks of heavily ionising particles such as fission fragments will be "developed" as slightly thinner regions on the surface of the flake. These can be seen under a high-power microscope or more easily with an electron-microscope. Where mica lies in close contact with uraniferous minerals, which will give a steady and calculable output of fragments from spontaneous fission, the age of the combined formation can be found by counting the tracks in a known region of the mica. The method is not yet fully developed and may never achieve great numerical accuracy, but it is unaffected by the presence of primeval lead or by possible loss of inert gases so that it may give a good check of other methods.

11.114 *Thorium* 230. Another decay product of uranium can be used for dating more recent deposits. This is thorium 230 (ionium) (80,000 y, α). The oceans contain traces of both uranium and the ordinary thorium 232 ($1 \cdot 39 \times 10^{10}$ y, α). The thorium, but not the uranium, is found in the sediments at the bottom. After the total thorium has been chemically extracted from perhaps a gram of sediment, the two isotopes are readily distinguished by their different α-particle energies which can be measured by an ionisation chamber or scintillation counter. The ratio of Th^{230} to Th^{232} in fresh deposits must be the same as in the sea. The rate of decay of the Th^{230} is, however, much greater so that in older ones the ratio will be less. If we assume that the ratio of uranium to thorium in the sea has been constant over the period concerned, the ratio can be used to give us the age of the sediment. The decay of Th^{232} is negligible over a few hundred thousand years so that, if the ratio of Th^{230} to Th^{232} in fresh sediment is r_0 and that in another sediment is r_t, then the age t of the older is given by

$$t = 80{,}000 \ln (r_0/r_t) \text{ years}$$

or

$$t = 184{,}000 \log_{10} (r_0/r_t) \text{ years.}$$

$$\left.\right\} 11.4$$

By examining cores from the bottom in this way, the rate of sedimentation in the Pacific has been shown to vary from about 1 mm per thousand years in the North to $\frac{1}{2}$ mm per thousand years in the South.

Not all old rocks contain enough uranium or thorium for age measurement. Several other long-lived radioactive elements exist which have also been used.

11.12 *Rubidium*

For example, rubidium-87 is β-active, with a half-life of 5×10^{10} years, giving rise to strontium-87. Hence the ratio of Sr^{87} to Rb^{87} in an old rock can be used to give its age. The present rate of strontium production is $3 \cdot 7$ μg per gram of rubidium per million years, and has not changed enough to matter during the life of any but the oldest rock formations. As in the uranium-lead method, the possibility of the presence of primeval strontium in the rock necessitates an isotope analysis by mass-spectrograph. Slightly longer ages tend to be given by this method. It is possible that the uranium-lead method gives systematically low results because a wrong value has been taken for the isotopic constitution of primeval lead or it may be that the half-life of Rb^{87} is in reality a little less than the accepted value.

11.13 *Potassium*

Another method which is increasingly used is the potassium-argon method. K^{40} has a half-life of $1 \cdot 27 \times 10^9$ years, 11% decaying by electron capture to A^{40} and 89% by β^--emission to Ca^{40}. Calcium is too common an element for the latter decay to be of much assistance but argon, like helium will certainly not have been one of the original constituents of the rock under investigation. A great advantage of this method is that potassium is a much commoner element than uranium or rubidium. The isotopic abundance of K^{40} is only $0 \cdot 012\%$, but the sensitivity of methods for detection of argon are great enough for this not to be serious. The amount of argon produced in less than a million years can easily be observed in suitable rocks. The important series of early fossil hominids found by Dr. Leakey at Olduvai Gorge have been dated by this method. One of the oldest, Kenyapithecus, has been shown to be 12 to 14 million years old, while others have been shown to be between one and two million years old although weathering of some of the minerals used has led to a considerable spread of the results.

The present rate of production of argon is $0 \cdot 0041$ cubic millimetres at S.T.P. per gram of potassium per million years or $4 \cdot 1$ cu mm^3 per

10^9 years, and the integrated total quantity produced since 10^9 T years ago has been $7 \cdot 4$ $(2^{T/1 \cdot 27} - 1)$ mm^3 per gm. As in the case of helium, it is essential that the rock should be impervious to gas. The method is particularly valuable for meteorites, since argon, unlike helium, is not produced in appreciable yield by the action of cosmic rays.

The decay of K^{40} is especially interesting, since it could give us an actual value, as first pointed out by Chackett, for the age of the earth. Whether the earth was formed hot or cold, a considerable part of its history, during which most of the primeval K^{40} decayed, was spent in a molten state, in which the argon produced would presumably have been liberated into the atmosphere. It is almost certain that nearly all terrestrial argon is radiogenic both because of the great preponderance of A^{40} and because of the anomalously great abundance on the earth of argon as compared to neon and krypton.

None of this argon, once formed, can have been lost from the earth as this would have required extended periods with atmospheric temperatures of the order of 60,000° K. Hence the 1 % of argon now in the atmosphere represents the total output during the molten period of all terrestrial potassium. An estimated quantity of the latter which is still given in reference books, however (2·6 % by weight of the earth; *American Institute of Physics Handbook*, pp. 7-9), leads to an age for the earth of a few hundred million years. This is obvious nonsense and shows that this estimate of abundance of terrestrial potassium must be seriously in error. If the hundred times smaller figure given by Suess and Urey is adopted, a very reasonable age is found. Their value for the potassium abundance is not certain enough, however, for this to be treated as a definitive value.

11.14 *Other long-lived elements*

There are a few more long-lived radioactive nuclides which could in principle be used for dating ancient formations, but for various reasons they seem unlikely to be as useful as those already described. Indium-115 $(6 \times 10^{14}$ y, $\beta^-)$, for example, has so long a half-life that even in a *thousand* million years only about a microgram of Sn^{115} would be produced per gram of indium and this would be too small a quantity to detect, let alone to distinguish from the background of aboriginal tin. Hafnium-174 $(2 \times 10^{15}$ y, $a)$ and vanadium-50 $(4 \cdot 8 \times 10^{14}$ y, e-capture) present a similar difficulty, aggravated by the fact that their isotopic abundances are only 0·18 % and 0·25 % respectively.

Samarium and neodymium are both rare earths and give rise to other rare earths. Sm^{147} $(1 \cdot 3 \times 10^{11}$ y, $a)$ forms 15 % of ordinary samarium and gives rise to Nd^{143}, which is 12.2% of ordinary neodymium.

Nd^{144} (5×10^{15} y, α) forms 24% of the normal element and gives rise to Ce^{140} which forms 88% of ordinary cerium. The rare earths are troublesome to separate chemically. More seriously, they tend always to be found together in very similar proportions so that it is extremely difficult to find rare-earth samples free of samarium or neodymium from which the "natural" abundance of the daughter isotopes can be determined.

Two other long-lived rare earths are known: La^{138} (2×10^{11} y, e-capture) forms only 0.089% of the parent element and, although it produces the chemically distinct barium, is not very promising since its "yield" is only about three micrograms per thousand million years per gram of lanthanum and these three micrograms will be the commonest isotope (72%) of primeval barium.

Lu^{176} (4.6×10^{10} y, β^-) forms 2.6% of ordinary lutecium and gives rise to Hf^{176} at the rate of 0.4 mg of hafnium per gram of lutecium per thousand million years; Hf^{176} is only 5.2% of ordinary hafnium so that this is a little more promising, but the yield is still some hundreds of times less than the yield of lead isotopes from uranium.

The most promising method of using any of the active rare earths would seem to be to neglect the heavy daughter-nuclide and to estimate the helium liberated by one of the α-active nuclides. Nd^{144}, like Hf^{174}, has too long a half-life, but Sm^{147} produces helium at the rate of 120 cubic millimetres at S.T.P. per gram of samarium per thousand million years. This is just a thousand times slower than the rate of evolution of helium from uranium but is not too slow to be usable in favourable circumstances.

11.2 Carbon-14

Some objects can be dated by measurements on short-lived radio-active materials which, though they could not possibly have survived from the time of the earth's formation, are continually renewed at a known rate. One such material, Th^{230}, has already been mentioned. A more important example is C^{14} (5700 y, β^-). This is continually being produced in the upper atmosphere, by free neutrons liberated by cosmic ray interactions, according to the reaction

$$N^{14} (n, p) C^{14}.$$

The carbon produced can diffuse down through the atmosphere in the form of carbon dioxide and becomes thoroughly mixed with the rest of the atmospheric carbon dioxide in a time short compared to the half-life of C^{14}. Atmospheric carbon dioxide is the ultimate source of the carbon in all living organisms. In consequence, all of these contain closely the same percentage of C^{14}. The amount is about one atom in 7.8×10^{11} atoms of stable carbon, giving 15 β^--disintegrations per gram of carbon per minute.

As soon as an organism dies, it stops taking in fresh carbon laden with the standard proportion of C^{14}. The radioactive carbon already assimilated decays away according to the ordinary laws of radio-activity. Measurement of the specific activity of a sample of carbon from the organism will therefore enable us to determine how long ago the organism died. If the specific activity of this carbon is n disintegrations per minute per gram, the age T years of the sample will be given by

$$n = 15 \times 2^{-T/5700}$$

or $$T = 5700 \, [\ln (15/n)]/\ln 2$$

$\therefore$ $$T = 8200 \ln (15/n) = 18{,}900 \log_{10} (15/n) \text{ years.}$$

We cannot measure the C^{14} activity of an organic sample simply by placing it below the window of a Geiger counter. Even with fresh material the counting rate will be only about 0·1 per minute, which requires a counter with anti-coincidence screening and there is a serious risk of contamination with other radioactive materials such as potassium, which is a normal constituent of organic matter. It is usual therefore to burn a suitable sample of the material to carbon dioxide which can be used directly as the counting gas inside a large proportional counter. The background can be considerably reduced by using a simple broad-channel discriminator to reject all counter pulses above 0·156 MeV (the maximum of the C^{14} β-spectrum) as well as the "noise" pulses at low energies. Further improvement can be obtained by using a surrounding set of anti-coincidence counters or a cage-type proportional counter of the kind shown in Fig. 16. Libby, who was the pioneer of this kind of work, and other recent workers have found it possible to establish dates of carbonaceous material back to 50,000 years ago (corresponding to a specific activity of about 0·04 counts per minute per gram of carbon) and this is certainly not the limit.

For the method to be valid it is necessary to suppose that the specific activity of atmospheric carbon dioxide has been constant throughout the relevant period. This implies the assumption that the intensity of cosmic rays has also been constant—at least when aver-aged over a few centuries. Both suppositions can be checked by determining the specific activity of carbon from materials of known ages; the results turn out to be consistent with the assumptions. The method has proved of great value in dating various events in human prehistory such as the appearance of hunters in North America and the establishment of various Neolithic settlements in the Middle East and elsewhere. Several possibilities of error exist. For example, when a sample of a wooden beam is used to determine the age of the building of which the beam formed part, it is of course the age of the wood, not of the building, that is given. Most of the carbon contained in the heartwood of large trees may have been incorporated hundreds of years—in the giant redwoods, even thousands of years—before the tree was cut down. Furthermore, the beam may have had a long and useful life in an earlier building before being used in the building currently under investigation. It is evident that a *maximum* value for the age of the structure is what will be obtained and that the apparently objective nature of the method does not enable the archaeologist to avoid the need for proper judgment of the meaning of the results.

Another source of error arises from the fact that in some biological processes an appreciable degree of isotope separation may occur. Thus the carbonate in the shells of shellfish contains a systematically higher proportion of C^{13} and C^{14} than does the carbonate in the seawater from which it is extracted. Modern seashells, therefore, show some 6% higher specific activity than would modern plant tissues. If this were not appreciated, the results of a test might suggest to that mythical creature, the unprejudiced observer, that he had got hold of a shell which had been induced by some scientist yet unborn to travel backwards through time from some 500 years in the future. It is wise when investigating any new type of material to use as control a piece of biologically similar material of known date.

In the last hundred years we ourselves have upset the constancy of the specific activity of biological carbon by burning immense quantities of coal and oil. These, being built from fossil carbon many millions of years old, contain no surviving C^{14} at all. Hence we have diluted the natural activity of the atmosphere with inactive material. It seems that as a result the wood of modern trees has a specific activity about 1 to 2% *lower* than that of the wood of similar trees 100 years ago. It thus could be mistaken, on the evidence of carbon-dating alone, for wood 200 to 300 years old. Doubtless this fact will in due course be noted and applied by the small but flourishing part of the antique-trade which is engaged in the manufacture of Jacobean furniture.

A newer complication still is the production of considerable quantities of C^{14} as the result of nuclear-weapon tests in the atmosphere. The normal quantity of C^{14} in the biosphere (the region within which carbon dioxide is thoroughly mixed in a period short compared with 5700 years) is probably some 80 tons. Only perhaps 3% of this, or 2½ tons, is readily available in the atmosphere and in the upper layer of the oceans. Nuclear weapons vary by a factor of 2 to 3 in their neutron-yield per megaton of explosive power, the fusion bombs producing more neutrons than the fission bombs, but a 20-megaton bomb must yield something close to 200 kg of C^{14}.

All atmospheric testing up to the end of 1962 must therefore have produced about 3½ tons of C^{14}. This has increased the total by over 4% but is made more important by the fact that it at least doubles the amount quickly available in the atmosphere and the upper layers of the ocean. Much is still high up in the stratosphere, where its residence half-time seems to be about five years, but mixing should be nearly uniform after a couple of decades. Mixing into the larger reservoir of the soil and the ocean deeps is slower and seems to have a half-time of about thirty years.

Measurements of the specific activity of carbon in fresh vegetation show a rise of 2 to 5% per year between 1955 and 1957, before the main series of tests. This is naturally consistent with the figures given previously, since it forms the main datum from which they were derived.

On the basis of the measured and estimated behaviour of C^{14}, and assuming that no further tests take place, a forecast can be made of the level of C^{14} to be expected in living organisms, including ourselves, over the next thirty years. Such a forecast is shown in Fig. 50. The increase shown is unlikely to be wrong by more than about a third. If it is correct, the specific activity of carbon will rise to a little more than double the natural value in the years 1970-75 and will then fall slowly to within about 4% of the natural value over the next 250 years. It will take a further twelve thousand years to fall to within 1% of the natural value.

It will be seen from equation 11.5, above, that a 1% change in specific activity corresponds to a difference in age of 82 years. To obtain a standard deviation of 1% requires 10,000 counts even if the background is negligible. A one-litre proportional counter filled with carbon dioxide at atmospheric pressure contains just over half a gram of carbon giving for new carbon only 8 disintegrations a minute. Counting would need to be continued for almost 24 hours to get a 1% standard deviation. With a practical background—accurately known—of 20 a minute, counting would need to be continued for nearly three days. Samples two half-lives old would need over a month. The samples are usually wood in some form, often charcoal from fires. It is then impossible to be sure of its "carbon age" at the time it was burnt. Hence an error of a hundred years or so in the radioactive dating is not usually of great consequence. Nevertheless, the advantage of large counters to take bigger samples is clear.

Some Soviet laboratories have used a liquid scintillation counter instead of a gas counter. Some 3 to 12 gm of carbon from the specimen are used to synthesise ethyl benzene which is then dissolved in 20 to 70 ml of scintillator. The larger quantity gives 57 counts per minute for contemporary carbon, with a background of 28 c.p.m.

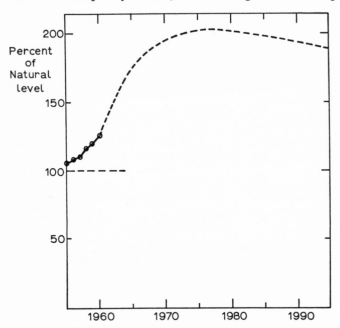

Fig. 50 Percentage variation of the concentration of C^{14} in living tissues observed (up to 1960) and estimated (after 1960) assuming that no further atmospheric tests of nuclear weapons are carried out after November 1962. (Data from Dr. C. E. Purdom, *New Scientist*, **15**, 255 (2.8.62).)

The standard deviation on a 48-hour measurement was then ± 35 years for a measured age of 5500 years.

11.3 Tritium

Traces of other radioactive materials are produced by cosmic rays in the upper atmosphere. The most important of these is tritium (12·3 y, β^-), which is a common product of the more violent nuclear disintegrations. Most of it doubtless decays before reaching the ground, but some is brought down with rain in the form of water. All recent rainwater contains detectable traces of tritium. The amount is exceedingly small, giving rise to about three disintegrations per hour per gram. Before measurement, therefore, the tritium must be concentrated at least a thousand-fold.

The electrolytic process used for separating deuterium can be used for this purpose. The separation factor T/H is about 14. This means that the proportional rate of loss of hydrogen during electrolysis is fourteen times as great as that of tritium. Then if the total hydrogen

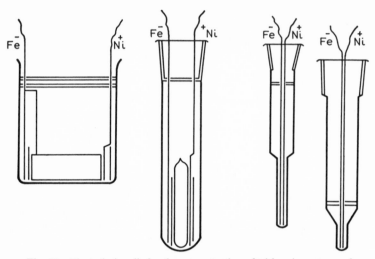

Fig. 51 Electrolysis cells for the concentration of tritium in water used by Grummitt and Brown. P-P, P.T.F.E. baffle-plates.

A. Capacity 1000 ml		Max current 25A.		Max electrolysis rate	7·5 ml/hr
B. ,,	140 ml	,,	10A	,,	3 ml/hr
C. ,,	10 ml	,,	0·86A	,,	0·26 ml/hr
D. ,,	1 ml	,,	0·18A	,,	0·054 ml/hr

is reduced N times in a single process of electrolysis, the tritium contained will be reduced by only $^{14}\sqrt{N}$ times. If N is 10,000 so that 1 l. is reduced to 0·1 ml, the tritium will be reduced by $^{14}\sqrt{10,000}$ or about a factor of two. Hence about half the tritium will still remain in the final 0·1 ml and an increase of concentration of about 5000 times will have been achieved. It is practically difficult to do this in one piece of apparatus as the cell and electrodes suitable for the original volume are quite unsuitable for the final one. A practical system is to use a series of cells in each of which the volume is reduced about ten times. In Fig. 51 are shown the cells used by Brown and

Grummitt to electrolyse 1 litre of water down to 50 mg. We quote their own description of the use of the system:

> "Each cell consists of a glass vessel suitably shaped to maintain a high liquid level around a nickel anode and iron cathode. Foaming and capillary action between the electrodes in the smaller cells permit electrolysis at a reasonable rate to a small volume. An initial electrolyte concentration of at least 1 % (w/w) potassium hydroxide was found necessary to prevent corrosion of the anode. This limits volume-reduction per stage to 10-20 fold, at which point the water is distilled into the next smallest cell, fresh potassium hydroxide added, and the electrolysis continued. Cells are cooled with running water to minimise evaporation losses and improve the separation factor, and current density is limited to 0·5 amp/in.² (75 mA/cm²) to prevent local overheating at the electrodes.

Following the suggestion of Kaufman and Libby we have used the natural deuterium in all water samples as a convenient tracer for tritium during the electrolytic concentration. It may be readily shown that

$$t/t_0 = (d/d_0)^{\alpha/\beta}$$

where d_0, d, t_0, t denote the initial and final number of moles of deuterium and tritium, and α and β the separation factors of deuterium and tritium, respectively, from protium. Thus, a measurement of the molar recovery of deuterium can be used to calculate the recovery of tritium, knowing the ratio α/β. While α and β are very dependent on experimental conditions they are likely to vary in a similar manner, hence the ratio α/β will be much less variable. In fact, Kaufman and Libby have shown α/β to be relatively constant over wide variations of α. Its value is close to 0·5 and it is sufficient for our purpose to replace the exponent α/β with a square root.

Deuterium analysis of the final concentrate requires only a few milligrams using an infra-red spectrometric method developed by Stevens and Thurston.

Reduction to Hydrogen

The remainder of the final concentrate is prepared for counting by conversion to hydrogen with zinc at 375°C. This reduction is rapid and complete, avoiding any isotopic fractionation and minimising cross-contamination between successive samples. The hydrogen is then passed through a liquid nitrogen trap and introduced into a Geiger counter with 6 cm argon and 5 cm ethylene."

It is essential to count the tritium, which gives a β-particle with an energy of only 18 keV, inside a Geiger—or proportional—counter. Larger concentrations of tritium can be converted to an aromatic compound and dissolved in a liquid scintillator, as has been done for C^{14}, but the background of this system is too high for use in the measurement of tritium in natural waters.

Rain and snow samples collected in Canada by Brown and Grummitt between 1951 and 1953 were found to contain from 17 to 42 tritium atoms per 10^{18} hydrogen atoms.

The H-bomb test in the Pacific in 1954 led to a rise of 10 to 100 times in Canadian samples taken later in the same year. The value then fell back towards normal with a "wash-out half-time" of 35 days. (The Bikini bomb was exploded near ground level; a high-altitude explosion would probably give smaller maximum levels but a longer "wash-out half-time".)

In samples of lake or seawater, the tritium content can be used as an indicator of the time that has elapsed since the water fell. This is of practical value in planning the proper use of underground water supplies. If the water can be shown to have arrived at the point of interest underground within a year or so of its deposition, there is little risk of permanent damage being done by drawing too much. On the other hand, if it has been underground for twenty years or so a similar period might be required for replenishment if it were seriously depleted.

Libby has confirmed the method by applying it to a 25-year-old vintage wine. It may be doubted whether such a use of vintage wine will help scientists to seem any more human to the general public.

11.4 Beryllium, Nickel and Manganese

Traces of the two beryllium isotopes Be^7 (53·6 d, e-capture) and Be^{10} ($2·5 \times 10^6$ y, β^-) are produced by cosmic rays from nitrogen and oxygen nuclei in the upper atmosphere. Little Be^7 will reach the ground before it decays and anyway much larger quantities are currently falling out from weapon tests. It is, however, possible that observable traces of Be^{10} might be extracted from deposits at various depths below the ocean-floor, making available a time-scale covering some millions of years.

Meteoric nickel and iron may carry observable quantities of Ni^{59} (8×10^4 y, e-capture) or Mn^{53} (2×10^6 y, e-capture) produced by cosmic rays outside the earth's atmosphere. Again, they may be recoverable from ocean deposits.

CHAPTER TWELVE

RADIOISOTOPES IN BIOLOGY

The field is so vast that only a few applications can be discussed. For a more complete list and a more detailed discussion the student must turn to one of the increasing number of volumes devoted entirely to biological applications of radioactive materials. A bibliography is given on page 331 and will be found to lead to extensive further references.

The examples below are chosen for the variety of principles and methods which they illustrate rather than for their biological importance.

12.1 Labelling of Whole Organisms

A simple but important technique is really an extension of the system used in Switzerland, and elsewhere, of locating cattle by tying bells round their necks. It is not practicable to keep track of a mosquito by tying a bell round its neck, but it is possible to label it with radio-active material. If a nuclide emitting a hard γ-ray is used, the labelled organism can be located even if it is inside a tree or a few centimetres underground.

For example, wireworms have been labelled by placing a small amount of metallic cobalt containing Co^{60} (5·27 y, β^-, γ) in their body cavities. Their movements underground could then be traced under natural conditions. It was found that when they were at a distance from food they followed a random search pattern, keeping within a limited range of depth, but when they came within a centimetre or so of a piece of potato their behaviour changed and they burrowed directly towards it, showing that they could detect it through an appreciable thickness of soil.

An alternative and often more convenient procedure is to label the food of the organism. In this case an element must be used which will be retained in the organism. Phosphate containing P^{32} (14·3 d, β^-) has been added to the water in which mosquito larvae swim so as to make the adults radioactive. It is then possible to investigate the distribution of the adults away from the breeding site, their mean expectation of life in the wild and so forth. It is important not to use so large a quantity of activity as to affect the growth or health of the insects. As little as 0·05 μCi/ml of P^{32} in water has been shown to affect the early stages of mosquito larvae and if there is more than 5 μCi/ml, hardly any reach the adult state at all. A very much higher concentration is, of course, reached in the body of the insect than exists in the water. Nevertheless, it is usually necessary for the radioactive adult insects to be caught or at least very closely approached before their activity can be detected.

The dispersion of bacteria from the human respiratory tract has been investigated by growing them first in a medium carrying a high specific activity of P^{32} in the form of phosphate.

12.2 Distribution of Substances within Organisms

After the investigations of movements of whole organisms we come naturally to movements of substances within organisms. For example, if phosphate labelled with P^{32} is supplied to the roots of a plant, the β-particles emitted can be detected as the phosphate reaches various parts of the plant and so the movement of the phosphate can be traced. An effective method for finding the distribution of freshly absorbed phosphate after any given time is to make an autoradiograph. The distribution of phosphate absorbed by a growing buttercup leaf is shown in Plate XVIII.

It may seem easy to make such a record at least roughly quantitative by measuring the degree of darkening of the film with a microphotometer. In a sense this is true, but the significance of the qualitative picture is easier to assess than is that of the numerical results obtained with a microphotometer.

In Plate XVI is shown an autoradiograph similar to that of Plate XVIII but of an older cut leaf of which the stalk was placed for a few hours in a solution of labelled phosphate. This shows the leaf-structure equally well but shows no concentration near the edge. In Plate XVII is shown the negative of a radiograph of the same leaf made with soft X-rays. There is no important difference between Plate XVII and Plate XVI. It is clear, therefore, that the main quantitative information obtainable from Plate XVI is simply the mass- and hence sap-distribution of the leaf, which can be obtained without the use of labelled phosphate and a microphotometer.

Even in the case of Plate XVIII the results are not so far very instructive. The fact that phosphorus derived from the roots is incorporated into the growing leaf has been well known for a considerable period.

The rate of movement of phosphates is perhaps not so easily measured by other means and can be found directly by following the rate of increase of activity in different parts of the plant after a sudden application of radioactive phosphate to the roots or by injection into the stem. The interpretation of such measurements is difficult, however. The "radioactive front" which travels through the plant may represent very few atoms and may travel at a rate well above the average, which is usually of greater interest.

A way in which this kind of measurement really comes into its own is in the demonstration of the active rather than passive nature of transfer of compounds in the plant tissues. By conventional methods it would be quite difficult to show whether phosphate was simply carried along by a general flow of sap or whether it was actively absorbed at one end of a cell and preferentially excreted from the other end. The known fact that minerals and phosphates move on the average from the roots to the leaves while carbohydrates such as glucose move from the leaves to the roots could represent counter flows in parallel vessels.

Injection of a mixture of two labelled compounds half way up the plant makes it possible to distinguish between the two mechanisms. For example, a single solution can be injected containing both P^{32} (14·3 d, 1·7 MeV β^-) as phosphate and C^{14} (5700 y, 0·156 MeV β^-) as glucose. Then on the passive counter-flow hypothesis, the same fraction of each label will be carried upwards, the value of this fraction depending only on the proportion of upward-flowing streams that happen to have been breached by the injection.

It has been shown in a number of cases that in fact most of the radioactive glucose flows downwards and most of the radioactive phosphate upwards. This demonstrates conclusively that transfer of both compounds must be due to specific actions of the living cells.

In many cases of this kind transference of activity from one part of an organism to another is not sufficient to prove that the labelled compound which was introduced is itself transferred. There is often a possibility or even a probability that the compound of interest is changed to something else in the organism and that it is a secondary metabolic product which travels. This possibility can be checked by re-examining the active material after its journey. The activity available may then be very much lower so that a sensitive method is necessary. An extract from the tissue containing the activity can be mixed with an inactive sample of the original compound together with inactive samples of the likely metabolic products. A chromatographic separation of the mixture can then be made, using an eluting agent which will separate the original compound from all other likely ones though it is not necessary to separate these from each other. An autoradiograph of the chromatogram will then show at once whether the activity is still associated with the original compound.

If the activities are very weak strip- or thread-chromatograms should be used.

The methods by which two mixed activities may be distinguished are worth a brief mention. Often one or both of the active materials will emit γ-rays of known energy which can be distinguished by a scintillation counter followed by a kick-sorter or narrow-band discriminator. Then different parts of the plant can be counted, at intervals after the injection, for each nuclide separately.

In the particular case considered, this method is impossible, as neither C^{14} nor P^{32} emit γ-rays at all. Instead we can use either the difference of energy or the difference of half-life.

To use the difference of energy conveniently, we should inject a solution containing perhaps 100 times as much C^{14} activity as P^{32} activity. Then on counting a thin layer of tissue with a thin window counter, we shall observe little besides the C^{14}. With 200 μ of aluminium absorber between the tissue and the counter we can block all of the easily absorbed β-particles from C^{14} while losing only 50% of the more energetic particles from P^{32}. In this way we can count only the P^{32}. A much thicker layer of tissue can now be used if desired to increase the counting rate.

To use the difference of half-life, the relative concentrations are less important, though it may be better to have the initial P^{32}-activity a few times the larger. This time it is necessary to dry the tissue samples and mount them properly so that they can be counted repeatedly. Then a decay curve must be plotted for each sample. From this we can easily find the part of the activity due to C^{14} after most or all of the P^{32} has decayed and then by subtraction the activity due to P^{32}.

Discrimination by half-life is more accurate than by energy, the latter being affected a good deal by small variations of thickness of the specimens used, but it is a good deal more laborious.

If a considerable excess of the shorter-lived activity is used the half-life difference may make possible autoradiographs of the same sample showing the two separate distributions. The total activities must be such that a good autoradiograph can be obtained of the shorter-lived material before any appreciable registration on the film of the longer-lived one has taken place. After the decay of the shorter-lived material the specimen is again exposed to the film for the much longer period required for a good autoradiograph of the distribution of the longer-lived material.

An important point to notice is that the resolution obtainable is much better with a nuclide such as C^{14} which gives β-particles of low energy. This is due to the effect of the range of the β-particles. The mean range of the particles from C^{14} in nuclear emulsion is in the region of 50 μ while that of the particles from P^{32} is nearer 500 μ. A point source directly on the emulsion will give circular images of finite size depending on β-energy and falling off rapidly in intensity from the centre outwards. If the point source is embedded in plant or animal tissue a little away from the emulsion surface the spreading effect is much increased, partly because the range of the β-particle is two or three times as great in the tissue as it is in emulsion and partly because the intensity of the image-disc will build up towards the centre far less rapidly, for simple geometrical reasons. In Plate XIX are shown a series of autoradiographs produced by the same very small P^{32} source held first directly against the emulsion and then separated from it by different thicknesses of Polythene. The exposures were varied to give about the same intensity in the middle of the image. The advantage in resolution of having the source close to the emulsion is obvious.

When we have a thick section of tissue to examine, and wish to find by autoradiography the distribution of radioactive substances, some of the activity will be in the interior but some will be close to the surface. If there are sharply localised concentrations in some regions a surprising degree of resolution of detail can be obtained if one of these regions happens to be close to the surface and therefore gives a sharp image (cf. Plate XIX).

For really good results, however, it is necessary to have tissue sections thinner than the dimensions of the structures that it is hoped to resolve. The sections should be held against the emulsion by a material of low atomic number—polyethylene is very suitable—so as to reduce the back-scattering of β-particles. When using X-ray film it is of especial importance to use a single-coated variety such as Flurodak rather than one coated on both sides, as many such films are. When a double-coated film is used, the difference in sharpness of the images on the two sides is most striking. Unfortunately a thin emulsion will necessarily need longer exposure—simply because much of the energy of the β-particles is not being used. A choice has, therefore, to be made in each particular case as to how much sensitivity should be sacrificed for resolution and vice versa. The special autoradiographic emulsions which are commercially available are made in different thicknesses.*

When an investigation is started in which high-resolution auto-

* For example, Kodak A.R.50 (50 μ thick) and A.R.10 (10 μ thick). These are strippable emulsions, i.e. they can be floated off their glass support in water and deposited directly on the surface to be examined, to which they adhere firmly, thus eliminating any gaps between emulsion and specimen. Detailed directions are given by manufacturers.

radiographs will be required it is useful to begin by considering what nuclide to use for labelling. For example, for a biological investigation of glucose distribution, a compound labelled with tritium would be much better than one labelled with C^{14}. The mean range in emulsion of the β-particles from tritium is only about a couple of microns and no γ-rays are produced. There is no advantage in making thin sections of the specimen since only particles from the top micron or so will be effective in producing an image anyway. Evidently there is also no need to bother about the thickness of the emulsion though a very fine grain will still be desirable.

Even when it is essential that the labelled nuclide should be an isotope of a particular element there may be alternatives available; for example for phosphorus we could in principle use P^{33} (24·4 d, 0·25 MeV β^-) instead of the usual P^{32} (14·3 d, 1·7 MeV β^-). Unfortunately P^{33} is a good deal more expensive as it must be made either in a pile from the rare (0·75 %) isotope S^{33} or in an accelerator using tritium as bombarding particle.

Even when a tracer nuclide with high β-energy is being used, high resolution can be obtained although at the cost of a large increase in effort. The specimen can be directly coated with an emulsion such as Ilford G5, in which the tracks of individual β-particles can be seen, the method of coating being described in Appendix 4. It is then left long enough for a suitable density of tracks to build up. The time required may be estimated by counting the particles from the specimen beforehand, close to the window of an end-window counter. Since the "background" of tracks in the emulsion is likely to be of the order of 50 per day per mm^2, the specimen should give at least 500 tracks per mm^2 and preferably 5000 to avoid troublesome corrections. If we assume 24 hours' exposure of the specimen to the plate, even 5000 counts per mm^2 correspond to only 300 counts per minute per square centimetre, which is easily produced. While a day's exposure would probably be suitable for a trial run for a specimen giving 300 c.p.m./cm^2, it is not possible to give an exact recommendation until the degree of variation of the activity over the specimen is known.

The film is developed and fixed still in contact with the specimen which is preferably on a transparent base. The whole can then be examined together under a high-power microscope. Tracks of individual electrons can be traced through the emulsion to their point of entry. This can be located to about half a micron and related to the underlying structure of the specimen itself. It is easier to identify and follow the tracks if the emulsion is not too thin; 50 μ is a suitable thickness.

Since the origins of the tracks can be found to half a micron, detail of this order of size can be detected so long as the thickness of the specimen is comparably small.

In some cases the results of the experiment can be obtained simply by focussing the microscope on the boundary between specimen and emulsion. Then the top of the specimen and the starting points of the β-tracks will both be in focus, and concentrations of tracks over particular regions can be seen. Even when the specimen is itself opaque or is mounted on an opaque backing, this method may be used if a microscope with epi-illumination (by reflection instead of transmission) is available. The resolution, however, is usually inferior in this type of lighting.

Where quantitative information is needed, the number of tracks per unit area from different regions can be counted, but this is a laborious process which should be undertaken only if someone with considerable patience and good mental stability is available to do the counting.

When thin sections of the active tissue can be cut, these can be deposited on a nuclear plate instead of depositing emulsion on the specimen. To do this the tissue may be freeze-dried and embedded in paraffin wax before sectioning. The paraffin ribbon leaving the microtome should then be spread with anhydrous glycerine instead of water to avoid loss of water-soluble radioactive compounds. When the glycerine-spread sections are placed on G5 plates a close and firm contact is achieved. If the wax can be removed by a solvent without damage to the section, development and fixation can usually be carried out with the section in place, so that again both section and autoradiograph can be examined together.

The ultimate limit of this technique was probably reached in an investigation of the distribution of cobalt in tumour-cell nuclei by the late Mrs. Liquier-Milward. She used Co^{60} (5·27 y, 0·31 MeV β^-) as tracer nuclide. Her procedure was as follows. Microscope slides selected for a low degree of radioactive contamination were covered with a thin layer of collodion or formvar. Cell-nuclei from tumour-tissue were labelled *in vitro* with Co^{60}. They were then placed on the collodion and covered with a layer of Ilford G5 nuclear emulsion only a micron or so thick. This was obtained by melting the gel at 55°C as described in Appendix IV and diluting with water to a suitable consistency before pouring it over the specimen. The slides were then dried and exposed in the dark, shielded as far as possible from other sources of radiation. The exposures necessary are several times as long as with a thicker G5 emulsion, since a small proportion only of the β-particles will be emitted in the plane of the emulsion.

Slides were developed normally but were fixed in two minutes. After washing, the double layer of collodion and gelatine was lifted from the glass slide without difficulty and a small area mounted on a grid of an electron-microscope. Electron-micrographs could then be taken showing both the cell structures and the silver grains of β-tracks. Fig. 52 shows a tracing from a typical result. By means such

Fig. 52 Tracing of electron-micrograph of a G5 nuclear emulsion showing β-particle tracks emanating from tumour nuclei labelled *in vitro* with Co^{60}.
(Original photograph by J. Liquier-Milward, *Cancer Research*, **17**, 844 (1957).)

as this the active part of a specimen may be located to a fraction of a micron. The technique is an exacting one which is not always practicable—the labelled compound, for example, must not be soluble in water or it will dissolve out into the emulsion when this is poured on. The section must also be able to stand the temperature of 55°C during the application of emulsion, which rules out some types of wax-embedded sections. The method is worth mentioning, however, as it will give information that can hardly be obtained in any other way.

12.3 Exchange of Elements and Compounds between Tissues

Apart from the demonstration of the direction and rate of movement of specific compounds, radioactive isotopes can be used to determine the rate of exchange of ions or compounds of the same kind. This can also be looked at as a discrimination between a static and a dynamic equilibrium. For example, it has been known for a long time that the concentration of potassium inside the living cells of animal tissues is normally greater than that in the plasma bathing the cells. The concentration in each region is practically constant. This would be consistent with the hypothesis that the extra potassium was taken into the cells during their initial development and then kept there by a cell wall impermeable to potassium ions. Alternatively the cell wall may be permeable but the permeability may be direction-dependent. We shall describe an experimental examination of these hypotheses for the potassium in human red blood cells or erythrocytes. The concentration of potassium in the cell fluid is nearly thirty times that in the plasma.

The procedure was as follows. K^{42} (12·5 h, β^-) as KCl was added to a known volume of blood and incubated for 24 hours.

Blood samples were taken at roughly hourly intervals after injection, the corpuscles separated from the plasma by centrifugation, and the activity of each determined separately. An hour after mixing, nearly 30% of the active potassium had left the plasma for the erythrocytes and after 21 hours' equilibrium had nearly been reached, 93·5% of the activity then being in the erythrocytes. The specific activity of the potassium in the cells was then within 20% of that of the plasma. This showed that in fact the difference of potassium concentration must be maintained by a continuous process involving a difference of permeability of the cell wall to potassium ions depending on the direction of their passage; a potassium ion striking the outside must have several times the chance of transmission possessed by one striking the inside.

Similar observations have been made on several other ions. The concentration of calcium in blood plasma, like the concentration of potassium, is constant. If Ca^{47} (4·9 d, β^-) is injected into the plasma, the initial fall of activity over successive samples shows that active calcium is leaving the plasma at a rate of 75% per minute. Since the total calcium content is constant, this implies that calcium is re-entering the plasma from other tissues, presumably the skeleton, at the same rate. With the help of Na^{24} (15·0 h, β^-) and H^3 (12·3 y, β^-)

it has similarly been shown that sodium ions and water in the plasma are exchanging with extravascular sodium and water at rates of 78% per minute and 105% per minute respectively.

Such experiments must be performed and interpreted with care as several different processes may be involved and may be taking place at different rates. Two points need particular attention. First, if the radioactive material is injected quickly into a vein, activity may be observable in most parts of the body in two or three seconds, but it may not be uniformly mixed with the whole blood-volume for considerably longer, especially in a sedentary subject or one in which cold is limiting the circulation rate in peripheral regions. On the other hand, if the experimenter waits too long after the injection, a serious amount of activity may leave the blood for other tissues before the first observation is made.

Second, there may well be a rapid exchange between, for example, blood and lymph, which reaches a quasi-equilibrium in a matter of minutes, followed by a slower exchange with bones or other organs.

These experiments will yield further information besides the rate of exchange. The proportion of activity which has been lost by the plasma when equilibrium is reached will give the relative amount of exchangeable material in the tissue which is taking up the activity. If the rate of loss of activity is measured by a series of observations, it may be possible to deduce this relation without waiting for equilibrium to be reached. This is often important when a short-lived nuclide is being used. It is worth while to discuss the process quantitatively.

Suppose that we wish to study the exchange of a particular ion or compound, X, between two tissues, A and B. Let the volumes of the two tissues be V_A and V_B litres and the normal concentration of X be $[X_A]$ and $[X_B]$ moles/litre respectively. Suppose that a labelled sample of the substance X, containing X with a known total activity of n_{A0} curies and of negligible mass (compared with that of the total X present) is added to tissue A. To avoid unnecessary complexity we shall assume that X is long-lived so that no change in total activity occurs during the course of the experiment. When this is not the case, allowance for decay can be made in the usual way by referring the activity of each sample observed back to some standard time.

The rate of mixing in A is not suitable for mathematical analysis as the details of the mixing process are rarely even roughly understood. It may well be necessary to carry out several subsidiary experiments to find out how best to achieve thorough mixing and to find when it has been achieved. We shall assume that thorough mixing occurs in A before appreciable exchange of activity with B begins.

I*

At the beginning of the experiment, at time $t=0$, then we have n_{A0} curies in A and no activity in B.

The total quantities of inactive substance in the two tissues are evidently $[X_A]V_A$ moles in A and $[X_B]V_B$ moles in B and the initial specific activities in the two tissues are therefore

$$S_{A0} = n_{A0}/[X_A]V_A \text{ curies per mole in } A \qquad 12.1$$

and S_{B0} = zero in B. Suppose now that the rate of transfer of total X from A to B is w moles per minute. Since the system is in equilibrium this must be identical with the rate of transfer of X from B to A.

At time t let the total activity in the two tissues be n_A and n_B curies respectively. Then

$$n_A + n_B = n_{A0} \qquad 12.2$$

(since we are neglecting the radioactive decay of X). The specific activities in the two tissues are then

$$S_A = n_A/[X_A]V_A \text{ and } S_B = n_B/[X_B]V_B \qquad 12.3$$

curies/mole respectively.

Then the rate of transfer of activity from A to B at time t is $n_A . w/[X_A]V_A$ curies per minute and from B to A is $n_B . w/[X_B]V_B$ curies per minute.

Then the *net* rate of transfer to A is

$$dn_A/dt = -n_A w/[X_A]V_A + n_B w/[X_B]V_B$$

$$= w\{-n_A/[X_A]V_A + (n_{A0}-n_A)/[X_B]V_B\} \qquad \text{(from eq. 12.2)}$$

$$\therefore \qquad dn_A/dt + n_A w(1/[X_A]V_A + 1/[X_B]V_B) = n_{A0}w/[X_B]V_B \qquad 12.4$$

This is a linear differential equation in n_A and t of which the solution is

$$n_A = \frac{n_{A0}}{[X_A]V_A + [X_B]V_B} \{[X_A]V_A + [X_B]V_B \exp [-w(1/[X_A]V_A$$
$$+ 1/[X_B]V_B)t]\} \qquad 12.5$$

(see for example Milne—Thomson and Comrie *Standard Four-Figure Mathematical Tables*, p. 240, 1931 edition).

When $t=0$ this, of course, gives $n_A = n_{A0}$ and when $t=\infty$ it gives

$$n_{A\infty}/n_{A0} = \frac{[X_A]V_A}{[X_A]V_A + [X_B]V_B} \qquad 12.6$$

i.e. the final activity in tissue A is the same proportion of the total activity as the amount of X in A is of the total amount of X. The specific activity is then

$$S_{A\infty} = n_{A\infty}/[X_A]V_A = n_{A0}/([X_A]V_A + [X_B]V_B).$$ 12.7

From equation 12.5 and equation 12.2 or by a similar piece of integration,

$$n_B = \frac{n_{A0}[X_B]V_B}{[X_A]V_A + [X_B]V_B} \{1 - \exp[-w(1/[X_A]V_A + 1/[X_B]V_B)t]\}$$ 12.8

This correctly gives $n_B = 0$ when $t = 0$. When $t = \infty$ it gives

$$n_{B\infty}/n_{A0} = \frac{[X_B]V_B}{[X_A]V_A + [X_B]V_B}.$$ 12.9

The specific activity is then

$$S_{B\infty} = n_{B\infty}/[X_B]V_B = n_{A0}/([X_A]V_A + [X_B]V_B) \text{ curies/mole.}$$ 12.10

This is exactly the same as the value for $S_{A\infty}$ as given by equation 12.6 as it should be when equilibrium is reached. Then the w moles of X which are being transferred in each direction per minute carry just the same amount of X^* in each direction per minute.

The expressions that we have derived are complex and it may be worth while to discuss their applications to actual measurements. These will usually consist of the determination of the activities of a series of equal samples of one tissue, say A, taken at zero-time and at suitable intervals afterwards. The samples must be too small to alter appreciably the total volume of A or some difficult corrections will be necessary. The quantities of interest are the volumes of the two tissues, the concentrations of X in the two tissues and the rate of exchange of X between them. It is clear from the form of the expressions that, by measurements of radioactivity alone, we cannot hope to find both the concentration $[X_A]$ and the volume V_A or both $[X_B]$ and V_B, although we can find the ratio of the total quantities of X in the two tissues directly from the equilibrium-expressions 12.6 and 12.9. Then if we know the two volumes, we can find the ratio of the concentrations or if the concentrations of X in the two tissues can be found by ordinary chemical analysis the comparative volumes of the two tissues can be calculated.

If we know the concentrations in and the volumes of the samples taken for measurement, the absolute value of the specific activity S_A at each time can be found, but this is usually not necessary as we are generally concerned only with the *proportional* change in specific activity with time and the ratios of the specific activities of two equal samples of A will simply be equal to the ratio of their total activities.

Typically then our data are $[X_A]$, $[X_B]$, n_{A0} and a series of relative values of S_A at different times.* Knowing $[X_A]$, V_A can be obtained from equation 12.1 if the first sample can be taken before appreciable activity has passed into B.

Equation 12.6 can be rearranged to give

$$[X_B]V_B = [X_A]V_A(n_{A0}/n_{A\infty} - 1) \qquad 12.11$$

Remembering that $n_{A0}/n_{A\infty}$ is simply the ratio of the activity of the first sample of A to that of a sample after equilibrium has been reached, this gives us $[X_B]$ V_B. If $[X_B]$ is known, V_B is then found, though often $[X_B]V_B$, the total mass of X in B, is itself the quantity desired.

We then have all the data required to obtain w, the rate of exchange between the two tissues, from equation 12.5.

For this purpose equation 12.5 can be simplified by the use of equation 12.6 and its modified form equation 12.11 to eliminate $[X_B]V_B$.

Equation 12.5 then becomes

$$(n_A - n_{A\infty}) = (n_{A0} - n_{A\infty}) \exp \{[-wt/[X_A]V_A][n_{A0}/(n_{A0} - n_{A\infty})]\}$$

$$\text{or} \quad (wt/[X_A]V_A)[n_{A0}/(n_{A0} - n_{A\infty})] = \ln (n_A - n_{A\infty}) - \text{const}$$

$$= 2 \cdot 303 \log_{10} (n_A - n_{A\infty}) - \text{const.}$$

$$12.12$$

The quantities n_{A0}, n_A and $n_{A\infty}$ can be replaced by the measured activities of equal samples, taken at time zero, time t and when the final state has been reached respectively. Then by plotting log $(n_A - n_{A\infty})$ against t we shall get a straight line, from the slope of which we can find $w/[X_A]V_A$ since n_{A0} and $n_{A\infty}$ are known. As in the investigation of radioactive decay, much time can be saved by the use of semilogarithmic paper. Then we can plot the sample activity in tissue A minus the equilibrium sample activity directly against time to give a straight line. The easiest way to determine w from this is to find first the "half-time" $t_{\frac{1}{2}}$ in which $(n_A - n_{A\infty})$ drops by a factor of two just as one finds the half-life of a radioactive material from a similar plot.†

* The same results can be obtained by taking samples from B. Which is done is simply a matter of experimental convenience. It is better to introduce the radioactivity into the tissue in which it will mix most rapidly and this is also in general the best for sampling although when there is a large difference in the quantities of X in the two tissues there is an advantage in introducing the activity into the tissue with the smaller amount of X as this will show a larger change of specific activity.

† Although it is obviously desirable to find the final activity $n_{A\infty}$ by continuing the experiment for a sufficient time, this may be impossible either because the half-life of the nuclide used is too short or because the time of the experiment is limited for biological reasons. Even in this case a value for $t_{\frac{1}{2}}$ can be found by the trial-and-error method. A series of arbitrary values of $n_{A\infty}/n_{A0}$ are tried and the one which gives the best straight-line plot is adopted. This sounds crude but usually gives an acceptable value of $t_{\frac{1}{2}}$ if the experiment has continued for $2t_{\frac{1}{2}}$ or longer.

Then from equation 12.12,

$$w = [(\ln 2)/t_{\frac{1}{2}}][(n_{A0} - n_{A\infty})/n_{A0}][X_A]V_A$$

or

$$w = [0 \cdot 693 \ [X_A]V_A/t_{\frac{1}{2}}][(n_{A0} - n_{A\infty})/n_{A0}] \qquad 12.13$$

The units of w will be fixed by those of $[X_A]V_A$ and $t_{\frac{1}{2}}$. If these are moles and minutes respectively, as we have assumed during the analysis, w will be given in moles per minute. More frequently, $[X_A]V_A$, which it will be remembered is the total quantity of X in tissue A, will be given in grams; in this case equation 12.13 will give w in grams per minute.

It is often useful to quote the rate of transfer in terms of a percentage of the total material in one of the tissues. Since the steady total quantities of X in A and B are different, these percentages will also be different.

Thus the transfer of X from A to B per unit time is

$$100w/[X_A]V_A = \frac{69 \cdot 3 \ (n_{A0} - n_{A\infty})}{t_{\frac{1}{2}} n_{A0}} \ \% \text{ of } X \text{ in } A \text{ per unit time} \qquad 12.14$$

while the transfer from B to A is

$$100w/[X_B]V_B = \frac{69 \cdot 3 \ n_{A\infty}}{t_{\frac{1}{2}} n_{A0}} \ \% \text{ of } X \text{ in } B \text{ per unit time}$$

where as before the unit of time is the same as the units used for $t_{\frac{1}{2}}$.

A quantity which is often quoted in biological work is the *turnover time*. This is the average time it takes to exchange any particular atom, which is sometimes described as the time for one complete renewal of the substance of interest in the tissue concerned.* In the example considered, for tissue A it will be $[X_A]V_A/w$ and for tissue B it will be $[X_B]V_B/w$. Then from equation 12.14 we have

$$\text{turnover time of } X \text{ in } A = n_{A0}t_{\frac{1}{2}}/0 \cdot 693 \ (n_{A0} - n_{A\infty})$$

$$\text{and turnover time of } X \text{ in } B = n_{A0}t_{\frac{1}{2}}/0 \cdot 693 \ n_{A\infty} \qquad 12.15$$

If the total weights of X in tissue A, tissue B and both tissues together are W_A, W_B and $W = W_A + W_B$, we may change equation 12.15 by the use of equation 12.11 to the alternative form

$$\text{turnover time of } X \text{ in } A = Wt_{\frac{1}{2}}/0 \cdot 693 \ W_B$$

$$\text{and turnover time of } X \text{ in } B = Wt_{\frac{1}{2}}/0 \cdot 693 \ W_A \qquad 12.16$$

Many of the expressions above could be simplified in appearance by thus replacing $[X_A]V_A$ and $[X_B]V_B$ by W_A and W_B respectively

---

* It should be noted that not every atom originally present will have exchanged in this time; on the other hand many will have crossed the boundary between the tissues more than once.

and these weights are often the interesting quantities. For example, when we consider the transfer of calcium between blood plasma and bone, it is of little interest to discuss the concentration of calcium in bone but of great interest to know what is the total quantity with which the plasma-calcium is freely exchanging. Measurements of the activity remaining in the plasma, when the final steady state is reached, show that the quantity of freely exchanging calcium in human bones is some hundred times the total in the plasma, or between a tenth and a fifth of the whole of the calcium in the skeleton. Further investigation would be required to find out whether the remaining fraction is in a different, more stable, chemical form or whether it simply represents the interior parts of water-tight crystals, which are not in contact with the plasma.

In other cases the total quantity of material is of little interest but the relative concentration is important. The difference of concentration between the potassium in the fluid inside individual cells and that in the fluid bathing them, which has already been cited, is a good example of this. It is best, therefore, to have the complete expressions available.

It is worth while to give a warning to those who wish to refer to the biological texts on this subject. We have been careful to distinguish between the *concentrations* $[X_A]$ and $[X_B]$ measured in moles *per litre* or grams *per litre* and the *total quantities* $[X_A]V_A$ and $[X_B]$ V_B measured in moles or grams. In some biological texts, even in so good a one as Comar's *Radioisotopes in Biology and Agriculture*, this distinction is not always made and the word "concentration" may be used to denote either kind of quantity indiscriminately. This does not lead to erroneous results so long as the word is used consistently in any one discussion but is clearly undesirable and can lead to serious confusion if it is not recognised.

When more than two tissues are involved, and the rates of transfer are all comparable, the quantitative analysis becomes very complex. The methods used are similar to those used in the analysis of the decay of radioactive series but since the analysis of even a three-component system requires a lot of paper and since no new principles are illustrated, it will not be included.

We have now considered the use of radioactive isotopes to measure (*a*) the rate at which a particular substance is *transferred* from one tissue to another and (*b*) the rate at which a substance is *exchanged* between one tissue and another. It will have been noticed that the observations made are qualitatively the same in each case. Radioactive material is lost from one tissue and appears in the other.

12.4 Discrimination between Exchange and Transfer

It is obviously of great importance that we should be able to tell which of the two processes is occurring, or whether both are taking place together.

Often the system concerned is well enough understood to know beforehand that only one of the processes is possible. For example, in the case of transfer of active potassium between the plasma and red cells of the blood, it is known that no net transfer can be taking place. Red blood cells survive for some weeks without serious change in their potassium content. The rapid transfer of K^{42} from the plasma to the cells can therefore be only by exchange with potassium already present.

The transfer of radioactive calcium from blood to bone must likewise be mainly an exchange. The blood of a normal individual contains some 300 mg of calcium. Radioactive calcium added to the blood is transferred at a rate of 70% per minute. If this showed a *net* transfer of calcium, it would represent a skeletal build-up of calcium of 200 mg per minute or over half a kilogram of calcium phosphate per day. Skeletons do not grow at this rate, especially in adults.

On the other hand, the transfer of radioactive carbon dioxide from the air to a growing plant in bright sunlight can equally simply be shown not to be the result, to any appreciable extent, of exchange. Very little activity is lost by the plant when transferred to air containing only ordinary carbon dioxide and little of the active carbon in the plant is in the chemical form of carbonate which might be capable of such exchange.

Difficulties may arise in complex systems or where only trace-quantities of the interesting substances are present. No useful general statement can be made about complex systems, each of which must be separately considered, but the second kind of difficulty is worth some discussion. The problem that arises when the quantities of material and the rates of take-up are small is that it is difficult to tell whether or not there is any net transfer of material. If the *whole* of the radioactive material in tissue A is transferred to a tissue B, clearly we have a transfer rather than an exchange. The fact that a saturation value is reached does not however prove the reverse. Tissue B may have a limited capacity to assimilate the substance of interest—at least during the length of the experiment. If suitable long-lived isotopes are not available we cannot simply lengthen the experiment. Discrimination is usually still possible, however.

As an example we will consider the uptake of fluorides by dental enamel. The enamel of normal sound human teeth consists mainly of the mineral hydroxy-apatite. This is a particular crystalline form of calcium phosphate, $3 Ca_3(PO_4)_2 . Ca(OH)_2$. A trace of fluoride is present, rising in the outermost layers to perhaps one part per thousand or more, probably as fluorapatite, $3Ca_3(PO_4)_2 . CaF_2$. When a clean fresh tooth is immersed in a dilute fluoride solution labelled with F^{18} (110 mins, β^+) it rapidly becomes radioactive. The total quantity of stable fluoride involved is too small to measure. Activity will be taken up from solutions containing one part per million or less of total fluoride. A single measurement gives no indi-

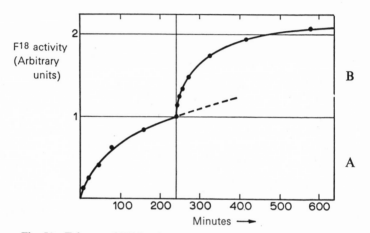

Fig. 53 Take-up of F^{18} by dental enamel from a solution containing 1 p.p.m. of fluoride. At 240 minutes the solution was relabelled with fresh F^{18}. In both parts of the curve the decay of the F^{18} has been allowed for and the activity of the solution normalised to the same strength at time zero.

cation as to whether fresh fluoride is being taken up permanently by the enamel, perhaps by exchanging with hydroxyl groups in the hydroxyapatite, or whether fluoride in the solution is exchanging with that already present in the surface layer of the enamel.

The two processes can be experimentally distinguished by making a succession of measurements of the activity taken up from the labelled fluoride solution by an enamel surface and recording them as a function of time. After allowing for the radioactive decay of the 110-minute F^{18}, a curve of absorption of the kind shown in Fig. 53A is obtained. Take-up clearly approaches a nearly constant level after a few hours.

When the activity of the original label has almost died away the

1 p.p.m. fluoride solution is relabelled with fresh F^{18}. The strength of the second labelling is "normalised" to that of the first by an adjusting factor which reduces the activity of the second solution to the same specific activity as the first at the same standard time. On plotting the results, a curve of the type B of Fig. 53 is obtained.

Now, apart from experimental errors, if the fluoride were being irreversibly absorbed by the enamel, the second part B of the curve should simply be a smooth continuation of the first part A, as indicated by the dashed curve. It is quite clearly nothing of the sort; the take-up of radioactivity has started again at approximately the original rate. In other words, the apparent take-up of fluoride to a near-saturation value as shown by curve A has in no way reduced the capacity of the enamel to take up fresh radioactive fluoride. This means that active fluoride from the solution must be exchanging with fluoride already in the surface. The method cannot of course prove in this or any other case that *all* of the radioactivity taken up corresponds to exchange and not to irreversible addition of new material. Its limit of capacity is to show that the permanent take-up rate is less than the experimental uncertainty in the exchange-rate.

Similarly, if the second part B of the curve *were* a continuation of the first, it would show only that the exchange-rate was less than the error in the permanent take-up rate.

Good evidence for a moderate permanent take-up rate occurring *together* with a large exchange-rate can sometimes be obtained with the help of a second longer-lived isotope of the same element. This is not possible in the case described since no radioactive isotope of fluorine with a half-life greater than that of F^{18} is known. Chemical analysis at appropriate long intervals can usually show easily enough whether permanent take-up of the element has occurred but has not the ability of the radioactive method to trace out the route by which the material is absorbed.

12.5 Determination of Metabolic Pathways

This brings us to the next and possibly the most important application of radioactive isotopes in biology. Many important biological reactions take place through a long and complicated series of separate steps. For example, one of the simpler compounds produced by a growing plant is glucose, $C_6H_{12}O_6$. The raw materials required by the plant are carbon dioxide and water. The overall result of the plant's activity can be represented by the equation

$$6CO_2 + 6H_2O \rightarrow C_6H_{12}O_6 + 6O_2.$$

The actual process, however, involves dozens of intermediate compounds, some of them highly complex. The life of intermediate compounds in the process is often very short and so they may be present only in very small quantities. Sometimes they are chemically very unstable. It is of basic importance for the understanding of living processes that we should be able to follow such trains of reaction in detail.

Several different methods are available. Suppose that we want to find the steps by which a compound A is transformed by a plant or animal, through a whole alphabet of intermediate compounds, into a new compound Z. The first step is to supply compound A in a radioactive labelled form, A*, to the organism of interest. Analyses are then made at intervals to find as many as possible of the radioactive compounds that have been produced. This can best be done by means of paper chromatography, which makes possible the simultaneous identification of a considerable number of compounds. An autoradiograph of the paper will show which compounds are radioactive. These analyses are repeated until appreciable quantities of radioactive compound Z* have appeared. One can be certain that all the intermediate compounds B, C, D, . . . etc. must have been made in radioactive form, but there is a risk that some of them persist for so short a time that there is too little to observe at any one instant. Furthermore, some may be so unstable as to be destroyed by the analysis. The chain found may therefore be incomplete, and it is then necessary to infer the existence of an unobserved compound, for example (Q*) intermediate between the observed compounds P* and R*. Careful work will usually leave few obvious gaps.

More important, there will usually be a number of compounds, Σ^*, $\mathcal{H}^*$, etc. which are also derived from A* but play no part in the series from A to Z. The first qualitative investigation gives no certain way of distinguishing these, though a hint may be obtainable from the rate at which each compound builds up after the beginning of the supply of A*.

Further information can be obtained by comparing the results of maintaining a steady supply of A* to the organism with the results of supplying A* for a period short compared to the time over which analyses are to be made. In the latter case, each intermediate between A and Z will show a build-up of radioactivity followed by a fall and if Z is a stable end-product the activity of the intermediate products will disappear as the total activity of Z reaches its saturation value. Many of the irrelevant compounds will show a quite different pattern of variation of activity with time, some persisting or even continuing to increase after Z has reached saturation. If Z itself is metabolised to further compounds, by a similar argument, none of the activity of B, C, D, . . . etc. should survive the disappearance of activity from Z.

This procedure will show up some of the irrelevant compounds but not all. Some of them may be intermediate in another chain and thus may persist no longer than the intermediates we want. Others, which can be found when labelled A* is continuously supplied, may be present in quantities too small for detection when A* is provided only for a brief period.

Confirmation of the compounds actually involved in the chain A-Z is more laborious. It consists of a series of experiments on separate organisms or groups of organisms which are supplied with suspected intermediate compounds in labelled form. Thus if we introduce D* into an organism we shall in due course find that a sample of Z extracted from the organism is radioactive. The chromatograms from the D* experiments will also show the intermediate substances E*, F*, . . . and hence by a series of such experiments the order of the intermediate substances can be worked out as well as their identify. (Many biological processes are reversible, so traces of C* or even B* may appear following the introduction of D*. Their proportion to the rest of the active compounds will, however, be markedly different from that in the original experiment using A*.)

It is not always easy to get the desired intermediate products in labelled form. It will very frequently be necessary for the investigator to make his own. Fortunately, the first experiment which has already been described will automatically provide labelled samples of all of them except for the unstable or short-lived. The appropriate pieces of the chromatogram can be cut out and the material re-extracted. In practice the procedure of the first experiment may need to be repeated several times to get adequate quantities of material in this way.

An alternative method is to supply extra quantities of a particular suspected intermediate in *unlabelled* form while a steady supply of labelled A* is being maintained. If the added compounds are true intermediates, the specific activity of Z will be reduced. This method is, however, less sensitive and also less satisfactory in that it may need enough added material to disturb normal metabolism.

Put in these general terms, fundamental research in biology looks very simple. Doubtless research in nuclear physics could similarly be made to look very simple by a biologist. In real life, a great many difficulties arise, of which we shall mention only two fairly obvious ones. First, many metabolic chains have two or more alternative routes, the extent of use of which may vary with circumstances.

Second, it is easier to describe the addition of active intermediate compounds than to carry it out. Many of such compounds may be used inside the cells in which they are produced and may not pass readily through the cell walls. It is not easy to inject labelled compounds through a cell wall at all, let alone to do this without upsetting the behaviour of the cell—which could wreck the experiment. Squirting them in a general way into the organism may be quite useless.

Radioactive materials do not make original research easy. At best they may make a particular investigation possible, although still very difficult, which without them would have been impossible.

As an example to illustrate the complexity of a real problem we shall give a sketch of work with radioisotopes on photosynthesis.

The whole of the higher organisms on our planet depend ultimately on the use of light energy by the plant kingdom to build up a variety of complex compounds from atmospheric carbon dioxide and water. Many different species have been used in investigating this. The mechanism is, so far as is known, essentially the same in all cases and it is often convenient to use a unicellular alga such as Chlorella rather than a flowering plant. Chlorella can be easily grown in continuous cultures from which samples with uniform characteristics can be taken as required. Radioactive carbon dioxide (labelled with C^{14} (5700 y, β^-)) can be bubbled through the culture and quick changes can be made from stable to labelled carbon dioxide and back again. Finally, the cells are very quickly penetrated by alcohol, which deactivates the important enzyme systems and can be used to block metabolic processes at a well-defined instant.

In Plate XX are shown autoradiographs of two two-dimensional chromatograms (produced as described in Chapter 10) showing the different compounds containing C^{14} made by Chlorella. In the top picture are shown the compounds produced within five *seconds* from the beginning of administration of $C^{14}O_2$. In the lower picture are shown the compounds produced after 30 seconds. Both pictures show a large number of compounds, especially as the patches due to "sugar phosphates" could themselves be further separated by more detailed analysis.

It can be seen that some extremely complex compounds are made very quickly. Comparison of the pictures shows that the compound PGA (phosphoglyceric acid) and the sugar phosphates, are made in good quantities in 5 seconds or less while others such as aspartic acid and the amino acids, alanine, etc., are formed later. The results after only five seconds, however, give plenty of scope for guesswork as to the first compound formed from the CO_2. Much further work has been needed to show that the first produced of the compounds shown, and in fact the first stable compound so far known, is phosphoglyceric acid.

12.6 Effects on Living Tissues

We shall now go on to mention some applications of radioactive isotopes to *change* biological processes rather than to investigate them. Ionising radiations have a variety of effects on living tissues, nearly all harmful. The removal of an electron from an organic molecule usually takes up a good deal more energy than is required to break the molecule up into simpler ones, so that when the molecule is neutralised by the return of an electron and this energy is liberated again, the molecule is likely to be destroyed. Furthermore, the charged ions derived from organic molecules—which are not produced in ordinary chemical processes—are exceedingly reactive and the molecule which remains after neutralisation may be quite different from the original one. Indirect processes may also occur; for example H_2O^+ ions and OH radicals may be produced from water and these may react with physiologically important compounds or may—especially in the presence of oxygen—give active chemicals, for example hydrogen peroxide, which then proceed to react with such compounds. Since water is the main constituent of most living tissues, these indirect effects are often of greater importance than the direct ones.

The first effects of radiation on living cells, therefore, are to destroy some useful compounds and to produce unwanted ones, the latter effect being the more important. This can affect or disorganise the delicately balanced series of reactions by which the cell lives. It is not unlike a type of poisoning in which the poisons are produced directly inside the cells.

12.61 *Units of measurement and physical basis of biological effects*

Radiation can produce effects only as a result of the energy actually absorbed so that if hard γ-rays pass through a tissue it is only the small fraction which stops in the tissue that needs to be considered. The biological effects, however, may also depend on the distribution of energy along the path of the radiation or particle. The absorption of equal amounts of energy from different radiations may therefore produce different biological results. This phenomenon is generally expressed quantitatively in terms of the *relative biological effectiveness* (R.B.E.) of the different kinds and energies of radiation. The R.B.E. varies not only with the different radiations but also with the biological effect considered. It generally increases with increase of the number of ion pairs produced per unit length of particle track; thus the effectiveness of α-particles is usually greater than that of β-particles. Some effects, such as the stopping of bacterial division or the inactivation of viruses, however, are produced more efficiently by β-particles.

In Fig. 54 are shown the variations of R.B.E. with linear ion density for various biological effects; the R.B.E. of radium γ-rays is arbitrarily taken as unity in every case.

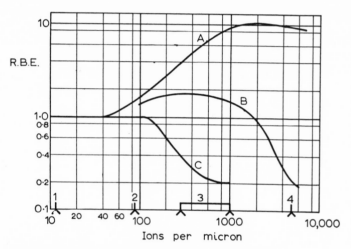

Fig. 54 Variation of relative biological effectiveness (R.B.E.) with linear ion density in ions per micron.
A. Growth inhibition (mouse tumours, wheat seedlings).
B. Inhibition of mitosis (Vicia Fabia).
C. Inactivation of tobacco mosaic virus.
   (Following D. E. Lea and L. H. Gray.)
Region 1. Radium γ-rays.
       2. 200-kV X-rays.
       3. Neutron recoil particles.
       4. α-particles.

For any particular radiation the effects will depend on the dose received, so that the measurement of this is important. The basic quantity is the energy absorbed per gram of tissue and correspondingly the fundamental unit of dose received is defined in terms of energy. It was laid down by the International Commission on Radiobiological Units in 1953 that this unit should be the *rad*. A material has received a dose of 1 rad when 100 ergs of energy have been absorbed in each gram of the material. The rad thus measures the integrated total of dose received whatever the time taken; in a particular flux of γ-rays the *rate* of absorption will be given in units such as rads per hour.

In practice it is very difficult to measure directly the total energy received per gram until this reaches some thousands of rads. Smaller doses are found indirectly by measuring the ionisation produced in a gas. The original unit, the *roentgen*, was in fact defined in terms of this; one roentgen was the radiation dose that would produce one electrostatic unit of charge of each sign in 1·293 mg of standard air. Unfortunately we cannot give a simple numerical factor to relate the roentgen to the rad, as the ionisation produced in air does not bear a constant relation to the energy dissipated in tissue although for β-particles and γ-rays up to 3 MeV the ratio is nearly constant and 1 roentgen is within a few per cent of 1 rad. For α-particles or very high energy radiations, however, the ratio is seriously changed either because of recombination of ions before they can be detected or because of the different behaviour of secondary radiations in gases and in solids or liquids. Fortunately β-particles and γ-rays of moderate energy are by far the most important radiations in biological applications so that various kinds of gas-ionisation chamber can usually be used to make an adequate estimate of tissue dose.

Having measured the radiation dose, one allows for the unequal effects of different types of radiation on the biological system of interest by multiplying the physical dose in rads by the appropriate R.B.E. Although not officially listed as a unit by the International Commission, the derived unit the *rem* can be used when it is necessary to add the doses of different radiations received by the same tissue. The rem is defined in terms of the rad by the relation:

$$\text{Dose in rem} = \text{Dose in rad} \times \text{R.B.E.}$$

It is a disadvantage that in older texts the unit rem (originally roentgen-equivalent-man) was defined somewhat differently, giving a unit about 16% smaller than that defined above. It is only rarely, however, that this small difference matters.

Neither rad nor roentgen is of any use in the measurement of neutron doses. Slow neutrons in particular produce neither direct ionisation nor energy dissipation. Secondarily they may produce both when absorbed in nuclei, but the energy will appear as a mixture of prompt γ-rays and delayed β-particles from any radioactive nuclei which are produced. The amount of γ-energy liberated and the half-lives of the compound nuclei differ widely with different nuclei and the cross-sections for absorption of neutrons in the first place may vary by a factor of 10^6 or so. Consequently neutron doses are normally recorded numerically in neutrons per square centimetre and are measured with boron trifluoride proportional counters (slow neutrons) or hydrogen-filled ionisation chambers (fast neutrons) as described in Chapter 3.

12.62 *Specific effects of radiation*

The phenomena produced by radiation on the main part of living cells, i.e. the parts outside the cell nuclei, are known as *somatic* effects. Large doses of radiation are required to produce them and they are not of great value as a tool in biological research although they are of considerable interest in themselves and may later become of practical value.

The use of radiation for low-temperature sterilisation is already established. Some drugs, such as penicillin, cannot be heated to destroy undesirable bacteria but can be irradiated, if necessary while refrigerated to low temperatures. Doses of the order of millions of rads are required.

If γ-rays or fast electrons of moderate energies are used, no radio-activity is left after irradiation. Raw foodstuffs can also be treated. Rats and dogs fed for up to two years on such irradiated foods have remained perfectly healthy. Unwanted changes of flavour often occur, but these can sometimes be disguised although research in this field uses reagents rarely employed in nuclear physics. For example, we may quote a paper by V. J. Kubala on the sterilisation of pork for army rations: "Best results, and a product which had acceptable characteristics, were obtained when a mixture of barbecue sauce and 40% base brine solution was injecto-pumped into the pork loins. The loins were processed by cooking, chilling, shining and packing with barbecue sauce prior to irradiation."

12.63 *Genetic effects*

Of much more importance to research are the genetic effects. The hereditary structure of every organism, and of every cell of every organism, is mainly determined by a group of highly complex molecules in the nucleus of each cell. This nucleus contains a number of thread-like bodies, the chromosomes, which in turn contain or consist of the genes which represent the elementary units of heredity. When a cell divides in the course of normal growth, the most important part of the process is the production of a duplicate set of chromosomes so that each daughter-cell can have a full complement. Each gene carries the information required to enable the cell to carry out one specific operation. Usually this consists of the production of a particular kind of protein molecule which itself can catalyse one of the reactions on which the life of the cell depends. If a gene in a fully grown cell is altered or destroyed it may not seriously affect the life of the cell, but when the cell divides, the new ones will find that their supply of the protein concerned is cut off. It may be replaced by some new kind of protein provided by the altered gene, but this is unlikely to be able to carry out the same function. This may often lead to the death or lack of development of the cell. When a less-essential protein is involved or where the changes in it still allow it to function, the cell may live and grow in a modified manner. Most of our knowledge of genetics has been obtained from study of the less important changes of gene for the obvious reason that one can learn more from an organism which develops abnormally than from one which simply does not develop at all. Thus in humans we know a lot about the genes controlling eye-colour or curliness of hair but not much about those controlling our basic biochemistry.

In the higher plants and animals each cell nucleus usually contains two sets of chromosomes, each with a full set of genes. In sexually reproducing organisms, one set will have been derived from each parent. Thus there will be two of each gene. These will normally be identical. When they differ, both may be active, producing some kind of average overall effect on the organism, or one may be dominant, the other being recessive and having little or no observable effect. The organism then develops just as it would have if both genes were of the type of the dominant one.

Ionising radiation can damage the molecules forming the chromosomes just as it can damage the other complex molecules in the cell. The present view of the chromosomes is that their essential constituents are long thread-like molecules of desoxyribonucleic acid or DNA. These consist of an enormously long chain composed of tens of thousands of alternating phosphate and sugar groups. To each sugar group is also attached one of the four bases adenine, guanine, cytosine and thymine. It is the order of arrangement of these bases that carries the genetic information, just as the order of dots and dashes in a Morse message determines the words which have been transmitted. Each gene is then represented by a particular set of bases in a particular order and the genes are strung together along the nucleic acid molecule in a way reminiscent of the sentences on ticker tape.

If ionising radiation modifies any of the bases comprising the DNA, the hereditary message will be changed. Such a change is known as a mutation. Once made it will be inherited—so long as the change is not so drastic as to prevent growth and reproduction of the organism. Mutations do occur in Nature as a result both of background radiation and of the action of heat and chemical agents, but they are rare. The natural rate of mutation may be only of the order of a few times per million reproductions for any one gene. Use of ionising radiations has enabled us to increase the mutation rate by thousands of times and hence to make it a practical field for experiment. The vast majority of changes are unfavourable, but improved characteristics of one or two cereals have already been obtained and doubtless further improvements will emerge as time goes on. X-rays, γ-rays or neutrons are the most convenient form of radiation used to produce the ionisation. The results of this general irradiation are entirely random in character and there would seem to be scope for attempts to develop more specific methods. For example, one might supply to the organism of interest a particular one of the important bases, in labelled form, with a high specific activity. When the radioactive atom in the base decayed, the latter would very reliably be changed. It would be possible either to label it in one of the "ring" nitrogen or carbon atoms, or in one of the attached hydrogen or other groups. In the first case the ring would be destroyed or a gap left in the regular series of bases. In the second case a smaller change might be achieved. A favourable case might be to label the carbon of the methyl group of thymine*: if this decayed it would probably recoil out of the molecule and might in a fair proportion of cases be expected to leave behind the base uracil* instead of the original thymine. Uracil is characteristic of another genetically important substance, ribonucleic acid, and its substitution for thymine in DNA might well modify the genetic message carried by the DNA. If so, then each decay of an active thymine incorporated into a chromosome should produce a mutation and, apart from the possibility of an instructive narrowing of the range of kinds of mutations produced, the actual rate of mutation ought to increase a good deal for the same amount of general radiation damage.

So far the main contribution of the applied physicist in this field has been the rather humdrum one of calculating and supplying appropriate dose rates of any desired kind of radiation to seeds or fruitflies. The techniques for doing this are covered in other chapters. The actual value of a suitable dose rate will depend on the organism. The number of mutations produced will be strictly proportional to the dose, so that the larger this can be made the better. On the other hand, if it is too large, the organisms will all be killed or made sterile. A dose rate that will kill or sterilise from 50 to 90% of the organisms

---

* The formula of thymine is $CHNHCONHCOC.CH_3$ and of uracil $CHNHCONHCOC.H$.

concerned would usually be the most economical. This may vary from tens of thousands of rads for bacteria or seeds to a few hundred rads for mammals.

Many miscellaneous biological uses for nuclear radiations are known. Only one will be mentioned.

12.7 Pest Control

The method to be described depends on the fact that the dose of radiation which will sterilise an organism is usually a good deal less than the lethal dose. It has been applied with great effect to the elimination of the screw-worm fly from certain areas of the United States. This fly has a limited climatic range, but within that range can do great harm to livestock. The eggs are laid in any minor scratch or injury and the resulting maggots feed on and extend the wound. The enlarged wound attracts further flies which lay their eggs in it and as a result, in a fly-infested country, almost any minor wound results in death. The navel of a newly born animal may also be attacked. The female of this fly mates only once before laying her eggs so that if the male with which she mates is sterile she will not be fertilised by a second male and the eggs will not hatch. Hence if sterile males can be liberated in any infested area, they will indirectly sterilise the wild females with which they mate. If they are more numerous than the wild males, the majority of the wild females will lay infertile eggs and the numbers of the next generation will be considerably reduced. If the same number of fresh sterile males are now liberated, they will outnumber the wild ones still more and reduce the population by a still larger factor. Complete extermination should be attainable in five or six generations.

Experiments have shown that irradiation of the pupae a couple of days before hatching can sterilise the flies without killing them or seriously reducing their period of active life. The males are sterilised by about 2500 rads and the females by about 5000.

After successful preliminary experiments on the islands of Sanibel and Curaçao, a major "factory" for the production of irradiated screw-worm flies was set up, in an aeroplane hangar in Florida. Six 500-curie Co^{60} γ-ray sources were used to give each pupa 8000 rads, to ensure complete sterility for both sexes. In full production, the factory was able to produce some 50 million irradiated flies per week. These were distributed by 20 planes over 180,000 sq.km (70,000 sq. miles) of the South Western United States, which was separated by adequate natural barriers from any other areas infected by the fly. After six months it was impossible to find living maggots anywhere in the area. The programme was continued for a further six months, finishing in November 1959.   No screw-worm flies have been seen in the area since.   There was no observable effect whatever on any other species, insect or otherwise, besides the screw-worm fly itself and any specific parasites that it may have had.   Since the fly was not merely controlled but exterminated, the cost was a capital one and the operation was almost certainly the most financially rewarding application of nuclear physics which has yet occurred.

CHAPTER THIRTEEN

MISCELLANEOUS APPLICATIONS

This chapter will be little more than a list of some of the general applications of nuclear physics. Most of these are non-specific, i.e. an isotope of any element can be used so long as it gives the appropriate kind of radiation and has a suitable half-life.

13.1 Observational Uses

13.11 *Pressure and thickness measurement*

α-particle emitters are not very extensively used because of the short range of the α-particles. They are, however, useful in one or two cases.

A commercial pressure-gauge has been made in which the gas whose pressure is to be measured is allowed to enter an ionisation chamber containing a fairly strong α-particle source. The ionisation and hence the current in the chamber is directly proportional to the gas pressure. Its sensitivity to different gases is proportional to the cross-section for ionisation of the α-particles in each gas. If the dimensions of the chamber, the strength of the source and this cross-section are known it can be regarded as an absolute instrument, though its accuracy is greater when used for relative measurements. The sensitivity is not very high, but it suffers from few systematic errors. A β-particle source would be much less satisfactory for this instrument because of the much lower ionisation produced in the gas at a given pressure.

A thin, flat α-particle source and a counter can be used for detecting holes or thin patches in continuous strips of thin metal or plastic. The α-particle source is placed close to and parallel to one side of the foil which is to be checked and the counter is put on the other side. The path of the particles is arranged so that they just fail to enter the counter, either by means of thin absorbers or simply by adjusting the distance for which the particles are travelling through air. Then a very slight reduction of the thickness of the foil which is being checked will allow large numbers of the particles to reach the counter. β-particles or the characteristic X-rays from a suitable target element can also be used for this kind of checking over a wide range of foil thickness, but, for thin foils, α-particles have a great advantage in their well-defined maximum range.

To measure a varying thickness, α-particle emitters are less suitable than they are when detecting divergence from a standard value. For measurement a continuous change of counting-rate with thickness is necessary so that uncollimated sources must be used. This limits the thickness that can be measured to about half the maximum range of α-particles. The absorption characteristics of β-particles make them much more suitable for such measurements. Tl^{204} (4·1 y, 0·76 MeV $β^-$) forms a suitable source for thicknesses from 0 to 160 mg/cm². For thinner films, say 0 to 20 mg/cm², C^{14} (5700 y, 0·156 MeV $β^-$) or S^{35} (87·1 d, 0·167 MeV $β^-$) are better, while a source of higher energy particles such as $Sr^{90} + Y^{90}$ (28 y, 0·61 + 2·24 MeV $β^-$) is needed for measurements up to 400 or 500 mg/cm². A very good arrangement employs a second source and counter, identical with the measuring source and counter. The outputs from the two counters are then fed to a mixing circuit set to subtract one from the other. Thus if a piece of material of the correct thickness is placed between the second source and counter, a zero-signal will be recorded when the correct thickness is passing between the measuring source and counter. The system can then be used for a variety of thicknesses simply by inserting the appropriate standard and of course several pairs of sources with different electron-energies can be kept in stock.

A large uniform plane source can be used to produce a map showing contours of equal thickness. For example, such a radiograph of a currency note or postage stamp will show the watermark but not the printed picture or design.

In some special cases very thin layers can usefully be measured by the back-scattering of β-particles. The back-scattering is roughly proportional to atomic number so that a thin film of a heavy metal on a light one can be measured. The thickness of silver or gold electroplating on copper or nickel could be found in this way and an application which has been made on a larger scale is the determination of the thickness of tin on tinplate. β-particles from a 16 mCi source of Tl^{204} were used. A roughly linear increase of counting rate occurred as the thickness of tin increased from 0 to 2·5 μ; fluctuations were about 0·15 μ. This method can also be used to measure somewhat greater thicknesses of light-element coatings on heavier elements, such as paint or varnish on a metal.

For much thicker materials an X-ray or γ-ray source is necessary. A γ-ray source is much the more convenient although it usually involves using a scintillation counter rather than an ionisation chamber.

For a centimetre or so of brass, for example, a very suitable source would be 100 mCi of Ir^{192} (74·5 d, β^-, several γ-rays 0·2 to 0·5 MeV) with a collimating lead diaphragm; this can satisfactorily be screened by about 2 cm of lead. For thicker materials still, Co^{60} (5·27 y, β^- γ 1·17 MeV) or Y^{88} (105 d, e-capture, γ 0·9 to 2·76 MeV) may be employed. Very strong sources of Co^{60} are available, making high-speed measurements possible.

Such sources can also be used for making radiographs of metal structures which cannot be conveniently brought to an X-ray set. For example, the checking of the lead jointing of cables or welds in pipe lines can be done very easily by placing a small strong source of Co^{60} behind or under the cable or pipe and an X-ray film on the other side. It is even better, when possible, to put a small symmetrical source at the centre of the pipe with film all round the outside. The source can be mounted on a rod projecting from the centre of a light umbrella-like structure to keep it located in the centre of the pipe as it is drawn through. Radiography by Co^{60} has also been used to check the position of the tungsten filament and molybdenum grid in a big radio transmitting valve during operation. (Such valves have tubular copper anodes which prevent direct visual inspection.)

13.12 Detection of leaks

A method of detection of leaks in underground water pipes is to run a solution of some γ-ray emitter slowly through the pipe and then, after it has been washed through, to search the ground for the activity left where active material has come through the pipe wall. Some tens or even hundreds of millicuries of a hard γ-emitter may be needed if the pipe is more than a few tens of centimetres below ground, so that a short-lived nuclide should be used which will not leave any long-term contamination. Na^{24} (15 h, β^-, 1·37 and 2·75 MeV γ's) is usually suitable. Even with this nuclide, of course, it must be remembered that the activity will remain at a serious level for over a week so that appropriate arrangements must be made for catching and storing the effluent from the pipe. A similar method has been used with a radioactive gas (methyl bromide, CH_3Br^{82}) to find leaks in gas-impregnated underground telephone cables. Some care is needed in the choice of a gas with suitable chemical properties as too inert a gas (e.g. Xenon) will diffuse too quickly out of the ground while too reactive a gas will interact with the insulation and stay in the cable.

13.13 Transport of fluids and solids

Radioactive sources from ten millicuries to one curie have been used to trace the movement of the boundary between two different oils sent successively along the same pipe. Such oils mix remarkably little in transit, but it is of course necessary to switch from one container to another when the second oil arrives at the receiving end with as little loss as possible. Admixture of a suitable activity into the leading edge of the second oil enables its approach to be detected.

"Go devils", the devices driven along pipes to sweep out accumulated debris, not infrequently get stuck part-way along. By making them radioactive it is possible to follow them in their journeys through underground pipelines so that when stuck they can be found and cut out as easily as possible.

In such cases it is not clear that the radioactive method is much better than a suitable battery-operated buzzer, switched on after appropriate times or by the build-up of pressure when the device sticks. However, radioactivity sounds much more scientific and is doubtless therefore preferred by Head Office.

An interesting application has been to the movement of pebbles in the Rhone. Ta^{182} (112 d, β^-; 1·12 MeV and several other γ's) was used, being waxed into holes in pebbles of various sizes from 2 to 20 cm diameter. Each contained about 0·5 mCi of activated tantalum wire. After 3 to 4 months the active pebbles were found to have moved $\frac{1}{2}$ to 1 km down the Rhone, being still quite easy to find on the bottom with a suitable γ-detector suspended from the boat. The spread of pebbles was, as might be expected, comparable with or greater than the mean distances travelled.

13.14 *Wear and abrasion*

Something nearer to chemical tracer work is illustrated by work on frictional wear and on the lubrication of bearings. This can be done in two main ways. First, a piston or cylinder, or an axle or its bearing, can be activated in a reactor. Then after a period of wear the amount of activity in the lubricating oil, washed out with a suitable cleansing agent, is measured and the amount of metal removed is deduced. This procedure has the advantage that it is easy to distinguish between the metal from the outer and from the inner member of the system. It has the disadvantage that large pieces of metal must be activated to high specific activities. This difficulty can be alleviated by inserting plugs of active material in suitable places or by introducing active sleeves or liners. The activity of even these can be reduced by screening unimportant regions with cadmium foil from the slow-neutron flux during irradiation.

The irradiation of such awkward objects is best done in a "swimming pool" reactor and even then a large object must be rotated during irradiation to give a uniform distribution of activity.

The second method is to collect the oil or washings from the system after a suitable working period without previous activation of either member. Then the quantity of metal in the oil or washings is found by activation analysis as described in Chapter 10. This has the advantage that very much smaller total activities need to be handled and, since the inner and outer members will almost always be of different alloys, an intelligent choice of elements for activation will normally allow identification of the material. It is not, however, so convenient when a large series of tests is to be made on the same test pieces for different times or with different lubricants.

In such investigations it is essential to estimate in advance the activity required of the test pieces to give conveniently measurable activity in the sludge. A factor of a million difference between weight of pieces and weight of absorbed material is all right; that is to say if a measurable activity of say 0·01 μCi in the sludge is given by an activity of 10 mCi in the pieces it is reasonably safe. If a factor of 10^9 is likely, however, it is probably wiser to use the activation-analysis method.

13.15 Diffusion rates

There is a true tracer application which is of great importance in metallurgy. This is the use of radioactive isotopes to measure diffusion rates in metals. The method is equally good for the investigation of self-diffusion which can hardly be observed in any other way. The active material may be introduced by welding an active piece of metal to a second similar inactive one, by electroplating of active atoms from solution on to an inactive metal surface, by evaporation or by several other methods. A fast rate of diffusion from the surface of a thin sheet can be measured by observing the rate of growth of activity on the far side of the specimen; a slower one by observing the rate of loss of activity (preferably using a soft β-emitter) from the active face. A more direct method is to take off a series of slices, parallel to the surface, and measure their activity. This is not easy for very thin layers, though in some cases a chemical polishing technique can be used in which successive layers as thin as 1 μg/cm^2 are removed for investigation. An extremely sensitive method developed by Frauenfelder to find the rate of diffusion of cadmium into silver used the detection of recoil atoms of Ag^{107} following the decay of Cd^{107} (6·7 h, e-capture, 0·85 MeV γ) atoms originally deposited on a silver plate. The recoil range of these was only in the region of two or three atoms of thickness so that exceedingly small diffusion rates could be measured.

K

13.16 *Prospecting for minerals*

Uranium and thorium ores have long been sought with the help of counting instruments. Search for such materials can be carried out by lowering a counter down the drill-pipe used in boring operations. This method can be extended to cover several other elements. A neutron generator (conveniently a Po-Be source) is lowered down the drill-pipe at intervals and, when it has been withdrawn, the counter is used to follow the decay of any radioactive nuclides produced by neutron absorption; the amounts and species of nuclides encountered are then inferred from the observed levels of activity and half-lives. This system is, however, usefully applicable to only a few elements; a much superior system is to use a scintillation counter, with a discriminator, to count the characteristic prompt γ-rays liberated by particular nuclides following neutron capture. In this way a continuous record may be obtained of the distribution in depth of any particular nuclide. The characteristic γ-ray (2·2 MeV) from the reaction H (n, γ) D can be used to detect oil or water and estimate its abundance. If a pulsed neutron source, such as an accelerating tube with a tritium or deuterium target, can be employed the characteristic γ-rays from short-lived radioisotopes can be used for identification, counting being carried out between the neutron pulses. Some indication of mineral content may also be found by lowering close together a fast-neutron source and a BF_3 counter. When the surrounding material contains an efficient moderator, particularly hydrogen and to a lesser extent carbon, the counting rate is increased.

13.2 Use of Radioisotopes for Operational Purposes

Most uses mentioned so far are for observational purposes. As the amount of radioactive materials available increases, other large-scale uses become possible.

13.21 *Static elimination*

Very large electrostatic voltages can be developed when insulators such as dry rubber or plastic are handled in bulk and a fire danger may result.

If the air can be kept slightly conducting by continuous ionisation, this risk is abolished. α-emitters or soft β-emitters, giving ionisation over only a limited range are usually used for this purpose. Tritium, which gives β-particles with a range in air of only about 4·2 mm, has been successfully used in some applications. The gas can safely be used when absorbed into thin evaporated layers of titanium. Of the α-particle emitters, giving particles with a four to five centimetres range in air, radium[226] (1622 y, α) is one of the most widely used. It is incorporated into a thin metal foil which is safe to handle in small quantities.

The main difficulty is that quite large activities, of the order of tens of curies, may be needed, and these form a serious health hazard unless great care is taken. Tritium is not so bad in this respect, being of low toxicity and having a biological half-life of only a few days. Strontium^{90} (28 y, β^-) on the other hand, is one of the most dangerous nuclides because of its long half-life and its tendency to stay in the bones. It and its daughter Y^{90} (64 h, β^-) are otherwise very suitable, being pure β-particle emitters, as no γ-ray screening is required.

It should be noted that it is quite unnecessary for α- or β-particles from the source to reach the surface to be discharged. The charge actually carried by each particle itself is negligible in comparison with that produced by ionisation in the air through which it passes. An electrostatically charged surface will itself attract from the ionised air whatever ions are needed to reduce its charge to a low level, so long as the ionised air lies within the range of its effective electric field.

13.22 *Induction of chemical reactions*

Many energetically possible chemical reactions do not occur at room temperature because one or more atoms require an energy of several times kT (which is only 0·02 eV at room temperature) to pass a potential barrier, before the reaction can proceed. It is not always practicable, especially in organic chemistry, simply to raise the temperature. The comparatively high energies which are transferred to atoms as recoil when nuclear radiations, even β- or γ-rays, pass through matter are more than enough for this; a head-on collision between a 1 MeV electron and a carbon nucleus will transfer a hundred or so electron-volts to the latter, which is tens of times as much as is needed to disrupt any kind of chemical bond.

A high radiation-density can therefore produce chemical reactions quite as easily as heat and, while the yield is much more liable to be reduced by irrelevant side-reactions, the prospects of use at least for research purposes seem to be bright. The prospect is opened up of producing whole classes of compounds which are unstable at high temperatures and which heretofore could not be made at low temperatures. Included in this is the possibility of the production of significant quantities of free radicals.

Another promising field is the large-scale manufacture of compounds which require a lot of energy for their production. The value of the chemicals produced by an industrial country is many times that of the electric power so that the direct application of nuclear energy for such purposes is very attractive.

Pilot studies have been made on the fixation of nitrogen by passing air directly through a nuclear reactor. The effective region of the reactor core was occupied by fine (5 μ) glass threads containing 10% of U^{235}. The energy of the fission products from these was then dissipated in the air traversing the core, producing a wide variety of reactions. The air was recirculated several times, 1 to 2% of NO_2 being removed on each cycle. Removal of active fission products does not appear to be difficult and the process should be useful as an adjunct to power production if not as an end in itself.

The polymerisation of simple organic compounds is another use for high-density radiation. Polymerisation by this method is carried out in the cold and it can be stopped conveniently at any point by removing the source of radiation, while polymerisation by heat is often self-accelerating and correspondingly difficult to prevent from going to completion.

Alternatively, radiation can be used to induce cross-linkages in simple, long-chain, thermoplastic polymers such as polyethylene. The presence of these cross-linkages between the chains makes the material more rigid and raises the softening temperature. Thus polyethylene vessels can be made which will not deform in boiling water but will still be free of the brittleness associated with most of the naturally cross-linked thermo-setting compounds. The cross-linking also reduces solubility and swelling in organic solvents. It requires intense irradiations; of the order of ten millions rads. For reasonably short irradiation times, this would involve using millions of curies of a γ-ray emitter such as Co^{60}, which is as yet hardly practicable, a source of 500,000 curies being the largest currently available. Such doses, however, are not difficult to supply by means of a high-energy electron beam. (This has nothing to do with the applications of nuclear physics, but its similarity to such applications makes it necessary to mention it.) The irradiation also improves the electrical properties and reduces fatigue-cracking. Polyethylene insulation for wire is now irradiated in this way commercially.

Work is being done on the chlorination of such polymers as polyethylene and polystyrene and on the production of mixed polymers. Both of these processes can be carried out at low temperatures with the help of γ-radiation or electron bombardment and should soon be of commercial interest.

13.23 *Power supplies*

The last application that will be mentioned is the use of radioisotopes for power supplies in units such as undersea repeater-amplifiers which are difficult to supply with power by normal methods. Electrical energy can be produced either directly, as by coating one electrode of a vacuum diode with an α- or β-emitter and collecting the charged particles on the second electrode, which is insulated, or indirectly, as by using the radiation to supply energy for some solid-state device analogous to a photo-voltaic cell.

The former system suffers from the drawback that it produces very small currents at very high voltages. 100 curies of polonium[210] coated in a thin layer on the inner of a pair of concentric spheres would deliver a maximum current of only 0·6 μA at a maximum of some $2\frac{1}{2}$ million volts. (The α-particles are doubly charged so that each will lose 5 MeV in overcoming a retarding potential of $2\frac{1}{2}$ million volts. Half of them will be wasted by passing down into the base on which the polonium is placed.) Thus 100 curies will give only about $1\frac{1}{2}$ watts in a very inconvenient form. The weight of the polonium would be only 20 mg so that it is not quite impossible that some application may be found in some unusual place, such as underground on the moon where the maintenance of a good vacuum and good insulation would be no difficulty.

The solid-state devices are rather more promising, as they deliver power at conveniently low voltages. Efficiencies of the order of 10 to 15% have been obtained but even so, curies are required to produce milliwatts of output. α-particle emitters or some β-particle emitters such as $Sr^{90} + Y^{90}$ are the materials most likely to be used to supply the power and the $Sr^{90} + Y^{90}$ pair is particularly promising with its half-life of 28 years and average β-particle energy of some $1\frac{1}{2}$ MeV. A small but increasing number of specialised applications may be expected of the type where a supply must be left unattended, but operating, for decades at a time. The cost of the solid-state diodes per kilowatt-hour is astronomical, but it must be remembered that the humble torch-battery continues to be produced in hundreds of millions although its cost per kilowatt-hour is upwards of £10 against the mains-supply cost to the consumer of little over a penny.

Many further applications will doubtless be found as time goes on, but it should be clear from even this brief survey that there will be an increasing market for the radioactive by-products which will be made at an increasing rate by the rapidly growing nuclear power industry.

CHAPTER FOURTEEN

SAFETY PRECAUTIONS

It is essential that anyone who works with radioactive materials should understand clearly the kinds of hazards involved and have a fair idea of their magnitude. In any laboratory in which such materials are used there should be a person who is responsible for checking the safety of any proposed experiment and who has the authority to ensure that suitable precautions are taken. This is of course additional to the individual responsibility of any civilised person to avoid action which will injure or seriously inconvenience anybody else. Details of the proper precautions for industrial establishments are prescribed by the appropriate section of the Factory Acts, and for University laboratories by the *Code of Practice for University Laboratories* issued by the Committee of Vice-Chancellors and Principals of the Universities of the United Kingdom (referred to below as *C.P.U.*) The author is extremely grateful to the Committee for permission to quote verbatim from several sections of *C.P.U.* (1961 edition) in this chapter. Further references are given in the Bibliography.

The level of danger from radioactivity permitted is much lower than for injury by, for example, industrial solvents or electric shock. This is particularly the case in University laboratories where precautions against the latter risks are in practice left to the good sense of the individual concerned.

This emphasis on radiation risks is unfortunate in so far as it tends to strengthen the already unreasonable depth of public concern about such things as the building of nuclear power plants, which are much less of a danger to public health than are existing power plants with their sulphurous effluent. On several other counts, however, the emphasis on radiation risks is highly desirable. Firstly, they are still unfamiliar and are less likely to be continuously and subconsciously borne in mind than are risks such as those of electric shock which have now been known by most people in this country since childhood. Secondly, radiation danger is not directly appreciable to the senses. This is, of course, true of the voltage on a live wire, but at least the wire can be seen and recognised, while any familiar object can become radioactively and invisibly contaminated if dangerous materials are carelessly handled. Thirdly, the use of radioactive materials is spreading very rapidly and is already much more general than, for example, research on pathogenic bacteria, which may represent an even more serious and very similar hazard. Finally, there is little doubt that many of the other hazards mentioned are in fact inadequately controlled. It is an excellent thing that the newest, and to the public the most impressive, new set of hazards should be treated in a way which sets a worthy example to those concerned with the more traditional risks.

The object of practical precautions is not to prevent entirely the absorption of radiation by human beings. We are already subject to steady radiation, from internal sources such as potassium and from the general background radiation of our environment. The object of precautions is to reduce the risks to such a level that they are no greater than many others commonly met in daily life and voluntarily accepted.

14.1 Radiation Measurement

The first requirement for this is evidently that radiation exposure should be properly monitored and recorded and that the data should be available as quickly as possible to the workers concerned. Some of the effects of exposure are cumulative over long periods so that a record must be kept of all significant doses received by each member of the laboratory.

Monitoring can be divided into two main kinds; in one, measurements are made of the radiation level in different regions of the laboratory to determine for what periods they may safely be used and in the other measurements are made of the radiation doses which have actually been received by each individual worker.

14.11 *Laboratory Monitors*

Various sorts of battery-operated ionisation chamber with convenient pistol-grips are commercially available. These give a direct reading of dose rate in milliroentgens per hour on a small meter, several ranges of sensitivity being available. Hand monitors based on counters are also used.

"A thin window Geiger-Muller counter monitor should be available at all times in every laboratory in which active materials are used. Where work with soft beta-emitters (e.g. C^{14}) is in progress, a mica window counter (e.g. EHM2/S or MX123) is essential. Such a counter cannot, however, be used for tritium. For the estimation of tritium a gas flow counter must be used or the tritium containing compound may be mixed with a liquid scintillator. Work with alpha-emitters requires a suitable alpha-monitor and facilities for measuring air-borne active dust." (*C.P.U.*, para. 3.3.7.1.)

For slow neutrons a similar device using a boron-trifluoride counter is available; this can be converted into a fast neutron detector by placing a cylinder of paraffin wax over the counter. The readings are usually given in terms of neutrons/cm²/sec. The effectiveness of neutrons in producing injury is shown in Table 14.1.

TABLE 14.1*

Neutron Fluxes, corresponding to a Dose of 0·1 rem† in Soft Tissue in a Forty-hour Week

Neutron Energy	R.B.E.†	Neutron Flux (neutrons/cm²/sec)
Thermal	3	670
5 keV	2·5	570
20 keV	5	280
100 keV	8	80
500 keV	10	30
1 MeV	10·5	18
5 MeV	7	18
10 MeV	6·5	17
14 MeV		10
20 MeV		10

* Table II, *C.P.U.*   † See section 12.61.

14.12 *Personal monitors*

Laboratory monitors measure the dose *rate* at the position of the monitor. Personal monitors are required to measure the *total dose* received by the worker carrying them. When work must be done in regions with a high radiation-level, it must be possible to observe the total dose at any time. For this the ionisation chamber described in Chapter 3 and shown in Plate I is very suitable. It can be clipped into a pocket and it is light and small enough to be taped to the back of the hand in critical manipulations.

The background fall of the fibre in such instruments is usually appreciable over a day, so that it is more appropriate for monitoring the doses received during particular jobs than for finding the dose received over weeks or months.

For this purpose the film-badge mentioned in Chapter 5 is usually used by everyone who regularly faces radiation hazards. It consists of a small piece of X-ray film which can be carried in a plastic case in which half of the film is covered by a thin sheet of lead. The fogging of the covered part then shows the exposure to γ-rays only, while the uncovered half will show γ-rays plus β-particles over about 100 keV (enough to penetrate the black paper which protects the film from light). No indication will be given of exposure to α-particles, which are not usually able to penetrate the black paper. If the badge—and therefore, presumably, its owner—has been exposed to radiation the developed film will appear darker, or fogged. The extent of the fogging depends on the amount of radiation absorbed and can be measured with a photometer. This is not an absolute method. Quantitative measurements are made by comparison with similar badges which have been exposed to a series of known doses.

Such badges are usually worn for a fortnight before development. They should be worn for the whole time that the person concerned is at work and a permanent record should be kept for each individual.

It is possible for such badges to be made up and developed by the laboratory using them but proper standardisation is both very important and rather difficult so that it is usually both cheaper and better to use the regular service provided by, for example, the "Radiological Protection Service".* This organisation not only supplies fresh badges, develops the used ones and sends regular reports of the results, but telephones immediately if any film badge is found to show a dose of more than three rads.

14.2 Effects and Permissible Doses of Radiation

The possible consequences of excessive exposure to radiation are severe, including skin injuries which in serious cases may fail to heal, interference with blood-cell formation, leukemia and cancer production, cataract formation, reduction of fertility and genetic effects. Much of the quantitative evidence on these injuries is still poor and individuals may well differ a good deal in, for example, their susceptibility to cancer. Consequently it is desirable that all doses should be kept as low as is practicable and that efforts should be made to keep regular exposures well below the maximum permitted values. This is particularly important for young people who will later have children, as mutations are almost always undesirable and are cumulative for the whole reproductive life of the individual. Most mutations are recessive, which means that their effects may not be apparent for several generations.

The value of the maximum permissible doses are periodically

* Clifton Avenue, Belmont, Sutton, Surrey.

K*

reconsidered as more data become available. The values currently recommended by the Medical Research Council are based on those proposed by the International Commission on Radiological Protection (1959).

Some organs are more sensitive than others so that the permitted dose will depend on how much, and which parts, of the body are exposed. For whole-body irradiation the generally accepted maximum permitted dose is 0·2 rem per fortnight or 5 rem per year for persons of over 18 years of age. Between 16 and 18 it should not exceed 1·5 rem per year and persons under 16 should not be permitted to work with radioactive materials. Even when a dose of radiation is uniformly received by the whole body, it is not entirely satisfactory to put the permitted maximum in such simple terms. This is because some effects, such as the genetic ones and (probably) the induction of cancer, are entirely cumulative; that is to say, the same effect will be produced by 5 rem per minute for 20 minutes as by 5 rem per year for 20 years. On the other hand, some other effects such as anaemia or skin damage are not entirely cumulative; 5 rem per minute for 20 minutes would certainly produce some degree of anaemia, whereas 5 rem per year for 20 years certainly would not. Hence it is not possible simply to say that no one should ever accumulate a total dose of more than say 100 rem. Over 40 years he could accumulate twice as much without any observable effect, although if he took it over a few minutes he might be ill for months. In other words, to avoid non-cumulative effects one must not exceed a certain dose *rate*, while to avoid cumulative effects one must not exceed a certain total dose, whatever period this takes to acquire. Actually no known effects are completely non-cumulative, i.e. depending only on dose *rate*. The damage produced by a dose of 100 rem in the blood-forming tissues would be little changed if instead of being absorbed over 20 minutes as in our example above it was absorbed over 20 seconds or 20 hours and no harm would be done by a rate of 100 rem per second if applied for only a millisecond. Hence dose rates may be averaged over moderate periods and in a complete statement of the permissible maximum doses these periods should be specified. This is not clearly done in any of the published recommendations. For example, we quote from *C.P.U.*, Appendix I:

14.21 "*Whole body, gonads, blood forming organs, eye lenses*

"The maximum permissible total dose accumulated over several years for any age over 18 shall not exceed that given by the relation

$$D = 5 (N - 18)$$

where D is the tissue dose in rem, and N is the age in years. This relationship implies an average dose rate of 5 rem per year or 0·1 rem per week. Over shorter periods of time this latter dose rate may be exceeded provided that it is not greater than 3 rem during any period of 13 consecutive weeks or 12 rem in any one year."

This may reasonably be interpreted as meaning that the dose rate must not exceed 5 rem per year but may be averaged over periods up to 2·4 years. It might, however, be interpreted as meaning that an individual starting radioactive work at say 28 years of age could be allowed 12 rem per year until he had reached the total given by the formula $D = 5 (N-18)$ (which would be after about a further 7 years). Until this point is clarified it is probably wise to take the more cautious interpretation that, if 12 rem are absorbed in one year by anyone at any age, his dose in the following years should be kept below 5 rem per year until his average dose measured from the beginning of the offending year has fallen to 5 rem per year.

It is likely that future editions of *C.P.U.* will discuss this point further; it would seem reasonable that they might also make some distinction between doses permitted during and after the reproductive period.

The recommendations given in *C.P.U.*, Appendix I, for the maximum doses for particular organs not listed above, are as follows:

14.22 "*Other internal organs*

"An average dose to the organ not in excess of 0·3 rem per week is permitted with short time relaxations such that the accumulated dose is not more than 4 rem in 13 weeks or 15 rem in one year.

14. 23 "*Skin (exceptions below) and thyroid*

"The long-term average dose should not exceed 0·6 rem per week, and, in addition, maximum accumulated doses of 8 rem in any consecutive 13 weeks or 30 rem in one year should not be exceeded.

14.24 "*Hands, forearms, feet and ankles*

"The long-term average dose to these members should not exceed 1·5 rem per week and, in addition, the maximum accumulated doses of 20 rem in 13 weeks and 75 rem per year should not be exceeded.

14.25 "*Accidental high exposure*

"An accidental high exposure that occurs only 'once in a lifetime' and contributes not more than 25 rem shall be added to the occupational dose accumulated up to the time of the accident. Higher exposures shall be considered as potentially serious and referred to competent medical authorities."

For those parts of the population not actually engaged in occupations requiring some exposure to radiation the recommendations given in *C.P.U.* are as follows:

14.26 "*Adults working in the vicinity of controlled areas but not occupationally exposed*

"The accumulated dose to the gonads, eye lenses, and blood forming organs or whole body shall not exceed 1·5 rem in any one year; the skin and thyroid may accumulate up to a maximum of 3 rem in any one year.

14.27 "*Members of the public living in the neighbourhood of controlled areas*

"This group differs from the above in that it contains children for whom it is considered that a lower figure, namely 0·5 rem per year (in the gonads, the blood-forming organs, and the lenses of the eye), should apply.

14.28 "*General population*

"The recommendations concerning the general population are based on an assessment of the genetic damage as this is considered to be the most important hazard occurring at low doses. It is suggested that, taking into account all sources of radiation except those due to medical procedures and the natural background, the total dose should not exceed 5 rem per generation.

"A detailed discussion of the permissible exposure of whole populations may be found in the Recommendations of the International Commission on Radiological Protection (see *Report of Main Commission*, Pergamon Press, 1959)."

The measures described so far might be described as passive ones, designed to find what has happened rather than to stop it happening.

14.3 Shielding

More active measures consist in shielding dangerous sources of radiation, shielding laboratories in which work involving radiation is done, use of protective clothing, use of remote-handling equipment and the safe disposal of radioactive wastes. All of these subjects have been discussed in great detail in books particularly concerned with health protection. Here we can give only a short summary of the main points involved.

The screening required by a dangerous source of radiation depends a good deal on the type of radiation. Thus a low-voltage X-ray set may be screened by quite a thin sheet of lead although it may easily be capable of delivering a dose of some thousands of rads per *second* to soft tissue close to the window of the tube.

In Fig. 55 is shown the percentage transmission of X-rays of various energies through lead sheet between 0 and 5 mm in thickness.

It is particularly important to see that the walls round an X-ray room are adequately screened with permanently fixed lead sheet so

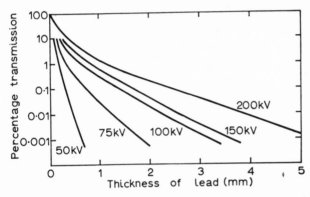

Fig. 55 Transmission through lead for X-rays excited at various constant voltages with initial filtration by 1 mm of glass (or 0·07 mm of copper).

This and Fig. 56 are derived from *Code of Practice for the Protection of Persons exposed to Ionising Radiations.* (H.M.S.O. 1957.)

The author is indebted to the Controller of H.M. Stationery Office for permission to reproduce the data contained.

that unsuspecting persons (and unsuspecting photographic material) cannot be affected whatever may be done by the operator of the machine. Local screening to protect the hands of the operator is also of special importance, but the need for means of access for servicing the equipment may make it impossible for this to be permanently fixed.

Large, fixed, γ-ray sources, such as the multi-kilo-curie sources of Co^{60} used for some radiochemical research, need much thicker screening. The half-thickness of lead for γ-rays around 1 MeV, as shown in Fig. 35, Chapter 6, is 12 to 13 mm, so that very large thicknesses of lead may be needed.

Furthermore, in the experimental use of such a source, it will often be surrounded by a large volume of material of low absorption, in which the effects of γ-rays are being studied or employed, so that the shield must be of considerable size. In such conditions concrete may easily form a cheaper screen, as may be seen from the discussion in section 6.532. In Fig. 56 are shown the absorptions in concrete for X-rays excited by electrons of various energies, together with those for the γ-rays of Co[60] and radium (Ra[226] in equilibrium with its decay products. These being nearly mono-energetic have a greater penetration than "white" X-rays with a considerably higher maximum energy but a lower average energy.)

The radiation dose rates at different distances from sources of γ-rays of 1 curie strength are given in Table 14.2 below.

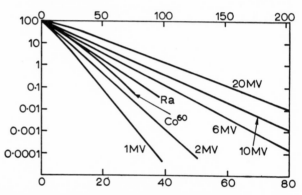

Fig. 56 Transmission of hard X-rays through concrete. As in Fig. 55, the figures give the exciting voltage. The transmission is given also for the nearly mono-energetic γ-rays fron Co[60] (1·17 and 1·33 MeV) and from Ra[226] in equilibrium with its decay-products (0·2 to 1½ MeV). The top scale is in cms and the bottom scale in inches.

Electron accelerators present much the same shielding problems as high-energy X-ray machines. A beam of high-energy electrons is, however, capable of giving hundreds of times the dose rate of the largest X-ray set so even the shortest exposure to it must be regarded as quite lethal. Furthermore, the electrons are scattered by air much more readily than are X-rays so that human access to any part of the room in which the beam is used must be made impossible during operation. Luckily the electrons themselves are easily stopped though protection may have to be provided against the secondary X-rays which they produce.

A high-current cyclotron presents a more serious problem owing to its production of a large flux of fast neutrons as well as high-energy γ-rays. Screening walls of one and a half to two metres of concrete are needed and again access to the working region during operation must be prevented by a suitable system of door-operated

TABLE 14.2

Gamma-ray Dose Rate at one metre from a curie* source of Various Isotopes

Isotopes	Half-life	γ-ray Energy (MeV)	Dose Rate† in roentgens per curie-hour at 1 metre
Na-22	2·6 years	1·3	1·32
Na-24	15·0 hours	1·38, 2·76	1·89
K-42	12·4 hours	1·5	0·15
Cr-51	27 days	0·32	0·02
Mn-52	5·7 days	0·73, 1·46	1·93
Mn-54	300 days	0·84	0·49
Fe-59	47 days	1·1, 1·3	0·67
Co-58	70 days	0·50, 0·81	0·56
Co-60	5·3 years	1·17, 1·33	1·35
Zn-65	250 days	1·11	0·30
As-74	19 days	0·6	0·46
As-76	27 hours	0·5-2·05	0·33
Br-82	36 hours	0·55-1·35	1·50
I-130	12·6 hours	0·42-0·74	1·25
I-131	8·0 days	0·08-0·72 (mainly 0·36)	0·225
I-132	2·3 hours	0·69-2·00	1·21
Cs-137	30 years	0·66	0·33
Tm-170	129 days	0·08	0·005
Ta-182	111 days	0·15-1·22	0·61
Ir-192	74 days	0·13-0·61	0·50
Au-198	2·7 days	0·41-1·09 (mainly 0·41)	0·24
Ra (B+C)	Filtered through 0·5 mm Pt.		0·83

* "Curies" are parent curies, and γ-branching ratios have been allowed for.
† Dose rates in milliröntgens per millicurie-hour at 1 foot may be obtained by multiplying the values in column 4 by 10·8.

switches, lock-switches or similar devices. (Lock-switches are key-operated switches, like those of car ignition systems, which prevent operation of the machine until all access-doors are locked.) All doors should be operable from the inside whether locked from the outside or not in case anyone should be shut in accidentally. A valuable additional safety factor is provided if on closing the final door a klaxon horn or similar device is sounded in the working region and if there is a 15-second delay in the relay circuit following the final door switch. It is the author's experience that one can extract oneself from a very complicated position and travel a considerable distance in the 15 seconds after hearing such a klaxon.

Very high energy accelerators such as synchrotrons often have small enough beam-currents for no screening to be necessary except for the direct extracted beam. It is worth remembering, too, that if a screen of one metre of concrete is placed in the way of scattered multi-GeV protons, it will actually increase the danger considerably. This is because such a screen will enable the primary protons to produce quite a number of secondary ionising particles of comparable penetrating power but will not be thick enough to stop any of them. Here then one must have a screen equivalent to several (4 upwards) metres of concrete or else no screen at all.

The problem of how to screen a nuclear reactor does not usually face the experimentalists working on it. Adequate screening will have been provided by the reactor builders and all that is necessary is to obey the rules.

In radiochemical or radiobiological work it is usually impossible to build a big screen round the source of radiation and then to go away and leave it. Much may be done, however, by building a screen of lead bricks (interlocking types of which are now available) up to chest height to protect the more sensitive parts of the body, and then to work over the screen. In such work, however, the main hazard is not so much the direct radiation from the source as the risk that part of the active material should contaminate the hands of, or be ingested by, the operator.

14.4 Permissible Doses of Ingested Radioactive Materials

Generally speaking, long-lived materials are more dangerous in this respect than short-lived ones, and elements which accumulate in the body are more dangerous than those which pass through quickly. The speed with which a given material is eliminated from the body can be put in terms of the time it takes for half of the material present to be removed. This is the *biological half-life*. The *effective half-life* is the time it takes for the total number of decays per second of the nuclide within the body to drop to half; this combines the effects of the biological and radioactive half-lives. With a short-lived nuclide which is only slowly eliminated, such as Sr^{83} (36 h, β^+) the effective half-life will be essentially identical with the radioactive half-life,

With a long-lived nuclide which is fairly quickly eliminated, such as Na^{22} (2·6 y, β^+), the effective half-life will be practically identical with the biological half-life.

The degree of hazard presented by any particular nuclide therefore depends largely on its biological half-life. A further factor must, however, be taken into account. This is the extent to which the active nuclide is concentrated in particular regions of the body. Thus radio-iodine is quickly concentrated in the thyroid which then receives a radiation dose 50 or more times greater than it would have if the active iodine were spread generally over the body. Elements which are concentrated near sensitive organs such as the blood-forming tissues are particularly dangerous in this respect. Finally, the kind of radiation emitted is important. (Naturally the decay products of the nuclides concerned must also be considered if they too are radioactive.) This is not only a matter of radiobiological effectiveness. Concentration of a nuclide which emits only hard γ-rays is not important, since few of the γ-quanta will be absorbed near the point of emission. On the other hand, the whole of the damage done by an α-particle will be done within one or two tenths of a millimetre of the emitting atom and hence the exact location of α-particle emitters will matter a great deal.

On the basis of these considerations, radioactive nuclides can be graded according to their different degree of toxicity.

A list of the toxicities of some of the more useful radioisotopes is given in Table 14.3* below, together with the maximum total quantities which may be permitted in the body, and the maximum concentration permissible in drinking-water and in the air of the laboratory. The values refer to the isotope in soluble form and are related to the most critical organ of the body; air concentrations refer to occupational exposure for 40 hours per week. More complete lists are given in *C.P.U.* and in the reports cited in the Bibliography. It should be noted that nearly all of the heavy α-particle emitters are in the class of very high toxicity.

* Data are derived from *C.P.U.*, Table III.

TABLE 14.3

Isotope	Half-life	Energy of Radiation (MeV)			Maximum Permissible Body-burden— micro-curies	Maximum Permissible Concentrations— micro-curies per ml	
		α	β	γ		Water	Air
Class I. Very High Toxicity							
Sr^{90}	25 y	—	0.54	none	2	4×10^{-6}	3×10^{-10}
Y^{90}	65 h	—	2.24	none	3	6×10^{-4}	10^{-7}
Pb^{210} (RaD)	22 y	—	0.029	0.047	0.4	4×10^{-6}	10^{-10}
Bi^{210} (RaE)	5.00 d		1.16		0.04	10^{-3}	6×10^{-9}
Po^{210}	138 d	5.303	—	0.803*	0.03	2×10^{-5}	5×10^{-10}
Ra^{226}	1620 y	4.795 4.611 4.21	—	0.188	0.1	4×10^{-7}	3×10^{-11}
Pu^{239}	2.41×10^4 y	5.147	—	{ 0.035 0.050	0.04	10^{-4}	2×10^{-12}

* Only 1·2 quanta to 10^5 α-particles.

Isotope	Half-life	Energy of Radiation (MeV)		Maximum Permissible Body-burden— micro-curies	Maximum Permissible Concentrations— micro-curies per ml	
		β	γ		Water	Air
Class II. High Toxicity						
Ca^{45}	152 d	0.255	none	30	3×10^{-4}	3×10^{-8}
Fe^{59}	46 d	0.46	1.1	20	2×10^{-3}	10^{-7}
Sr^{89}	54 d	1.5	none	4	3×10^{-4}	3×10^{-8}
Y^{91}	57 d	0.555 1.56	none	5	8×10^{-4}	4×10^{-8}
I^{131}	8.04 d	0.60 0.32	0.364 0.284 0.080 0.638	0.7	9×10^{-5}	10^{-8}
Ba^{140}	12.8 d	1.022 0.48	0.54 0.306 0.160	4	8×10^{-4}	10^{-7}
Th-nat.	1.39×10^{10} y	various	various	0.01	3×10^{-5}	2×10^{-12}
U-nat.	4.498×10^9 y	various	various	5×10^{-3}	5×10^{-4}	7×10^{-11}

Isotope	Half-life	Energy of Radiation (MeV)		Maximum Permissible Body-burden—micro-curies	Maximum Permissible Concentrations—micro-curies per ml	
		β	γ		Water	Air
Class. III. Moderate Toxicity						
Na^{22}	2·6 y	$0·54\beta^+$ $1·8\beta^+$	1·28 0·511*	10	10^{-3}	2×10^{-7}
Na^{24}	14·9 h	1·39	1·38 2·758	7	6×10^{-3}	10^{-6}
P^{32}	14·3 d	1·718	none	6	5×10^{-4}	7×10^{-8}
S^{35}	87·1 d	0·167		90	2×10^{-3}	3×10^{-7}
K^{42}	12·4 h	3·6 2·4	1·5	10	9×10^{-3}	2×10^{-6}
Mn^{52}	6·2 d	$0·582\beta^+$ $2·66\beta^+$	0·511* 0·734 0·94 1·46	5	10^{-3}	2×10^{-7}
Mn^{54}	310 d	1·0	0·835	20	4×10^{-3}	4×10^{-7}
Mn^{56}	2·59 h	2·86 1·05 0·73	0·845 1·81 2·13 2·7 3·0	2	4×10^{-3}	8×10^{-7}
Fe^{55}	2·94 y	K		10^3	0·02	9×10^{-7}
Co^{58}	72 d	$0·47\beta^+$	0·511* 0·81	30	4×10^{-3}	8×10^{-7}
Co^{60}	5·25 y	0·31	1·33 1·17	10	10^{-3}	3×10^{-7}
Cu^{64}	12·8 h	0·57 $0·65\beta^+$	0·511* 1·34	10	0·01	2×10^{-6}
Zn^{65}	250 d	$0·32\beta^+$	0·511* 1·14 0·201	60	$3·10^{-3}$	10^{-7}
As^{74}	17·5 d	0·69 1·36 $0·92\beta^+$ $1·53\beta^+$	0·511* 0·635 0·596	40	2×10^{-3}	3×10^{-7}
As^{76}	27·6 h	3·12 2·56 1·4 0·4	0·58 1·20 1·76 2·02	20	6×10^{-4}	10^{-7}
Br^{82}	35·7 h	0·465	0·547 0·787 1·35 1·7	10	8×10^{-3}	10^{-6}
I^{132}	2·4 h	1·5 2·2	0·6 1·4	0·3	2×10^{-3}	4×10^{-7}
Cs^{137} Ba^{137}	33 y 2·6 m	0·518 1·17	0·663	30	4×10^{-4}	6×10^{-8}
Class IV. Slight Toxicity						
H^3	12·5 y	0·018	none	10^3	0·1	2×10^{-5}
Be^7	54·5 d	K	0·479	600	0·05	6×10^{-6}
C^{14}	5700 y	0·155	none	300	0·02	4×10^{-6}
F^{18}	1·83 h	$0·6\beta^+$	0·511*	20	0·02	5×10^{-6}

* Annihilation radiation.

14.5 Precautions in handling Radioactive Materials

With the help of this table it is possible to make a further table showing the amounts of the various nuclides that can be handled safely in ordinary chemical operations without special precautions. These are given in Tables 14.4 and 14.5* which give also the modifying factors

TABLE 14.4

Class of Isotope	Grade of Laboratory		
	C	B	A
I	10 μCi max.	1 mCi max.	> 1 mCi
II	100 μCi max.	10 mCi max.	> 10 mCi.
III	1 mCi max.	100 mCi max.	>100 mCi
VI	10 mCi max.	1 Ci max.	> 1 Ci

TABLE 14.5

Procedure	Multiplying Factor for Figures in 14.4
Storage (stock solutions)	× 100
Very simple wet operations.	× 10
Normal chemical operations	× 1
Complex wet operations with risk of spills . . .	× 0·1
Simple dry operations	× 0·1
Dry and dusty operations	× 0·01

Example of application of Modifying factor

In a Grade-B laboratory normal chemical operations may be carried out with quantities up to 10 mCi of Fe^{59} (Class II isotope). If, however, the work is concerned with dry grinding of metallic samples containing Fe^{59}, then the maximum permissible quantity is 100 μCi (modifying factor × 0·01).

required for operations other than ordinary solution-chemistry. Thus storage in solution presents little hazard while dry dusty operations involve greater ones.

A Grade-C laboratory is simply a good chemical laboratory with standards of care and cleanliness equivalent to those of a microchemical laboratory, with good fume cupboards which have surfaces

* From *C.P.U.*, Tables VI and VII respectively.

covered with a washable gloss paint and with disposable waterproof coverings on benches.

Grade-A and Grade-B laboratories need specially fitting out and equipping. This is a job which requires expert design and should not be undertaken without the advice and supervision of an experienced radiochemist. No recommendations will be given here, therefore, except that such advice *must* be sought if quantities of activity greater than those given in Column 1 of Table 14.4 are to be handled.

Even within the limitations fixed by Table 14.4, as modified by Table 14.5, a number of precautions must be taken by each individual experimenter. The quantities permitted of the safer long-lived nuclides, such as C^{14}, are quite large. If these substances are carelessly handled, it is very easy to get enough spread around to make accurate low-level counting entirely impossible in the laboratory concerned, even if there is no particular danger to health.

Protective clothing should be worn in the laboratory but, to avoid the risk of carrying material into other laboratories, should not be worn outside it. In a Grade-C laboratory an ordinary laboratory coat is sufficient and, even when contaminated, can usually be laundered in the ordinary way.

It is advisable to get used to wearing rubber or disposable plastic gloves regularly; they *must* be worn to handle amounts of active material as much as one tenth of the levels given in Column 1 of Table 14.4, or for any amount if the worker has any open cuts or abrasions on the hands. No food, drink, eating utensils, tobacco or cosmetics should be brought into the laboratory. No equipment should be operated by mouth: flexible Polythene wash-bottles and bulb-operated pipettes should always be used.

It is valuable to acquire the normal medical habit of washing the hands thoroughly after every experiment, preferably with the help of a soft nailbrush. It is essential to wash thoroughly on leaving the laboratory and before meals.

When handling strong β-particle sources goggles should be worn; ordinary spectacles are, however, usually sufficient.

Several nuclides which are classed as highly or very highly toxic for ingestion emit little or no penetrating radiation. Quantities of these which would be dangerous to handle in the open laboratory can be safely manipulated in a *glove box*. This is in effect a small hermetically sealed laboratory, perhaps a metre wide, 75 cm high and 50 cm deep, with a large Perspex or glass window in front. Below or through this are sealed a pair of long-sleeved stout rubber gloves which can reach any part of the interior. These are so placed that the operator can put his hands into them from the outer laboratory and can then use his gloved hands freely within the box. Highly toxic materials can be introduced through a lock in their original containers which are not unsealed until they are safely inside the box. Such boxes usually form part of the equipment of Grades A and B laboratories and as such will not be discussed in detail here. It is, however, worth pointing out that they can be very useful in handling dusty or otherwise troublesome materials which might be a nuisance to counting equipment even at levels well below those which could be any danger to health.

14.6 Waste Disposal

Safe disposal, as already indicated, must be considered before starting the experiment. For high levels of activity again no advice will be offered except that a specialist *must* be consulted before any experimental work is begun. For most Grade-C operations, however, no special arrangements are needed so long as the following instructions are followed.

These are quoted from *C.P.U.*, section 3.3.9:

"(i) Short-lived materials should be stored in some unfrequented place until their activity has decayed to a sufficiently low level for disposal as inactive waste. The containers should be adequately shielded and labelled. It is useful to remember that in a period of 10 half-lives the activity is reduced almost exactly to 1/1000 of the original.

"(ii) Solid waste of low activity may be disposed of in the ordinary laboratory waste, provided that the total activity put into the waste by one worker in any day does not exceed 10^{-3} μC for isotopes in Class I, and 1 μC for isotopes in Class II, III and IV (see section 3.3.1).

"(iii) Putrefiable solid waste, e.g. animal carcasses, may conveniently be preserved without decomposition by means of refrigeration or cocooning for a sufficient time for radioactive decay to take place. (For further details of cocooning see Appendix III.) (C.P.U.)

"(iv) While the occasional incineration with normal refuse of combustible waste containing small amounts of radionuclides (say 10^{-2} μC per charge for Class I isotopes or 100 μC per charge for Classes II, III, or IV) would not be objectionable, there might well be reservations if this were done regularly. For example, if the ash from several incinerations were allowed to accumulate, the disposal of the ash might present a difficult problem. Whether active wastes should be incinerated depends on the design of the incinerator and the relative volumes of active and inactive waste.

"(v) Before any carrier-free material or material of high activity in solution is discharged into the drains it should be mixed with a large quantity of its non-radioactive counterpart (carrier) which should, wherever possible, be in the same chemical form as the active material. In general, the maximum amount of any single isotope in dilute solution which may be discharged into the drains by each research worker in any one day is as follows:

Class I materials 10^{-3} μC
Class II materials 1 μC
Class III materials 10 μC
Class IV materials 100 μC*

Volatile liquids or liquids immiscible with water should not be put into the drains.

"(vi) Radioactive liquids should only be poured into sinks which are directly connected to the main drainage system of the building by closed drains. Sinks which discharge into open gulleys or acid traps should not be used for radioactive work. In places where liquid may accumulate, e.g. traps, the dose-rate at the surface of the trap should not exceed 10 mrad/8 hr day.

"(vii) Glassware, etc., should be washed only in sinks such as those described in section (vi) and sinks so used should be marked with a Radiation Symbol. (Fig. 57.)

"(viii) As far as possible, solid waste should be separated into separate radioisotopes and into combustible and non-combustible forms.

"(ix) Where it is necessary to store liquid waste of specific activity too high for discharge into the drains, a considerable meas-

* These general figures are given as they are easy to refer to and remember. They refer to the most toxic isotope in each class and, if necessary, may be increased for less toxic members of each class to amounts which are given on the tables on pp. 28-33. In certain circumstances the University Radiation Protection Officer may give permission to exceed these latter amounts, as for example, when it is known that no other quantity of a given isotope has been, or will be discharged into the drains from a certain department on a certain day."

ure of concentration may be obtained by mixing the liquid with aluminium hydroxide floc, when most of the activity is adsorbed on to the floc, or by the use of suitable ion-exchange resins.

"Maintenance work on sinks or fume cupboards should not be carried out in laboratories or drains leading from laboratories unless such work has been declared safe by the U.R.P.O. or D.R.S.

"Emergency procedure

1. In the event of an accident involving any personal injury, however minor, the Department Radiation Supervisor should be informed immediately.

Fig. 57 Radiation symbol for marking containers, etc., of radioactive materials. The symbol is usually coloured yellow or red.

2. In the event of a minor spill involving no radiation hazard to personnel, the operator (wearing rubber gloves) should proceed as follows:

"Wet spill: the liquid should be absorbed by blotting paper or similar material.

"Dry spill: the material should be carefully wiped up with absorbent tissue moistened with a 5% solution of glycerine in water. In the case of very active materials, which also present an ingestion hazard, a suitable face mask should be worn.

"All papers used in cleaning should be placed in a suitable waste receptacle and the affected areas monitored. Decontamination should be carried out until no further reduction is being achieved, provided that the contamination level is then below the maximum permissible."

Any other accident should be reported at once to the responsible authority who should investigate immediately in case radioactive material should have been involved.

The whole of this chapter must be regarded as an introduction to the subject. Anyone who works at all seriously with radioactive isotopes must expect from time to time to be regarded as an authority by various juniors and laboratory technicians. He or she should regard it as important to study at least one of the references given below which are wholly devoted to the subject and henceforth to keep informed of new advances in methods of protection.

Nuclear radiation detectors work by generating electrical signals which convey information about the incident particles. There are several points of information that we may wish to have; e.g. type of radiation, average intensity, energy per particle or quantum, mutually related occurrence of nuclear events, etc. To convey the desired kind of information electronic circuits are very widely used, with a degree of complexity ranging from a single-transistor indicator to a highly sophisticated multi-channel analyser or "on-line" computer. The number of different types of circuit now used in nuclear physics is growing very fast. Only a small sample of the commonest and most important can be mentioned here. The main object of showing even this sample is to give the experimenter or his electronics technician an idea of the trouble and cost of building equipment for various specific purposes if cost precludes the far more convenient method of buying a finished commercial unit. The reader who wishes to make a serious study of such circuits should refer to more detailed sources of information such as: R. L. Chase, *Nuclear Pulse Spectrometry*, McGraw-Hill Book Co. Inc., 1961; I. A. D. Lewis, F. H. Wells, *Millimicrosecond Pulse Techniques* (2nd ed.), Pergamon Press, 1959; or periodicals such as *Nuclear Instruments and Methods*; *Review of Scientific Instruments*; *Journal of Scientific Instruments*; *Electronics*.

Ionisation chambers

An ionisation chamber used for monitoring the average level of radiation may be connected directly to an electrostatic electrometer. Sensitive models are, however, very delicate so that it is usual to use some form of electronic amplification of the ion-chamber current before feeding it to a more robust but less sensitive instrument.

In the simple arrangement of Fig. 58 the electrometer measures the

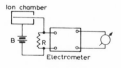

Fig. 58

voltage drop across resistance *R*. For very small ionisation currents *R* tends to be very large ($\sim 10^{12}$ ohm), so that the circuit of Fig. 59

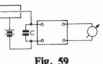

Fig. 59

is more useful, in which the current is measured by the rate of accumulation of charge on a known capacitance C. Initially, C is discharged and after time t the voltage across the capacitor is

$$U = I_{ion} \cdot t / C \qquad (U_B \text{ must always greatly exceed } U)$$

In Fig 60 a simple valve electrometer ("Cutie Pie") is shown. The electrometer valve 5886 forms one arm of the Wheatstone bridge. A

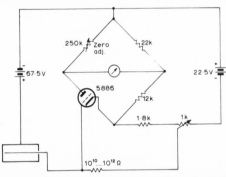

Fig. 60   "Cutie Pie" valve electrometer with an ionisation chamber

more modern ion-chamber monitor using solid state devices is shown in Fig. 61. The MOS (metal-oxide-semiconductor) transistor used in the first stage has an input resistance (10^{13}—10^{18} ohm) of the same order as or even larger than the commercially available electrometer tubes. The second stage with a field-effect transistor is a DC amplifier and the third stage provides the negative feedback to stabilise the electrometer against a change of supply potentials and a drift in the second stage. The ion chamber potential (250 V) is supplied from an AC-DC converter, with a transistor as a blocking-oscillator, a silicon diode rectifier and a gas discharge tube voltage-stabiliser.

The most sensitive electrometer systems, employing the principle of conversion of the measured DC current into an AC signal by means of a vibrating capacitor, can measure currents as low as 10^{-17} A. If an ionisation chamber operates as a detector of individual events, then it represents a proportional detector and the electronic circuitry should be essentially the same as for proportional gas counters and scintillation detectors.

Proportional detectors

In a proportional detector the total charge collected is proportional to the primary ionisation or to the energy loss in the sensitive volume of the counter, but very often not the charge itself. The resulting current pulse in the detector circuit is amplified and further processed. Signals from proportional detectors vary in shape and magnitude.

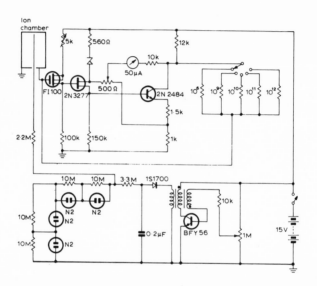

Fig. 61 MOS-FET electrometer for an ionisation chamber (SGS-Fairchild).

In many applications (e.g. spectrometry) it is convenient to work with very short pulses to prevent them from piling-up when counting rates are high. The pulses obtained directly from detectors have a length at least equal to the collecting time t_c but it is customary to shorten the pulses and make their shape more suitable for the amplifier. As the collecting time and therefore the shape of the initial pulse may change, depending upon particle energy, ionised track orientation, etc., the clipping (differentiating) time constant should be larger than the collecting time, with an ample margin left.

An amplifier for proportional detectors should have sufficient gain to activate the devices connected to the output (discriminators, analysers, etc.); a rise time better than the rise time of the detector pulse; good linearity and overloading characteristics. In some cases (e.g. spectrometry with solid state detectors) a very low noise level is also essential.

The simplest pulse shaping, as done by a single high-pass RC circuit (Fig. 62*a*) is not always sufficient because it is difficult both to satisfy the low noise requirements and to have immunity against high overload. Sometimes, therefore, two clipping RC circuits are used, located in different stages of the amplifier.

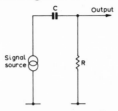

Fig. 62*a*   RC differentiating circuit.

Another method of clipping employs a delay line (Fig. 62*b*). If a pulse is applied to the short-circuited delay line, after twice the propagation time the signal will return to the input end and partially (because of attenuation in the line) cancel the input signal. If two

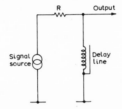

Fig. 62*b*   Delay line clipping circuit.

delay lines are used in the amplifier, one gets a bipolar signal, which has good overloading characteristics and is very suitable for accurate timing, by reference to the time of crossing the zero point (Fig. 63).

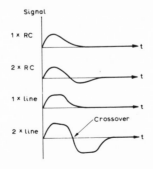

Fig. 63   Waveforms of a detector pulse after shaping.

Linear amplifiers are usually divided into a preamplifier and a main amplifier. The former can be placed close to the detector to avoid shunting effects of the cable capacitance and transmission-line loss of the initial very weak signal. The preamplifiers for scintillation or proportional gas counter can be of normal voltage-sensitive type (Fig 64a),

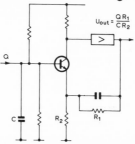

Fig. 64a Voltage-sensitive gain-stabilised amplifier.

but for semiconductor detectors a special type has been developed in which the output signal is proportional to the charge collected on the input capacitance (including the capacitance of the detector which varies with the bias voltage), (Fig. 64b). A circuit of a commercial

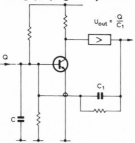

Fig. 64b Charge-sensitive gain-stabilised amplifier.

preamplifier (ORTEC, model 118) is shown on Fig 65. The first stage employs FET, due to its high input impedance and temperature stability. Transistors Q_1-Q_2-Q_3-Q_4-Q_5 form a DC amplifier with the required capacitive feed back to obtain charge sensitivity. Q_6-Q_7 is a voltage amplifier and in the output stage, four transistors are connected in a Darlington amplifier with complementary symmetry. The full diagram of the main amplifier is shown on Fig. 66 and the block diagram on Fig. 67 (ORTEC model 410). The amplifier provides two modes of differentiation (RC and delay line) and produces a unipolar signal, which can finally be converted into a bipolar signal in the Bipolar Amplifier, which is fitted with another clipping circuit (RC or delay line).

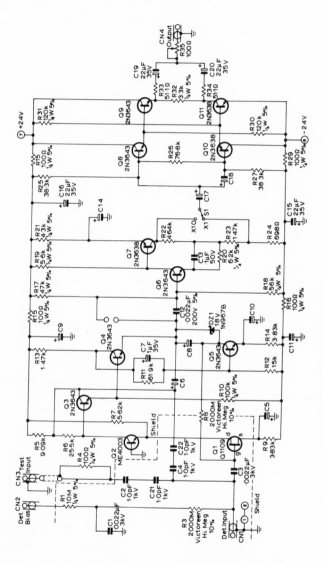

Fig. 65 Preamplifier for semiconductor detectors (ORTEC Model 118)

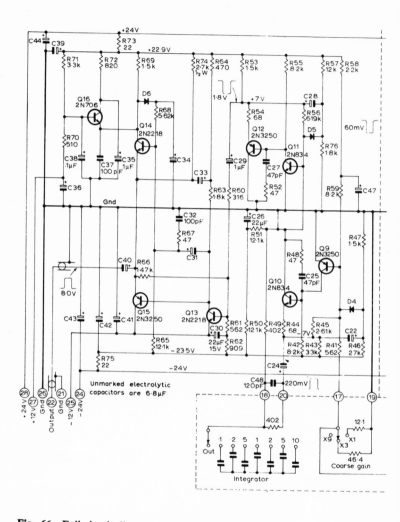

Fig. 66 Full circuit diagram of the linear amplifier (ORTEC, Model 410).

308

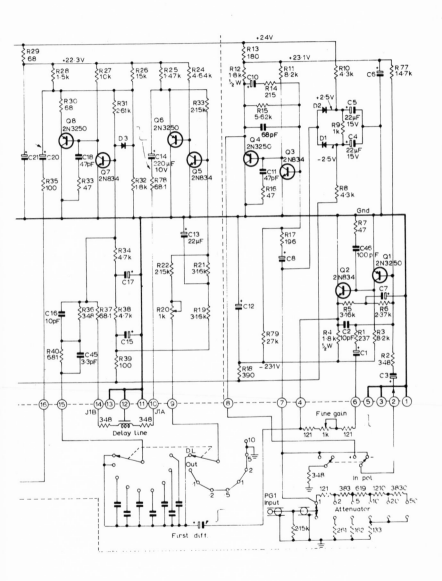

309

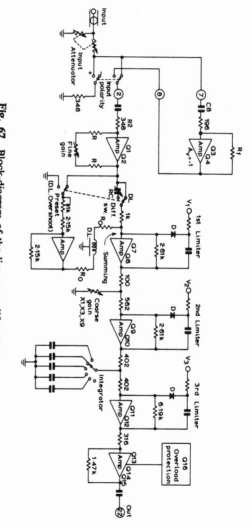

Fig. 67 Block-diagram of the linear amplifier (ORTEC, Model 410)

Sometimes a relatively simple emitter follower (e.g. Fig. 68) can be used as a preamplifier.

Coincidence and anticoincidence circuits

There are many cases in which it is necessary to select from all the signals generated in one detector only those which occur simultaneously (within the resolving time) with signals from a second detector. For this purpose coincidence circuits are used. A simple diode coincidence circuit is shown in Fig. 69. In the absence of input signals, both diodes are conducting. The pulse that appears at the output when both diodes are cut off by positive input pulses is many times bigger than when only one diode is blocked. Hence, an amplitude discriminating circuit can select coincidence pulses only. If the diodes and the power supply polarity are reversed, the circuit will accept two negative signals. If only one diode is reversed, the circuit can be used for anticoincidence (with different polarities of input signals), used for example for shielding against cosmic ray or laboratory background. A commercial (EG and G Model C 104) coincidence unit diagram is shown in Fig. 70. To understand the principle of operation let us first consider the upper (J1) channel only. If switch S1 is closed D5 conducts while D1 does not. Conduction of D5 keeps D10, a tunnel diode, in its low-voltage state. When a positive signal appears on J1, the D1 switches on and takes over the current from D5. If this process happens in all of the switched-on channels, the tunnel diode receives no current and changes its state to the high-voltage one; this change is sensed by a differential amplifier on Q1 and Q2. The fifth channel (VETO) is an anticoincidence channel. Transistors Q3 and Q4 are output line drivers.

Introduction of a delay line into one of the coincidence channels enables us to count only events separated by a predetermined time interval.

Amplitude discriminators

Amplitude discriminators permit us to select only those pulses which exceed a given amplitude. The simplest discriminator is a biased diode (Fig.71), but more reliable are Schmitt triggers. These simultaneously standardise the output pulse. One is shown in Fig. 72; it is an emitter-coupled univibrator. Potentiometer P biases Q1 so that the circuit will change state at the desired polarity and threshold level of the signal. Because of regenerative action this change takes place rapidly. The circuit remains in the new state until the input signal returns to the lower value, when the reverse change takes place. Usually the threshold on the way up differs slightly from that on the way down (hysteresis). Diodes in the input circuit protect the input transistor from overload and from bottoming. The circuit can be made current- or voltage-sensitive.

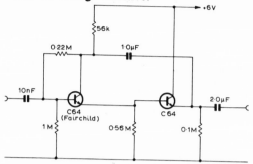

Fig. 68   A simple emitter follower as a preamplifier.

Amplitude analysers

A single amplitude discriminator can be used for an integral pulse-amplitude (spectrum) analysis. In this case only those pulses are fed into the output device (scaler or ratemeter) which exceed an

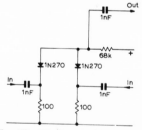

Fig. 69   Diode coincidence circuit.

adjustable level. Two amplitude discriminators and an anti-coincidence circuit make it possible to construct a single channel differential analyser, which registers only the pulses falling between the two preset amplitudes (Fig. 73). By change of the level of both discriminators or by introduction of a preceding variable gain amplifier it is possible to scan the whole range of amplitudes. (Fig. 74).

One of the most useful and versatile electronic instruments of the modern nuclear laboratory is the multichannel analyser. This selects and sorts the incoming signals into its memory channels (now available up to 4096 channels). Signals can be sorted according to their amplitude (Pulse Height Analysis), to the time of arrival (Time Interval Analysis) or the analyser may be used as a multiscaler, repetitively counting the whole range of pulses at regular intervals.

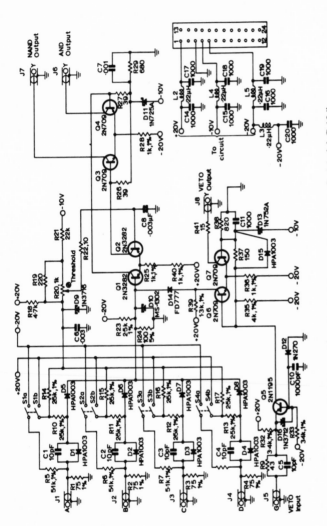

Fig. 70 Four-channel coincidence unit (E. G. & G. Model C 104)

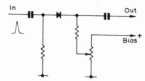

Fig. 71 Biased diode pulse-height discriminator.

The number of counts in each channel is stored in the memory and can be displayed on an oscilloscope and printed out on tape. The multichannel analyser is usually designed also to perform other

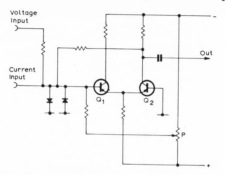

Fig. 72 Transistorised pulse-height discriminator

functions—the background can be subtracted from previously recorded counts, preselected channels can be used for stabilisation of the detector (mostly photomultiplier) gain, etc.

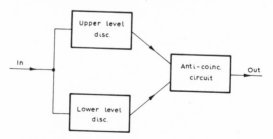

Fig. 73 Block diagram of a simple differential (window) discriminator for pulse-height analysis.

The most important part of the multichannel analyser is the Analog-to-Digital Converter which codes the amplitude of the incoming signal in the form of number code, so that it can be later processed by standard computer techniques. In Fig. 75 the block diagram of a relatively simple commercial multichannel analyser is shown (Baird-Atomic, model 554). The most common method of Analog to Digital Conversion employs a capacitor charged to the analysed voltage level (Fig 76). The capacitor is discharged by a train of pulses, each one removing the same small amount of charge. The number of pulses necessary to discharge the capacitor completely is thus proportional to the initial voltage.

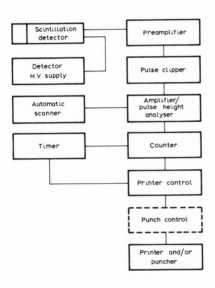

Fig. 74 Block diagram of an automatic differential pulse-height analyser for scanning the whole energy range in the spectrum (Philips).

Scalers

The commonest output device in a nuclear counting system is a scaler. Scalers are characterised by their capacity, resolution (or maximum counting rate) and by their method of display. The essential part of nearly every scaler is a binary counter; a bistable circuit, now usually using solid state devices. Several such units are connected through appropriate links to form a scaler. It is important

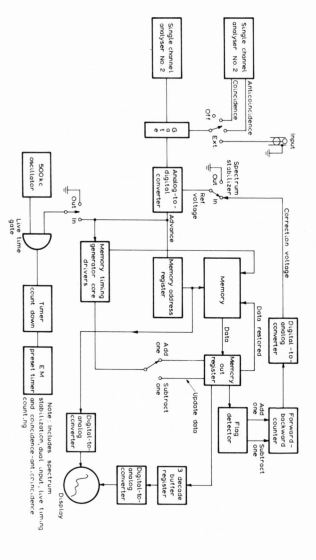

Fig. 75 Block diagram of the multi-channel analyser (Baird Atomic, Model 554)

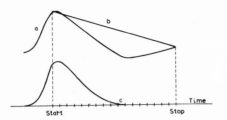

Fig. 76 Pulse-to-height converter of a discharge type. Input signal (*a*) charges the capacitor which is discharged linearly (*b*) Time needed for total discharge is being measured by counting the periods of known oscillator (*c*).

to display the state of each binary unit, so that the total number of counts can be found. A binary counting unit (Mullard) is shown on Fig. 77. As indicators of state, special glow tubes can be used.

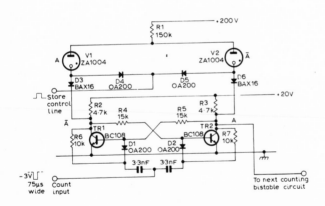

Fig. 77 Circuit of a binary counter with neon display (Mullard).

For experimentalists it is more convenient to have a decimal scaling system. It is possible, ·by use of a combination of binary counters to achieve true decimal counting. The results can be displayed on NIXIE indicators (glow tubes with electrodes in the form of figures). As the "decimalisation" usually reduces the resolution available from the components used, a Binary Coded Decimal system is often used. In a BCD (1248) system for each decade there are four indicators tagged "1", "2", "4", "8". The state of the decade is determined by summing the switched-on indicators. Very fast scalers (up to 350 MHz) are now made usually with integrated circuits. Fig. 78 gives a diagram of a single decade of BCD scaler using integrated flip-flops and gates (SGS-Fairchild). Scalers are usually provided with additional facilities such as print-out output, live-time correction, programmed stop etc.

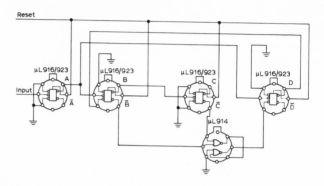

Fig. 78 Decade counter using integrated circuits (Fairchild). Display not shown.

Ratemeters

Ratemeters are used for the determination of the average number of pulses coming in a given period of time. The simplest, linear rate-meter is usually of Cooke-Yarborough type (Fig. 79). The diode D1 and capacitor C form a "diode pump" which injects at every pulse the same fixed charge to the capacitor C_s. Capacitor C_s is in turn discharged continuously through resistance R, and RC determines the time constant of the circuit. The voltage across the capacitor is in these conditions proportional to the average count rate.

Ratemeters are often used with Geiger Counters in simple radiation monitors, because the pulse from a GM counter is nearly stan-dardised and hence can be directly used to drive the diode pump. In more elaborate, commercial designs the diode pump is preceded by an amplitude discriminator.

Power supplies

Detectors usually need DC voltages higher than can be obtained by direct rectification of the mains and stabilised against variations

of mains supply voltage. When practicable, there is still something
to be said for using a well-insulated set of dry H.T. batteries in
series.   In this case resistances totalling some 100 kilohms per
battery should be connected in series between the batteries.The short-
circuited current of a new H.T. battery is several amperes. It is the
current, not the voltage, that kills you when you get a serious electric
shock and without suitable current-limiters, four innocent-looking
H.T. batteries in series are more dangerous than a 5000-volt T.V.
supply which can deliver only a small number of milliamperes.

For high voltages, supplies that must be continuously variable,
or where batteries are otherwise inconvenient, a smoothed stabilised
power supply must however be built.

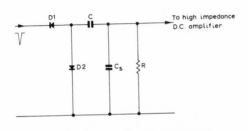

Fig. 79   Simple ratemeter integrator.

For GM counters, the simple gas discharge stabiliser is usually
sufficient, but solid state detectors and photomultipliers require
highly stabilised voltage supplies. A circuit for the EHT supply for a
photomultiplier is given on Fig. 80. Low voltage transistor circuits
usually also require highly stabilised voltages if reliable and stable
operation is to be expected. In the stabilised power supply of Fig. 81,
a Zener diode is employed as a reference voltage source.

Unit modules

A valuable development in nuclear electronics has been the production of a whole series of simple units of identical external dimensions
which can be used to build up more complex devices as required.
This system is known as modular construction and many firms now
produce compatible sets of modules of such individual items as
amplifiers, coincidence and anticoincidence circuits, power packs, etc.

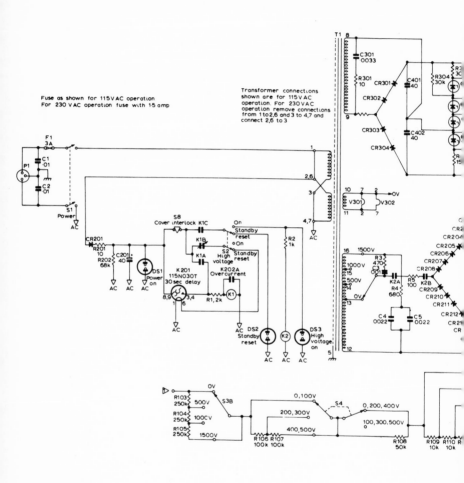

Fig. 80 High voltage power supply circuit diagram (Fluke, Model 412 B)

320

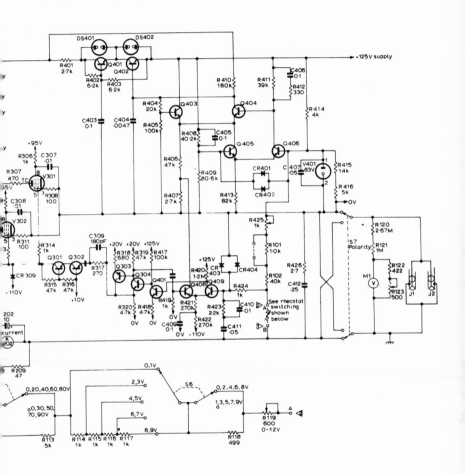

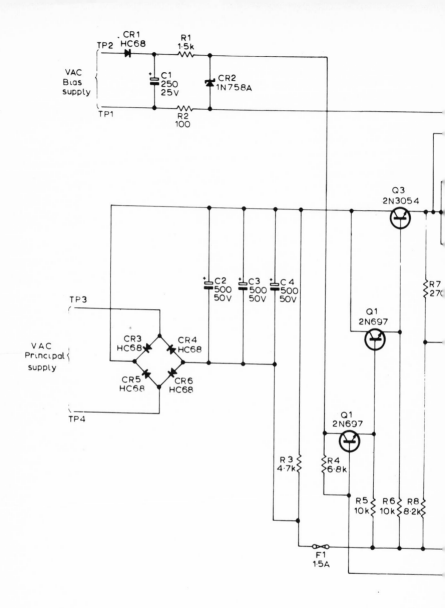

Fig. 81 Low voltage stabilised power supply for transistor circuits (E. G. & G, Model M 1

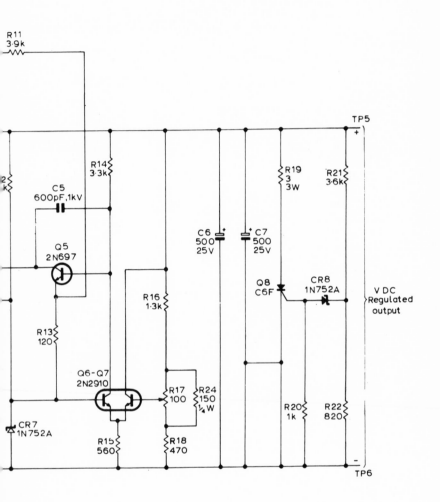

Counter Correction to be added for a Dead-time of 600 μ sec

for an observed counting rate of N per minute

COUNTS PER MINUTE

N	0	100	200	300	400	500	600	700	800	900
0	0	0	0	1	2	3	4	5	6	8
1,000	10	12	15	17	20	23	26	29	33	37
2,000	41	45	49	54	59	64	69	75	81	87
3,000	93	99	106	113	120	127	134	142	150	158
4,000	167	175	184	193	203	212	222	232	242	252
5,000	263	274	285	297	310	320	330	340	360	370
6,000	380	400	410	420	440	450	470	480	500	510
7,000	530	540	560	580	590	610	620	640	660	680
8,000	700	710	730	750	770	790	810	830	850	870
9,000	890	910	930	950	980	1,000	1,020	1,040	1,060	1,090
10,000	1,110	1,130	1,160	1,180	1,210	1,230	1,260	1,280	1,310	1,330
11,000	1,360	1,390	1,410	1,440	1,470	1,490	1,520	1,550	1,580	1,610
12,000	1,640	1,670	1,700	1,730	1,760	1,790	1,820	1,850	1,880	1,910
13,000	1,940	1,970	2,010	2,040	2,070	2,110	2,140	2,170	2,210	2,240
14,000	2,280	2,310	2,350	2,390	2,420	2,460	2,500	2,530	2,570	2,610
15,000	2,650									

The absolute accuracy of the correction is rarely much better than 5 to 10% but may remain constant over short periods to 1% or so. Where small changes in counting rate over short times are important, as is often the case, use of the number of significant figures shown in the table may be worth while.

Appendix III

Ranges of Charged Particles in various Materials

A distinction must be made between the *actual* range of a particular particle of energy E and the *average* range of particles each of energy E. Owing to "straggling" it is not useful to try to measure the range of a particular to a greater accuracy than about 1%. When many particles are known to have the same energy, the average range or extrapolated range can be measured to better accuracy than this.

For protons or heavier particles it makes little difference whether the range is taken as the distance between start and finish of the particle's path or whether it is taken as the actual distance travelled by the particle, following all twists and turns of its track. For slow β-particles, there is a very large difference as can be seen from the tables below.

Heavy Particles

Ranges in Emulsions

If l is the projected length of the track on the focal plane of the microscope, as measured for example with an eyepiece graticule, and z is the vertical difference between the ends of the track, the full range r is given by

$$r^2 = l^2 + (2 \cdot 3z)^2$$

owing to the contraction of 2·3 times in the thickness of the emulsion during processing. The energy of a known kind of particle at the start of its track can be deduced from its range using the following tables.

TABLE III.1

Ranges in Ilford C2 Emulsion

PROTONS, LOW ENERGY*

Range Microns	Energy in MeV										
	0	1	2	3	4	5	6	7	8	9	10
0	0	0·11	0·215	0·31	0·40	0·475	0·54	0·61	0·67	0·73	0·79
10	0·79	0·84	0·90	0·95	1·00	1·04	1·09	1·14	1·18	1·23	1·27
20	1·27	1·31	1·35	1·39	1·43	1·47	1·51	1·55	1·59	1·63	1·67
30	1·67	1·71	1·74	1·78	1·82	1·84	1·88	1·91	1·95	1·98	2·01
40	2·01	2·05	2·08	2·11	2·14	2·17	2·20	2·23	2·26	2·29	2·32
50	2·32	2·35	2·38	2·41	2·44	2·47	2·50	2·53	2·56	2·59	2·61
60	2·61	2·64	2·67	2·70	2·73	2·76	2·78	2·80	2·83	2·85	2·88
70	2·88	2·90	2·93	2·96	2·98	3·01	3·03	3·05	3·08	3·10	3·13
80	3·13	3·15	3·17	3·20	3·22	3·25	3·27	3·29	3·31	3·33	3·35
90	3·35	3·37	3·40	3·42	3·44	3·46	3·48	3·50	3·53	3·55	3·57

* 0-3 MeV from Rotblat, Catala and Gibson, *Nature* 167, 550-1, (1951).

PROTONS, MEDIUM ENERGY*

Range Microns	Energy in MeV										
	0	10	20	30	40	50	60	70	80	90	100
0	0	0·79	1·27	1·67	2·01	2·32	2·61	2·88	3·13	3·35	3·57
100	3·57	3·78	3·98	4·18	4·37	4·56	4·74	4·92	5·09	5·26	5·41
200	5·41	5·56	5·72	5·88	6·03	6·18	6·33	6·48	6·61	6·76	6·90
300	6·90	7·03	7·17	7·28	7·42	7·56	7·68	7·81	7·93	8·05	8·17
400	8·17	8·29	8·41	8·53	8·65	8·76	8·88	8·99	9·10	9·21	9·32
500	9·32	9·43	9·54	9·65	9·76	9·86	9·97	10·07	10·17	10·27	10·38
600	10·38	10·48	10·58	10·68	10·78	10·88	10·98	11·07	11·17	11·27	11·37
700	11·37	11·46	11·56	11·65	11·75	11·85	11·95	12·04	12·13	12·23	12·32
800	12·32	12·41	12·50	12·59	12·68	12·77	12·86	12·95	13·03	13·12	13·21
900	13·21	13·30	13·39	13·47	13·55	13·63	13·72	13·80	13·88	13·96	14·04
1000	14·04	14·12	14·20	14·28	14·35	14·43	14·51	14·59	14·66	14·74	14·82
1100	14·82	14·89	14·97	15·04	15·12	15·19	15·27	15·34	15·42	15·49	15·56
1200	15·56	15·63	15·70	15·77	15·84	15·90	15·97	16·04	16·11	16·18	16·24

* 3-16.24 MeV from W. H. Barkas, *Il Nuovo Cimento*, **8**, 201, (1958).

TABLE III.2

α-particles

Range Microns	Energy in MeV										
	0	1	2	3	4	5	6	7	8	9	10
0					1·18	1·50	1·79	2·08	2·34	2·60	2·84
10	2·84	3·07	3·30	3·52	3·74	3·95	4·15	4·34	4·52	4·69	4·87
20	4·87	5·04	5·21	5·38	5·54	5·70	5·86	6·02	6·17	6·33	6·48
30	6·48	6·63	6·78	6·93	7·07	7·21	7·34	7·48	7·61	7·75	7·88
40	7·88	8·01	8·14	8·27	8·40	8·52	8·65	8·77	8·90	9·02	9·14
50	9·14	9·26	9·38	9·50	9·61	9·73	9·84	9·96	10·07	10·18	10·29

The figures given are for vacuum-dried Ilford C2 emulsion. G5 and K2 emulsions give the same values within about 2%; ranges in emulsion at normal laboratory humidity are longer by 1 to 3%. Where absolute measurements of greater accuracy than this are required, the makers should be consulted as to the characteristics of the emulsions in current production.

All results are given as tables, which are more accurate and much quicker to use than graphs. The last column in each table is repeated to make interpolation easier. Owing to straggling, the accuracy can never be better than about 1 μ or 2%, whichever is the greater. The extra figures are put in to make it easy to find the rate of energy loss at any energy with reasonable accuracy.

The rate of loss of energy is proportional to the square of the charge and is otherwise a function of velocity only. The range-energy relation for any other charged particles can be deduced from the proton tables if we remember that a particle of atomic weight A and charge Z will have A times the energy of the proton at the same velocity and will be losing energy Z^2 times as fast. It will therefore have A/Z^2 times the range.

For example, suppose we see a deuteron ($A=2$, $Z=1$) track of length 1000 μ. Then we know it has a range twice as great as that of a proton with the same velocity. Such a proton would therefore

TABLE III.3

Percentage Excess of Ranges of Protons and of α-Particles in Aluminium over those in dry Ilford C2 Emulsion*

%	0	1	2	3	4	5	
E_p	0·97	1·02	1·08	1·13	1·20	1·35	1·55
E_a	3·9	4·0	4·3	4·5	4·8	5·4	6·2

%	6	7	8	9	10	11	12	13
E_p	1·8	2·1	2·7	4·8	9	11	14	18
E_a	7·2	8·4	11	19	35	44	54	72

This is a critical-difference table, i.e. it gives the range of energy over which the path in aluminium is $n\%$ greater than that in emulsion, to the nearest whole number. Thus for protons between 4·8 and 9 MeV, the path in aluminium is 10% longer than in emulsion or for α-particles between 5·4 and 6·2 MeV the path in aluminium is 5% longer than in emulsion.

* Figures for range in aluminium from H. Bichsel, *Phys. Rev.* **112**, 1089 (1958).

have a range of 500 μ. From Table III.1 the proton energy would be 9·32 MeV. But the deuteron has twice the mass, and hence twice the energy of a proton of the same velocity. Hence finally the energy of a deuteron of range 1000 μ is 9·32 × 2 or, closely enough, 18·6 MeV.

Similarly, a helium-3 nucleus with the range 300 μ has three times the energy of a proton with 4/3 times the range, or $3 \times 8 \cdot 17 = 24 \cdot 5$ MeV.

For the isotopes of helium, it is better to use Table III.2 when possible; at higher energies, when the proton tables must be used, a small correction should be made to compensate for the fact that, when moving slowly, they are not all the time doubly charged. They occasionally pick up an electron and carry it with them for a short distance, being for this time only singly charged and hence losing energy four times less quickly. This effect could be allowed for by subtracting 1·3 μ from the measured range of an α-particle or 1 μ from that of a He³ nucleus before carrying out the calculation above. This is rarely worth while, owing to straggling, unless the average energy of large numbers of particles is required. This effect is more serious with still more highly charged nuclei but we shall not be concerned with these.

Protons and Alpha-Particles in Aluminium

At an energy of 1·0 MeV for protons or 4 MeV for α-particles, the ranges in aluminium are identical with those in dry C2 emulsion. At higher energies the ranges in aluminium are somewhat larger. The difference does not justify a complete fresh table of ranges.

Table III.3 can be used to correct the emulsion tables to the value for aluminium.

To convert ranges in aluminium in mg/cm², to ranges in microns, divide by 0·270.

Protons and Alpha-Particles in Air and Argon

The range in gases is inversely proportional to pressure and proportional to absolute temperature. It is usually given for 15°C at atmospheric pressure. Detailed figures will not be given but approximate values can be obtained by multiplying the emulsion ranges by the figures in Table III.4 below. It will be seen that for

TABLE III.4

Ratio of Range in Air and Argon at 15°C and Atmospheric Pressure to Range in C2 Emulsion

Proton Energy	1	2	3	4	6	8	10	12	14	MeV
α-Particle Energy	4	8	12	16	24	32	40	48	56	MeV
Ratio air/C2	1640	1790	1880	1920	2000	2060	2100	2120	2130	
Argon/C2	1680	1790	1840	1840	1900	1960	2000	2010	2020	

Ranges in air are taken from *Experimental Nuclear Physics*, ed. Segré, Vol. 1, Part II, Sec. 1A. Ranges in argon from F. W. Martin and L. C. Northcliffe, *Phys. Rev.* **128**, 1166 (1962).

moderate energies one micron of emulsion is equivalent to 1·6 to 2·1 mm of air or argon.

Protons and Alpha-Particles in Miscellaneous Substances

Ranges in hydrogen, copper, silver, lead and glycerol tristearate are given in the section of *Experimental Nuclear Physics* already

L*

cited. A nomogram of fair accuracy usable for all elements is given at the end of the same section.

Beta-Particles in Aluminium

The ranges of β-particles have been measured much more extensively in aluminium than in emulsion. In Table III.5 are shown the extrapolated ranges of β-particles in aluminium as a function of the maximum energy; these are derived from the data given by Katz and Penfold, *Rev. Mod. Phys.* **24**, 28, 1952.

The range in mg/cm² is 0·27 times the range in microns. Above 2·5 MeV the range-energy relation is practically linear and the range in aluminium can be written as (1·96 E max − 0·39) millimetres or (530 E_{max} − 106) mg/cm² where E max is given in MeV.

TABLE III.5

Extrapolated Range of β-Particles in Aluminium

LOW ENERGY

Range Microns	Energy in keV										
	0	1	2	3	4	5	6	7	8	9	10
0	0	10	15	19	22	25	28	31	34	36	38
10	38	41	43	45	47	49	50	52	54	55	57
20	57	59	61	62	63	64	66	67	69	70	71
30	71	73	74	76	77	78	79	81	82	83	84
40	84	86	87	88	89	90	92	93	94	95	96

MEDIUM ENERGY

Range Microns	Energy in keV										
	0	10	20	30	40	50	60	70	80	90	100
0	0	38	57	71	84	96	106	117	127	137	146
100	146	156	165	174	182	190	198	206	214	222	230
200	230	238	245	253	260	267	274	281	288	295	301
300	301	308	314	320	326	332	338	344	350	356	362
400	362	368	374	380	386	392	398	404	410	416	421
500	421	427	433	439	445	450	456	462	467	473	479
600	479	485	491	496	502	508	513	519	525	530	536
700	536	542	547	553	559	564	570	575	581	586	592
800	592	598	603	609	614	620	625	631	636	642	647
900	647	653	658	664	669	675	680	685	691	696	701

HIGH ENERGY

Range mm	Energy in MeV										
	0	0·1	0·2	0·3	0·4	0·5	0·6	0·7	0·8	0·9	1·0
0	0	0·15	0·23	0·30	0·36	0·42	0·48	0·54	0·59	0·65	0·70
1	0·70	0·75	0·81	0·86	0·91	0·96	1·01	1·06	1·11	1·16	1·22
2	1·22	1·27	1·32	1·37	1·42	1·47	1·52	1·57	1·62	1·67	1·72
3	1·72	1·77	1·82	1·87	1·91	1·96	2·01	2·06	2·11	2·16	2·21
4	2·21	2·26	2·30	2·35	2·40	2·45	2·50	2·55	2·60	2·65	2·70
5	2·70	2·75	2·80	2·85	2·90	2·96	3·01	3·06	3·11	3·16	3·21
6	3·21	3·26	3·31	3·36	3·42	3·47	3·52	3·57	3·62	3·67	3·72
7	3·72	3·78	3·83	3·88	3·93	3·98	4·03	4·08	4·13	4·19	4·24
8	4·24	4·29	4·34	4·39	4·44	4·50	4·55	4·60	4·65	4·70	4·75
9	4·75	4·80	4·85	4·90	4·95	5·01	5·06	5·11	5·16	5·21	5·26
10	5·26	5·32	5·37	5·42	5·47	5·52	5·57	5·62	5·67	5·72	5·77
11	5·77	5·82	5·87	5·92	5·97	6·02	6·07	6·12	6·18	6·23	6·28

Range mm	0	5	10	15	20	25	30	35	40
Energy—MeV	0	2·7	5·3	7·8	10·4	12·9	15·4	18·0	20·5

From Katz and Penfold, *Rev. Mod. Phys.* **24**, 30 (1952).

The thickness of aluminium needed to reduce the counting rate of a β-particle source ten times is very close to one third of the extrapolated range if the source is placed close to an end-window counter. It is easy to measure and does not depend on knowledge of the thickness of the counter window or of the source. "Close to" means within say 0·2 of the window diameter. If γ-rays are also present and are not taken into account the energy will usually be overestimated by 10% or so, but this is about the error to be expected anyway.

To the accuracy obtainable, there is no difference between the ranges of β^-- and β^+-particles of a given energy.

Beta-Particles in Emulsions

The extrapolated ranges of β-particles in emulsion have not been measured very accurately, but are close to half of those in aluminium.

In Table III.6 is shown the range of β-particles in G5 emulsion, measured along the track of the electron. This is not easy to do, as will be appreciated by a glance at Plate XII, remembering that the vertical displacements are as complex as the horizontal ones and must be taken into account. Generally speaking, it is worth while only to measure a few sample tracks with the view to recognising the regions of energy in which they lie. When it is desired to find the maximum energy of a β-emitter by observation of tracks in emulsion, a much quicker method is to use the "radius" of the track. This is defined as the radius of the smallest sphere whose centre is the starting point of the track and which wholly contains the track. It is thus the maximum distance reached by the particle from its point of origin. It is never possible to be sure that all the tracks

TABLE III.6

β-Particles in Emulsion

(along the track)

Range in Microns	Energy in keV										
	0	1	2	3	4	5	6	7	8	9	10
0	0	5	10	15	19	23	26	29	32	35	37
10	37	40	42	44	46	48	50	52	54	56	58

Range in Microns	Energy in keV										
	0	5	10	15	20	25	30	35	40	45	50
0	0	23	37	48	58	67	75	84	92	100	107
50	107	114	121	127	134	140	146	153	158	164	169
100	169	175	180	185	190	195	200	204	209	213	217
150	217	221	225	228	232	236	239	242	245	248	250

Figures from Zajac and Ross, *Nature*, **164**, 311 (1949).

measured are due to the nuclide of interest. It is best therefore to use the median rather than the average radius. (The median radius is greater than the radii of half the tracks and smaller than the radii of the other half.) Then the effect of a few abnormally long tracks or tracks which leave the emulsion is negligible.

The relation between energy and mean radius of tracks in G5 emulsion has recently been measured by Levi and Rogers; using this and the distribution to be expected from an "average" spectrum, we obtain Table III.7.

Measuring the radii of a hundred tracks, one might hope to get an answer right to better than a factor of $1\frac{1}{2}$ from this table, even if

10% of the tracks observed were irrelevant. This would be quite good enough to show whether the tracks were derived from the expected nuclide.

TABLE III.7

β-Particles in G5 Emulsion

Median Radius Microns	1	1·5	2	3	4·5	6	10	15	20	30
Maximum energy of β-emitter (keV)	27	38	47	66	89	110	160	230	300	400

Beta-Particles in Air

Little accurate work is done with β-particles in air. The stopping power of air is somewhat greater than that of aluminium (weight for weight) and for approximate estimates of energy loss it is usually sufficient to say that the range in air is about 2000 times that in aluminium, so that 2 mm of air is equivalent to 1μ of aluminium.

Beta-Particles in Miscellaneous Materials

Few reliable results are available for other materials than aluminium. For β-particles with energies above about 100 keV the rate of energy loss is fairly closely proportional to the number of electrons per cc of the absorbing material. If this were the only factor, the range in microns would be proportional to $A/Z\rho$ and the range in mg/cm² to A/Z when A, Z, ρ are the atomic weight, atomic number and density of the absorber. Increased nuclear scattering and radiation loss reduce the ranges somewhat in the heavier elements but the expressions above can at least be used to find out whether self-absorption in a source is likely to be important.

Photographic Emulsions for Nuclear Research

1. *Storage*

Emulsions should be kept cool but not frozen and at a moderate humidity. 10°C and 50% R.H. is ideal for prepared plates; emulsion gel should be kept between 0 and 5°C.

2. *Lighting*

The emulsions are not colour-sensitive and an orange safelight may be used, though with discretion. Ilford S(902) for general illumination and Wratten No. 1 or Ilford F(904) for direct illumination are suitable. The emulsion may be examined as closely as desired to the latter for a short time but should not be exposed even to this for longer than necessary.

3. *Processing*

For emulsions of thickness 50 μ or less development follows conventional lines. Most dilute developers can be used. A very satisfactory one is I.D.19, diluted to 50% of its usual strength. The formula for I.D.19 is given below. It may also be bought ready mixed.

I.D.19 full strength

Metol	2.2 gm
Sodium sulphite, anhydrous . .	72 gm
Hydroquinone	8.8 gm
Sodium carbonate, anhydrous . .	48 gm
Potassium bromide	4 gm
Water to	1 litre

For plates up to 50 μ, development time at 23°C in 50% I.D.19 would be about 5 minutes for the α-sensitive emulsions and up to 15 or 20 minutes for the G5, L4 and K5 emulsions. Rinse in water, fix for 50% longer than the time it takes to clear the plates, wash for an hour and dry. For thicker emulsions it is necessary to presoak the emulsion in water and in cold developer so that the lower layers may be reached by the developing chemicals before much development has taken place at the top. Some workers prefer to use a very slow development for some hours at 10°C, others warm the emulsions for a short developing period after they are well soaked with cold developer. This procedure is best carried out with a neutral amidol developer and even with the thinner emulsions gives an appreciable improvement in track visibility. A good developer is

Brussels Amidol Developer

Sodium sulphite (anhydrous) . . .	18 gm
Potassium bromide	0·8 gm
Amidol	4·5 gm
Boric acid	35 gm
Water to	1 litre

The best times of development depend somewhat on the type of track to be observed, the amount of background fog and the preferences of the observer. Soaking times are approximately proportional to emulsion thickness and are given below for 100 μ. A satisfactory procedure for producing good electron tracks in G5, K5 and L4 emulsions is as follows:

Soak in distilled water, starting at 20°C and cooling to 5°C—20 minutes per 100 μ.

Soak in developer at 5°C—20 minutes per 100 μ.

Remove plates from developer and blot off surplus liquid gently with filter paper.

Warm plates in a moist atmosphere (for example, in a stainless steel dish in a water bath) to 25°C—50 minutes for any thickness.

Cool to 5 to 10°C in stop-bath of 1% acetic acid and keep in this for 20 minutes per 100 μ.

Remove surface fog by *very gently* rubbing with cotton wool.

Rinse and transfer to 30% plain hypo 5 to 10°C—50% longer than clearing time, perhaps 2 hours per 100 μ.

Exchange hypo solution slowly for water at the same temperature to avoid a sudden change of osmotic pressure.

Wash in running water not above 10°C—4 hours per 100 μ

Soak in 1% glycerol solution* to prevent cracking of the emulsion in dry conditions—5 to 10 mins per 100 μ.

Dry horizontally in clean air.

Handling of emulsion in gel form

All work must be done with a suitable safelight, such as Ilford S.902, or Wratten No. 1, which should not be brought closer than necessary.

---

* This will also reduce slightly the shrinkage of emulsion in processing. If 5% glycerol is used, the finished emulsion will be half as thick as the unprocessed material. Its thickness is however sensitive to the relative humidity.

Specially prepared glass, to which the emulsion will adhere well, may be obtained from Ilford. The emulsion may also be poured directly on to a specimen. If it should not stick well to this it may be floated off before processing and allowed to dry on to a piece of prepared glass before processing in the normal way.

The glass or specimen should first be set up carefully on a levelling table which, at least for the thinner layers, should be prewarmed. If a thick layer is required, the sides must be built up with strips of

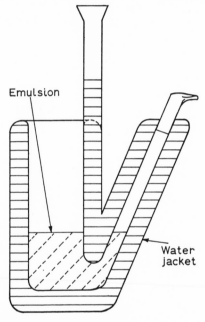

Emulsion

Water jacket

Fig. 82 Double-walled glass vessel for melting and pouring emulsion gel. A thermometer may be kept in the vertical tube of the water jacket.

glass or otherwise to prevent the emulsion from running over the edge.

An appropriate amount of emulsion should be weighed out, touching it only with glass or stainless steel, and allowing a little extra for loss in the heating vessel. Surprisingly little stays in this and a preliminary test will easily show whether any allowance is necessary. The density of the gel is only 1·3 gm/cm³ so that 10·4 gm will be needed to cover 100 cm² (16 in²) with a thickness when dry of 100 μ.

A very convenient double-walled glass "teapot", designed by Mrs. J. Mathieson, for heating the gel is shown in Fig. 82. This has its own waterjacket which prevents overheating, reduces the need for stirring and keeps the emulsion temperature constant while small quantities are being poured. The "teapot" in turn may be heated on a hotplate if its temperature is watched continually, or may be kept in a thermostatically controlled bath.

The gel should be warmed to 45 to 55°C. The thicker the layer desired, the lower can be the temperature without risk of premature solidification. A glass rod or stainless steel spatula may be used to help the emulsion to cover the required area, but should be used as sparingly as possible. The plates must be dried in the dark in dust-free air which must not be hot but may be very slightly warmed if desired.

Appendix V

X-ray Absorption and Emission Energies

Calculated from the *International Tables for X-ray Crystallography*,
Vol. 3, 1962

Z	Element	K Absorption Edge keV	Emission Energy, keV Kα mean
3	Lithium	0·055	0·052
4	Beryllium	0·114	0·110*
5	Boron	0·189	0·185*
6	Carbon	0·284	0·282
7	Nitrogen	0·400	0·392
8	Oxygen	0·532	0·523
9	Fluorine	0·694	0·677*
10	Neon	0·874	0·848*
11	Sodium	1·08	1·04*
12	Magnesium	1·30	1·25
13	Aluminium	1·56	1·49
14	Silicon	1·84	1·74
15	Phosphorus	2·14	2·01
16	Sulphur	2·47	2·31
17	Chlorine	2·82	2·62
18	Argon*	3·20	2·96
19	Potassium	3·61	3·31
20	Calcium	4·04	3·69
21	Scandium	4·50	4·09
22	Titanium	4·96	4·51
23	Vanadium	5·46	4·95
24	Chromium	5·99	5·41
25	Manganese	6·54	5·89
26	Iron	7·11	6·40
27	Cobalt	7·71	6·92
28	Nickel	8·33	7·47
29	Copper	8·98	8·04
30	Zinc	9·66	8·63
31	Gallium	10·37	9·24
32	Germanium	11·10	9·87
33	Arsenic	11·86	10·53
34	Selenium	12·65(4)	11·20

Z	Element	K Absorption Edge keV	Emission Energy, keV	
			$K\alpha_2$	$K\alpha_1$
35	Bromine	13·48	11·88	11·92
36	Krypton	14·33	12·60	12·65(0)
37	Rubidium	15·20	13·34	13·40
38	Strontium	16·11	14·10	14·17
39	Yttrium	17·04	14·88	14·96
40	Zirconium	18·00	15·69	15·78
41	Niobium	18·99	16·52·	16·62
42	Molybdenum	20·00	17·38	17·48
43	Technetium	21·05*	18·25	18·37
44	Ruthenium	22·12	19·15	19·28
45	Rhodium	23·23	20·07	20·22
46	Palladium	24·35	21·02	21·18
47	Silver	25·52	21·99	22·16
48	Cadmium	26·71	22·98	23·17
49	Indium	27·93	24·00	24·21
50	Tin	29·19	25·05	25·27
51	Antimony	30·49	26·11	26·36
52	Tellurium	31·81	27·20	27·47
53	Iodine	33·17	28·32	28·61
54	Xenon	34·58	29·49	29·81
55	Caesium	35·96	30·63	30·97
56	Barium	37·41	31·82	32·20
57	Lanthanum	38·94	33·04	33·44
58	Cerium	40·45	34·28	34·72
59	Praseodymium	42·00	35·55	36·03
60	Neodymium	43·58	36·85	37·36
61	Prometheum	45·20*†	38·16	38·66
62	Samarium	46·85	39·53	40·13
63	Europium	48·52	40·88	41·53
64	Gadolinium	50·24	42·29	42·99
65	Terbium	52·01	43·74	44·48
66	Dysprosium	53·80	45·20†	45·99

Z	Element	K Absorption Edge keV	Emission Energy, keV	
			$K\alpha_2$	$K\alpha_1$
67	Holmium	55·62	46·69	47·53
68	Erbium	57·49	48·21	49·11
69	Thulium	59·35	49·77	50·74
70	Ytterbium	61·31	51·34	52·37
71	Lutecium	63·31	52·97	54·07
72	Hafnium	65·32	54·62	55·80
73	Tantalum	67·40	56·28	57·54
74	Tungsten	69·51	57·99	59·32
75	Rhenium	71·62	59·72	61·14
76	Osmium	73·89	61·49	63·00
77	Iridium	76·13	63·29	64·90
78	Platinum	78·39	65·13	66·83
79	Gold	80·80	66·99	68·81
80	Mercury	83·08	68·91	70·83
81	Thallium	85·68	70·83	72·87
82	Lead	88·07	72·81	74·97
83	Bismuth	90·46	74·82	77·11
84	Polonium	93·07*	77·10	79·52
85	Astatine*	95·74	78·97	81·51
86	Emanation*	98·40	81·09	83·83
87	Francium*	101·2†	83·26	86·10
88	Radium*	104·0	85·50	88·49
89	Actinium*	106·8	87·68	90·89
90	Thorium	109·8†	89·96	93·35
91	Protoactinium*	112·6	92·25	95·81
92	Uranium	116·0	94·67	98·44
93	Neptunium*	118·6	97·01	101·1†
94	Plutonium*	121·8	99·50	103·7
95	Americium*	125·0	102·0	106·4
96	Curium*	128·2	104·5	109·2
97	Berkelium*	131·5	107·2	112·0
98	Californium*	134·8	109·7†	114·9
99	Einsteinium*	138·2	112·4	117·9
100	Fermium*	141·7	115·1	120·8

* Not based on direct experimental results; most of these are based on interpolation or on the energies of the atomic levels.

† Members of pairs for which the difference in energy is less than the experimental error.

Appendix VI

The Counting of Rapidly Decaying Sources

Suppose that at time $t=0$ we have N_0 atoms of the active substance with decay-constant λ and half-life $\tau_{\frac{1}{2}}$. Then at time t the number of atoms surviving will be $N = N_0 \exp(-\lambda t)$.

Then if the counter efficiency is ϵ, the counting-rate at time t will be ϵ times the rate at which atoms are decaying; this is

$$-\epsilon dN/dt = \epsilon \lambda N_0 \exp(-\lambda t) \qquad 6.1$$

(the negative sign merely indicates that N falls as t increases).

What we observe is not directly this counting rate, but the number $N_{t_1 t_2}$ of counts which occur between two times t_1 and t_2 say. This will be ϵ times the number of atoms which decay in this period, or ϵ times the change in N:

$$N_{t_1 t_2} = \epsilon[(N_0 \exp(-\lambda t_1) - N_0 \exp(-\lambda t_2)]. \qquad 6.2$$

The mean counting *rate* over this period is therefore

$$\epsilon N_0[\exp(-\lambda t_1) - \exp(-\lambda t_2)]/(t_2 - t_1). \qquad 6.3$$

This average rate must be equal to the actual rate at some particular time. We will suppose that it is equal to the counting rate at the particular time $(t_1 + t_2)/2 + \Delta t$. This, from equation 6.1 is

$$\lambda N_0 \exp\{-\lambda[(t_1 + t_2)/2 + \Delta t]\}.$$

The choice of the rather complex-looking assumption is governed by the realisation that for very short time-intervals, i.e. for very small values of $\lambda t_2 - \lambda t_1$, the *mean* counting rate must tend towards the *actual* counting rate at the middle of the interval. This mid-time is $(t_1 + t_2)/2$. We shall expect, therefore, that Δt will tend to zero as $\lambda t_2 - \lambda t_1$ tends to zero.

By definition, then,

$$\epsilon \lambda N_0 \exp\{-\lambda[(t_1 + t_2)/2 + \Delta t]\} = \epsilon N_0[\exp(-\lambda t_1) - \exp(-\lambda t_2)]/(t_2 - t_1).$$

Dividing both sides by $\epsilon N_0 \exp[-\lambda(t_1 + t_2)/2]$,

$$\exp(-\lambda \Delta t) = \{\exp[\lambda(t_2 - t_1)/2] - \exp[-\lambda(t_2 - t_1)/2]\}/\lambda(t_2 - t_1).$$

Hence, remembering that $\dfrac{\exp x - \exp(-x)}{2} = \sinh x$,

$$\exp(-\lambda \Delta t) = \sinh\left[\frac{\lambda}{2}(t_2 - t_1)\right] \Big/ \frac{\lambda}{2}(t_2 - t_1). \qquad 6.4$$

Now $(\sinh x)/x \to 1$ as $x \to 0$.

As expected, for very small values of $\dfrac{\lambda}{2}(t_2 - t_1)$, therefore,

$$\exp(-\lambda \Delta t) \simeq 1.$$
$$\text{or } -\lambda \Delta t \simeq 0.$$

For larger values of $\lambda(t_2 - t_1)$, i.e. for longer counting periods, from 6.4,

$$\Delta t = -\frac{1}{\lambda} \ln \left[\sinh \frac{\lambda}{2}(t_2 - t_1) \middle/ \frac{\lambda}{2}(t_2 - t_1) \right].$$

Replacing λ by $0.693/\tau_{\frac{1}{2}}$, since the half-life is more convenient to use than the decay constant, we have finally

$$\Delta t = -1.443\tau_{\frac{1}{2}} \ln[(\sinh 0.347\, \overline{t_2 - t_1}/\tau_{\frac{1}{2}})/(0.347\, \overline{t_2 - t_1}/\tau_{\frac{1}{2}})] \qquad 6.5$$

This may be written $\quad \Delta t = -\tau_{\frac{1}{2}} f(\overline{t_2 - t_1}/\tau_{\frac{1}{2}}) \qquad 6.6$

The function of $f(\overline{t_2 - t_1}/\tau_{\frac{1}{2}})$ is plotted in Fig. 41 as a function of $(\overline{t_2 - t_1}/\tau_{\frac{1}{2}})$, up to counting-intervals of $2\frac{1}{2}$ half-lives.

Lithium-drifted Detectors

In the four years since the above sections were written, a new technique has been developed for producing bulk silicon or germanium crystals with almost the properties of intrinsic semiconductors. This process is known as *lithium drifting*. It depends on the fact that lithium forms a donor impurity with the very low ionisation energies of 0·033 eV in silicon and 0·0093 eV in germanium. The resulting positive lithium ions, unlike positive phosphorus ions, can move through the crystal lattices of silicon or germanium with relatively high mobility under the action of an electric field. A suitable concentration of these positive ions can be made just to compensate the negative ions of an acceptor impurity such as boron in a p-type semiconductor. Electrostatic mutual repulsion of the corresponding free positive holes then drives these out of the region leaving a very high impedance region in which, however, electrons and positive holes liberated by fast charged particles can move freely.

The necessary accurate compensation is attained as follows; a typical process will be described for germanium but differs only in details for silicon. A polished single-crystal block of p-type germanium is heated for a few minutes in an inert atmosphere under a thin layer of liquid lithium at 450°C. Lithium diffuses in giving a low-impedance n-type region to a depth of perhaps $\frac{1}{2}$ mm. The block is allowed to cool and the excess lithium removed from the surface. A *"drift voltage"* of about 500 volts is then applied so as to drive positive lithium ions down further into the block and the temperature adjusted to give a fixed current of a few mA. The positive lithium ions then travel into the p-type region while the free negative electrons in the n-type region and the positive holes in the p-type region are driven away from the p-n junction where the Li^+ concentration equals the B^- concentration. Now the unchanged p-type region may have a specific resistance of only a few tens or hundreds of ohm-cm, the lithium-drifted n-type region will have a lower resistance still, while the approximately compensated region at the junction will have a very high resistance. Consequently, most of the potential difference between the two faces of the block will be concentrated across this nearly compensated region so that lithium ions are dragged rapidly across it into the p-type region, and the compensated region grows. Very complete compensation is obtained by an automatic self-adjustment process. If an excess of positive lithium ions should develop the effect of its space charge will necessarily increase the electrostatic field ahead of it and decrease the field behind it. Hence lithium ions will flow out of it forwards towards the remaining p-region faster than they flow into it from behind. Similarly, if a deficiency of lithium ions developed anywhere the resulting negative space charge of uncompensated boron ions would have the reverse effect.

Fully compensated thicknesses of the order of centimetres are now obtainable although the drifting process may take weeks, the final thickness obtained being proportional to the square root of the drifting time.

Once the desired depth is established the excess n-type and p-type regions can be removed by etching and electrodes evaporated onto the final surfaces. Further diffusion of lithium atoms must be prevented by cooling to liquid nitrogen temperatures and for full efficiency the detector must be kept at such temperatures for its whole life, whether or not it has voltages applied.

So far (1969) the thicknesses obtained are not large enough for lithium-drifted germanium detectors to be very efficient detectors of high-energy gamma rays and for the same reason the full-energy peak is a far smaller proportion of the escape peaks and Compton continuum than for a large NaI crystal. Almost the full theoretical resolving power is however obtainable and though less than 1% of counts may occur in the full-energy peak this is so narrow as to be actually higher than in the far more efficient sodium iodide case, and gamma rays whose energy differs by only 5 keV or so in the region of several MeV can be clearly resolved. The use of such detectors has completely transformed the analysis of complex gamma-spectra.

Fig. 83 shows a typical pair of spectra of Co^{60} observed by two different detectors for comparison.

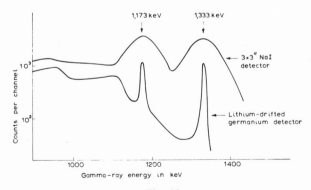

Gamma-ray spectra from Co^{60} showing superior resolution of lithium-drifted germanium over scintillation detector

Fig. 83

BIBLIOGRAPHY

In those cases marked *, where the publisher is not mentioned the book or report was published by the International Atomic Energy Agency, Vienna.

1. General

Nuclear and Radiochemistry. G. Friedlander and J. W. Kennedy. (Wiley, 1955.)

Radioisotope Laboratory Techniques. R. A. Faires and B. H. Parks. (George Newnes, 1958.)

Radioisotope Techniques. (2 vols.) Proceedings of the Isotope Techniques Conference, Oxford, 1951. (H.M. Stationery Office, A.E.R.E. 1952.)

Directory of Radioisotopes, revised edition. (1962.)

The Radiochemical Manual. 1. Physical Data. (The Radiochemical Centre, Amersham.)

Publications of the International Atomic Energy Agency (IAEA) *Vienna.* A classified cumulative list is issued periodically and may be obtained on request. Several of these publications (up to mid 1962) are separately included. (Marked *.)

Gamma-rays of radionuclides in order of increasing energy. D. N. Slater. (Butterworth, 1962.)

Elementary Statistics. H. Levy and E. E. Preidel. (Nelson, 1944.)

2. Nuclear Physics and Instruments

Nuclear Physics: an Introduction. W. E. Burcham. (Longmans, Green & Co., 1963.)

Introductory Nuclear Physics. D. Halliday. (Wiley, 1950.)

Experimental Nuclear Physics. (3 vols.) E. Segré. (Wiley, 1953-1959.)

Methods of Experimental Physics Vol. 5, part A, Nuclear Physics. Ed. Yuan and Wu. (Academic Press.)

Nuclear Physics and Instrumentation. Second United Nations International Conference on the Peaceful uses of Atomic Energy, Vol. 14. (U.N., Geneva, 1958.)

Radioactive Isotopes and their Production under Neutron Irradiation. (1961.)*

Electromagnetic Isotope Separators and Applications of Electromagnetically Enriched Isotopes. Ed. J. Koch. (North-Holland Publishing Co., 1958.)

Metrology of Radionuclides. (1960.)*

The Measurement of Radioisotopes. Denis Taylor. (Methuen Monograph, 1951.)

Radiation Dosimetry. Ed. G. J. Hine and G. L. Brownell. (Academic Press Inc., New York, 1956.)

Selected Topics in Radiation Dosimetry. (1961.)*

Nuclear Radiation Detectors. J. Sharpe. (Methuen, 1960.)

Ionization Chambers and Counters. D. H. Wilkinson. (Cambridge University Press, 1950.)

Energy measurement with Proportional Counters. D. West. (Progress in Nuclear Physics 3, 18, 1953.)

Organic Scintillators. F. D. Brooks. (Progress in Nuclear Physics 5, 252, 1956.)

Liquid Scintillation Counting. Conference Proceedings, 1957. Ed. C. G. Belland, F. N. Hayes. (Pergamon Press.)

Cherenkov Detectors. G. W. Hutchinson. (Progress in Nuclear Physics 8, 237, 1960.)

The Diffusion Cloud Chamber. M. Snowden. (Progress in Nuclear Physics 3, 1, 1953.)

The Bubble Chamber. D. V. Bugg. (Progress in Nuclear Physics 7, 1, 1959.)

On Bubble Chambers. H. Slatis. (Nuclear Instruments and Methods, 5th Jan. 1959.)

The Study of Elementary Particles by the Photographic Method. C. F. Powell. (Pergamon Press, 1959.)

Photographic Emulsion Techniques. J. Rotblat. (Progress in Nuclear Physics 1, 37, 1950.)

3. Electronics

Principles of Electronics. M. R. Gavin and J. E. Houldin. (E.U.P. 1959.)

Nuclear Electronics. (1959.)*

Nuclear Electronics. (1962.)*

Multichannel Pulse-height Analysers. (Nuclear Science Series, Nat. Acad, Sci., Washington, D.C. 20, 1957.)

Millimicrosecond Pulse Techniques. I. A. D. Lewis and F. H. Wells (2nd edit.). (Pergamon Press 1959.)

4. Radiochemistry and Analysis

Radioactive Tracers in Chemistry and Industry. Pascaline Daudel, translated by U. Eisner. (Charles Griffin and Co., 1960.)

Chemical effects of Nuclear Transformations, Vols. 1 and 2 (1961).*
Methods of Radiochemical Analysis. W.H.O. Technical Report Series No. 173. (Geneva 1959.)
Activation Analysis Handbook. R. C. Koch. (Academic Press, 1960.)
Chemical Dosimetry. Jerome Weiss. (Nucleonics **10**, 28th July 1952.)
Tritium in the Physical and Biological Sciences, Vols. 1 and 2 (1962).*
Monographs on the radiochemistry of individual elements: U.S. National Academy of Sciences, National Research Council; Nuclear Science Series.
Paper Chromatography and Paper Electrophoresis. R. J. Black, E. L. Durrum and G. Zweig. (Academic Press, 2nd edition, 1958.)

5. Biological and Medical Applications

Isotopic Tracers in Biology. Martin D. Kamen. (Academic Press, New York, 3rd edition 1957.)
Radioisotopes in Biology and Agriculture. C. L. Comar. (McGraw, Hill, 1955.)
The Application of Radioisotopes in Biology. (1960.)*
Mechanisms in Radiobiology. Ed. Maurice Errera and Arne Forssberg. (Academic Press 1960.)
Radiation Biophysics. Howard L. Andrews. (Prentice-Hall, Inc., 1961.)
Autoradiography in Biology and Medicine. G. A. Boyd. (Academic Press, 1955.)
Radiation in Agricultural Research and Practice. (1961.)*
Radioisotopes and Radiation in Entomology. (1962.)*
Effects of Ionizing Radiations on Seeds. (1961.)*
Effects of Nuclear Weapons. Ed. S. Glasstone. (U.S.A. Atomic Energy Commission, revised edition, 1962.)

6. Dating

Radioactive Methods for Determining Geological Age. L. H. Ahrens. (Reports on Progress in Physics **19**, 80, 1956.)
Geologic Dating. G. Wasserburg and G. Weatherill. (Wiley 1963.)
Progress in Isotope Geology. K. Rankama. (Interscience 1963.)
Lead isotopes in Geology. R. D. Russell and R. M. Farquhar. (Interscience 1960.)
Radiocarbon dating systems. G. J. Fergusson. (Nucleonics **13**, 18, 1955.)

7. Miscellaneous Applications

Radioisotopes in the Physical Sciences and Industry, (3 vols.) (Proceedings of the Conference held by the IAEA with the co-operation of UNESCO at Copenhagen, 6-17 Sept. 1960.) (1962.)*

Peaceful Uses of Atomic Energy. (Report on the Impact of the peaceful uses of Atomic Energy to the Joint Committee on Atomic Energy, United States Congress, 1956.)

Symposium on Applied Radiation and Radioisotope Test Methods. American Society for Testing Materials. Publication No. 268, 1959.

Large Radiation Sources in Industry. (1960.) (Proceedings of a conference on the applications of large radiation sources in Industry, and especially to chemical processes. International Atomic Energy Agency, Warsaw, Sept. 1959.)

Surveying and evaluating Radioactive Deposits. (1959.)*

8. Health Precautions

Radiation Protection. Recommendations of the International Commission on Radiological Protection. ICRP Publication 2. Report of Committee 2 on Permissible Dose for Internal Radiation. (Pergamon Press, 1959.)

Safe Handling of Radioactive Isotopes. Code of Practice for Protection against Ionising Radiation. National Bureau of Standards Handbook issued by the U.S. Department of Commerce, Nos. 42 and 48 onwards.*

Code of Practice for the protection of persons exposed to ionising radiations in University laboratories. (Assoc. of Universities of the British Commonwealth, 1961.)

Report of the Committee on Training in Health and Safety to the Authority Committee on Health and Safety, UKAEA. (H.M. Stationery Office, February 1960.)

Safe Handling of Radioisotopes—Health Physics Addendum. (1960.)*

Safe Handling of Radioisotopes—Medical Addendum. (1960.)*

The Packing, Transport, and related handling of Radioactive Materials. (1961.)*

Health Physics Instrumentation. John S. Handloser. (Pergamon Press, 1959.)

Use of Film Badges for Personnel Monitoring. (1962.)*

Whole body counting. (1962.)*

Medical Radioisotope Scanning. (1959.)*

Radiation Shielding. B. J. Price, C. C. Horton and K. T. Spinney. (Pergamon Press, 1957.)

Fundamental Aspects of Reactor Shielding. H. Goldstein. (Pergamon Press, 1959.)

Atomic Energy Waste. Ed. E. Glueckauf. (Butterworth, 1961.)

Radioactive Waste Disposal into the sea. (1961.)*

Report of the Panel on Radioactive Waste Disposal into Fresh Water. (1962.)*

Radioactive Substances in the Biosphere. (1961.)*

PLATES

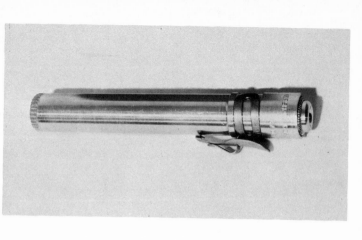

Plate I The Kershaw dosimeter, an ionisation chamber used as a personal monitor.
(*By Courtesy of Cinema-Television Ltd.*)

Plate II Counters for special purposes : (*a*) Liquid counter with built-on container for a standard volume of active liquid.

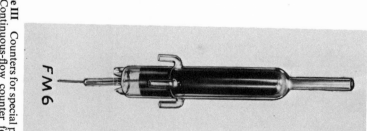

Plate III Counters for special purposes: (*b*) Continuous-flow counter for monitoring the variations of activity in a continuously flowing liquid.
(*By courtesy of 20th Century Electronics Ltd.*)

350

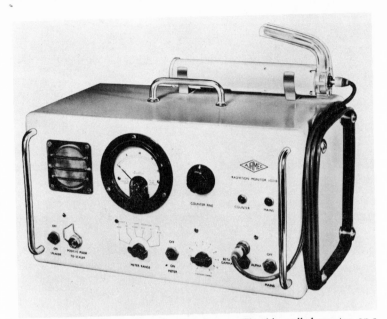

Plate IV Laboratory monitor for β- or γ-rays. The thin-walled counter, on a flexible lead, has a movable screen to stop β-particles with little attenuation of γ-rays of moderate energies. (*By Courtesy of Airmec Laboratories Ltd.*)

Plate V Photographs of two photomultiplier tubes; *a*—56 AVP, a high-definition 14-stage tube; *b*—54 AVP, an 11-stage tube for use with a large block of phosphor. (*By courtesy of Mullards, Ltd.*)

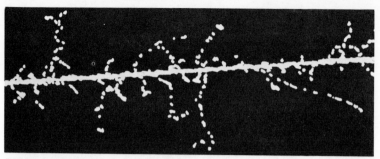

Plate VI Expansion chamber photograph showing the track of a 100 MeV carbon ion from the Birmingham University cyclotron. Numbers of δ-ray (knock-on electrons) tracks can be seen.

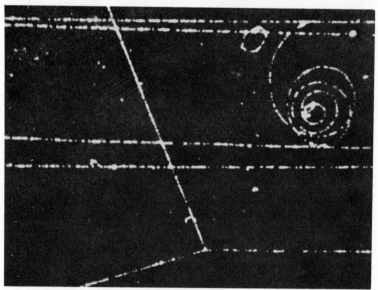

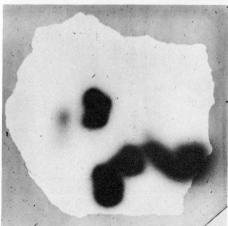

Plate VII (*above*) Tracks in a hydrogen bubble chamber in a magnetic field. Track A is that of a 975 MeV proton from the Birmingham University synchrotron which makes a collision with a stationary proton (hydrogen nucleus) in the chamber. The track of a fast secondary electron can also be seen, bent into a helix by the magnetic field.

Plate VIII (*left*) Autoradiograph of a piece of torbernite, a hydrated phosphate of copper and uranium. The distribution of uranium is clearly far from uniform.

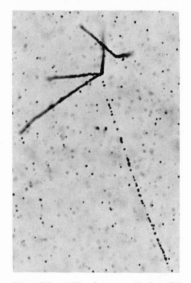

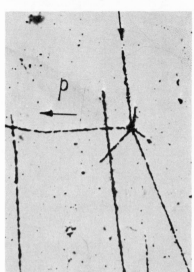

Plate IX "Thorium star" in G5 emulsion. α-tracks from the decay of the thorium decay-product Th^{228} and its descendants Ra^{224}, Em^{220} Po^{216} and Po^{212}. The slight separation due to diffusion of the gas Em^{220} can be seen. The β-track from Bi^{212} can also be seen.

Plate X Nuclear reaction in a C2 emulsion. A 25-MeV nucleas of He^3 from the Birmingham University cyclotron strikes a stationary nucleus of N^{14} in the emulsion, yielding four α-particles and a proton (track p).

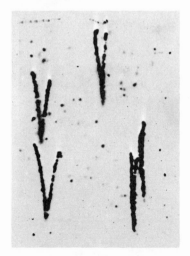

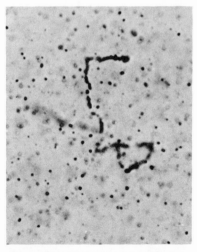

Plate XI Track of the molecular ions HeH$^+$ in a C2 emulsion.

Plate XII Track of a β-particle from C^{14} in a G5 emulsion.

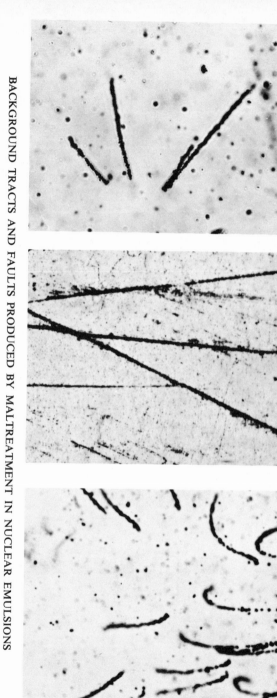

BACKGROUND TRACTS AND FAULTS PRODUCED BY MALTREATMENT IN NUCLEAR EMULSIONS

Plate XIII Truncated thorium star due to contamination of glass.

Plate XIV Surface scratches.

Plate XV α-tracks distorted by excessively rapid drying in hot air.

Plate XVI Autoradiograph of buttercup leaf which has absorbed phosphate labelled with P^{32} through the cut end of the stem.

Plate XVII X-radiograph of the same leaf.

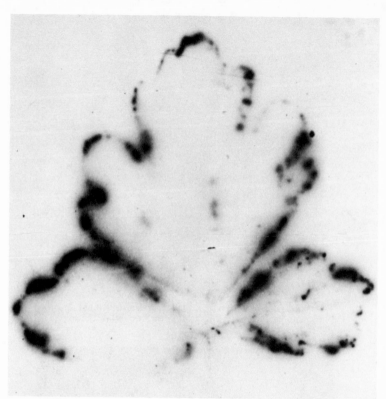

Plate XVIII   Autoradiograph of growing buttercup leaf which has absorbed phosphate labelled with P^{32}.

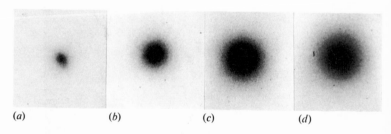

(a)   (b)   (c)   (d)

Plate XIX   Autoradiographs of small (not point) P^{32} source, $\times 20$ magnification.

(a) Directly against a fine-grain plate. (Ilford N 40.)
(b) Separated from plate by 80 μ (7·4 mg/cm²) of polyethylene.
(c) Separated from plate by 160 μ (14·8 mg/cm²) of polyethylene.
(d) Separated from plate by 320 μ (29·6 mg/cm²) of polyethylene.

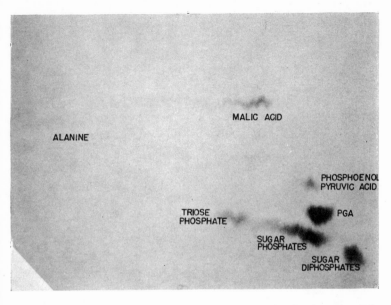

Plate XX Autoradiographs of two-way chromatograms showing the compounds produced by Chlorella from $C^{14} O_2$ by photosynthesis:

(*above*) in five seconds.
(*below*) in thirty seconds.

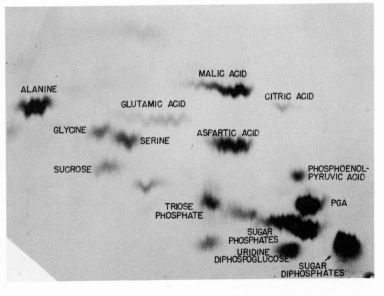

(*By the Courtesy of Dr. J. A. Bassham*)

INDEX

In order to eliminate the need for a glossary, references to definitions in the text are given in **heavy type**.

Abrasion, 276
Absolute counting, *see* Counters, 4π
Absorbing materials, 126 *seq.*, 132-3, 135, 289 *seq.*
Absorption, 119 *seq.*
  of radiation, 62, 128, 132, 289 *seq.*
  in air, 113
  measurement of, 128
  self-, 135 *seq.*
Absorption coefficient, **129**
Absorption edges (X-ray), 134
  Table, 333
Accelerators, 172 *seq.*, 178, 180, 209
Acceptor impurity, 83
Accidents, 287, 300
Activation analysis, 208 *seq.*, 219
Activation energy, 162
Actual range, 321
Adenine, 269
Adsorption, 198, 224
Age determination, 229 *seq.*
  see Radioisotopic dating
Along-the-track measurement, 99
  beta-particle table, 327
Alpha-particle counters, 32 *seq.*, 60, 74, 75, 87
Alpha-particles, 4, 273
  absorption of, 120
  emulsion tracks of, 97, 98, Pls. IX-XI
  energy measurement of, 34, 40, 89, 98
  ranges of (tables), 322 *seq.*
  reactions of, 178
  straggling of, 121, 314
Alpha-radioactivity, 14, 24
Aluminium hydroxide, 300
Amplifiers, 33, 39, 88, 304 *seq.*
Anaemia, 286
Analysers, amplitude, 312
Analysis, 207 *seq.*
  activation, 208 *seq.*
  inverse isotope dilution, 223
  isotope dilution, 221, 234
  neutron-absorption, 228
  radiometric, 226
Angular momentum, 7, 8
Anion exchange, 198
Annihilation, 10, 117
Anthracene, 70
Anticoincidence circuit, 57, 311
Antimony-beryllium source, 179
Anti-neutrino, 10
Arsenic, activation analysis for, 209, 213
Atom, structure of, 1
Atomic bomb, 165
Atomic number, **6**
Atomic weapon tests, 195, 238
Atomic weight, basis of scale, 16

Autoradiographs, 96
  in analysis, 209, 220, 225
  in biology, 245-7
  resolution of, 247
Avalanche, 45
Average life, **21**
Average range, 140, 314

Background, 51, 69, 100, 314
  current, 84
  in diffusion chamber, 93
Background rates in α-counters, 74
  in Cherenkov counters, 79
  in coincidence measurements, 118
  in emulsions, 100, 101
  in Geiger-counters, 52, 56
  in scintillation counters, 66, 69
Back-scattering, 114, 274
Back-scattering factor, **115**
Barn, **171**
Barrier detectors, 88
Bassham, J. A., Pl. XX
Beryllium, in neutron sources, 179
Beta counting, absolute, 58
  from thick sources, 139 *seq.*
  gas counters, 47 *seq.*
  scintillation counters, 74
Beta-particles
  bubble chamber tracks of, Pl. VII
  emulsion tracks of, Pls. IX, XII
  range energy tables, 325 *seq.*
Bikini, 243
Binding energy, 15
Biological half-life, **292**
Blood plasma, 252, 258
Bohr, N., 1
Bremsstrahlung, **124**
Brown, R. M., and Grummitt, W. E., 241 *seq.*
Bubble chamber, 94

Calcium in blood plasma, 252, 258
  in bone, 258
  as carrier for strontium, 195
Carbon, activation analysis for, 219
Carbon-14 dating, 236 *seq.*
  from weapon tests, 238-40
Carrier-free materials, 193
Carriers, 193 *seq.*
Cataract formation, 285
Cation exchange, 198
Cell nucleus, 268
Chackett, K. F., 29, 55, 235
Chain reaction, 164
Chemical reactions, 188 *seq.*, 279
  yield, 222

359